Computer-Mediated Communication in Personal Relationships

This books is part of the Peter Lang Media and Communication list.
Every volume is peer reviewed and meets
the highest quality standards for content and production.

PETER LANG
New York • Washington, D.C./Baltimore • Bern
Frankfurt • Berlin • Brussels • Vienna • Oxford

Computer-Mediated Communication in Personal Relationships

EDITED BY Kevin B. Wright & Lynne M. Webb

PETER LANG
New York • Washington, D.C./Baltimore • Bern
Frankfurt • Berlin • Brussels • Vienna • Oxford

Library of Congress Cataloging-in-Publication Data

Computer-mediated communication in personal relationships /
edited by Kevin B. Wright, Lynne M. Webb.
p. cm.
Includes bibliographical references and index.
1. Online social networks. 2. Telematics—Social aspects.
3. Internet—Social aspects. 4. Interpersonal communication.
5. Interpersonal relations. I. Wright, Kevin B. II. Webb, Lynne M.
HM742.C65 302.30285—dc22 2010040633
ISBN 978-1-4331-1082-5 (hardcover)
ISBN 978-1-4331-1081-8 (paperback)

Bibliographic information published by **Die Deutsche Nationalbibliothek**.
Die Deutsche Nationalbibliothek lists this publication in the "Deutsche Nationalbibliografie"; detailed bibliographic data is available on the Internet at http://dnb.d-nb.de/.

The paper in this book meets the guidelines for permanence and durability of the Committee on Production Guidelines for Book Longevity of the Council of Library Resources.

29 Broadway, 18th floor, New York, NY 10006
www.peterlang.com

Printed in the United States of America

DEDICATION

Kevin B. Wright: To my wife Melissa

Lynne M. Webb: To my husband Robert B. Moberly,
our children (Laura E. Moberly, Richard E. Moberly,
Reed J. Moberly, Laura A. McLeod),
and grandchildren (Luke M. Moberly and Henry M. Moberly)

CONTENTS

PART 3: Influences of CMC on Relational Contexts

PART 4: The Dark Side of Computer-Mediated Communication in Personal Relationships

PREFACE

It began with a simple email: "When you edit a book on CMC in close, personal relationship, let me know. I'd like to contribute a chapter." Then an email back offering a co-editorship for such a book, and the adventure began. From the beginning, we imagined that the edited volume would offer factual descriptions, based in research and theory, of how individuals employ computer-mediated communication (CMC) to effectively enact their online and offline relationships. Worth noting, we developed all our ideas and framing for this book in what seemed like record time: We were three months from the first email suggestion to a signed contract. Our 12-page proposal envisioned an edited volume of chapters that reviewed cutting-edge research examining how individuals use CMC to enact their personal relationships. The book you hold in your hand achieves that goal. Also worth noting: We communicated about our ideas and framing for the book exclusively via email, including developing the book proposal and the call for chapter proposals. We developed our book describing CMC via CMC.

Our vision for the book. We proposed an edited volume that focused on CMC issues of ongoing importance to interpersonal communication researchers, such as interpersonal perception, privacy issues, and social support. We wanted the volume to provide equal focus on issues that CMC scholars thought important in explaining interpersonal communication online, such anonymity, synchronicity, and channel limitations. In short, we wanted to gather essays that covered the intellectual ground at the intersection of these two important bodies of research. We noted that no textbook or book of readings existed on the interpersonal aspects of CMC. We designed our book to serve as the ideal textbook for advanced undergraduate classes as well as graduate seminars in the popular and fast-growing area of CMC in personal relationships.

Scholarly contribution. The body of research on CMC has grown by leaps and bounds, including studies of interpersonal issues in CMC. Nearly 38% of the manuscripts submitted to the *Journal of Computer-Mediated Communication* (an ICA journal) between March 2006 and March 2008 had an interpersonal focus (ICA, 2008). In the past decade, additional journals, including *Human Communication Research, Communication Monographs, and Communication Research;* have published an increasing number of articles reporting interpersonally focused CMC research.

Our book adds to the literature of the discipline by offering the first collection of essays on interpersonal CMC. The book brings together the multiple ideas, topics, theories, and authors addressing the interpersonal aspects of CMC. It focuses scholarly attention on an applied, pragmatic area of study in the field of communication. We envision our book inspiring and facilitating further growth of the CMC area of specialization within the discipline of communication, an area widely considered one of the fastest growing specializations within the communication discipline.

First of its kind. We wanted this book to be a first of its kind in multiple ways. We knew that a sufficient body of scholarship existed to warrant an edited volume of literature-review essays on the topic of CMC in close, personal relationships. We reasoned that because scholarship in CMC (or if you prefer, "communication and new technologies" or simply "new media") was growing at such an enormous rate, it was only a matter of time before specialty books and courses began appearing in narrower subject areas within the broad CMC area. We wanted a book on our interest area, interpersonal communication via CMC, to be among the first to appear on the scene. We believed that if we could develop a readable book, accessible to advanced undergraduates as well as graduate students and their professors, the book would encourage the development of upper-division specialty courses in CMC beyond simply the "general course" in CMC as well as further research on CMC in close, personal relationships. Thus, our goals were two fold: to inspire further research and teaching on the subject and to inform scholars unfamiliar with this body of research about its existence and its enormous potential to assist us in understanding communication within relationships in the 21st century.

Debunk the myths. The popular press emphasizes negative aspects of online interaction (e.g., cyberporn and pedophiles trolling for unsuspecting victims). However, we knew from reading the scholarship (see the final four chapters of this book) that such a depiction was simply misleading as it failed to describe how the vast majority of users employed new technologies to meet people as well as, develop, maintain, and end relationships. We wanted our book to be one of the first books to contradict these negative, popular notions, not once or with one study, but systematically, with chapter after chapter examining diverse bodies of research.

Theory-based chapters. We wanted to develop, for the first time, a multi-theoretical book on CMC. While various communication journals had published attempts by interpersonally oriented CMC scholars to better integrate traditional communication theories (and extensions of these theories) with empirical findings (e.g., See Walther's work linking empirical findings to social information processing theory), our new book provides more in-depth discussions of communication theory vis-à-vis empirical findings. Indeed, we designed our book to integrate recent CMC theory with recent CMC findings in new lines of research (e.g., relational maintenance, online health-care provider communication, social network sites).

To these ends, we wanted each chapter to interpret a line of research via theory. Thus, our book affords readers the opportunity to survey three distinct bodies of theory: (a) theories used in face-to-face research on relational communication that also provide useful explanations for CMC in relational contexts; (b) CMC theories that, although designed to describe CMC in general, also provide explanations of CMC in personal relationships; and (c) the native theories developed within the communication discipline to explain how relational partners communicate via new technologies.

Represent the diversity of the research. While we wanted our book to cover the intellectual territory which we staked out, we also wanted the chapters individually to display the diversity of topics and approaches that exist in the relevant lines of research. Indeed, we wanted each chapter to describe relevant theory and research regarding *one distinct aspect* of CMC in personal relationships (e.g., online dating). For this reason, we selected chapters that examine a diverse array of relation types (family members, romantic partners, friends, collaborators, colleagues, and other types of partners in personal relationships), that analyzed interactions at a wide variety of communicative levels (words, language, interacts, transactions, relational stages, communication patterns across relationships and groups of relationships), and that examined messages communicated via a wide variety of media (email, skype, cell phones, twitter, texting, www, virtual networks). In sum, we elected to examine the phenomenon under study in its complexity rather than limit our examination. Indeed, at every turn, we embraced opportunities to expand rather than limit our examination of CMC in close, personal relationships.

Organizing the diversity. We settled on an organizational scheme for the chapters that we believe provides a fair representation of the diversity of approaches undertaken by scholars who study interpersonal communication enacted via CMC. We organized the chapters into four units:

1. The Power of Technologies to Influence How Relational Partners Communicate via Computer
2. Processes and Goals in Computer-Mediated Communication in Personal Relationships
3. The Mutual Influences of the Medium and Relational Contexts on Computer-Mediated Communication
4. The Dark Side of Computer-Mediated Communication in Personal Relationships

Thus, we offer chapters centered on four factors: (1) *channel factors* (e.g., interaction via online social networking utilities), (2) *process factors* (e.g., relational maintenance via CMC), (3) *relational factors* (e.g., CMC in long-distance relationships), and (4) *task factors* (online deception and surveillance). This format offers the reader perspectives from a variety of angles, thus providing insights into connections and diverse phenomenon. Each chapter provides a comprehensive review of relevant theory and research findings. Together, the chapters present knowledge obtained via a variety of research methods, theoretical orientations, and relational foci. These views are united by a common perspective—the perspective of the communication discipline, specifically from an applied communication perspective.

A word about using this volume as a textbook. Faculty members find relatively few choices for undergraduate and graduate CMC books. No previous book focused on CMC in personal relationships, an avid interest of undergraduate students and many college professors selecting textbooks for their courses in CMC. Despite increased interest in interpersonally oriented CMC research, a brief scan of available CMC textbooks and edited volumes reveals relatively few books written at the advanced undergraduate or graduate level. While previous CMC books were innovative at the time of their publication and did an outstanding job covering the breadth of CMC research, we saw a clear need for a new undergraduate/graduate book that provides an in-depth look at the growing area of interpersonal CMC research, on such topics as online dating, self-disclosure, and surveillance.

To undergraduate instructors. Our book addresses this need and offers an ideal selection for an instructor looking for a text for an upper-division, undergraduate course focused on the interpersonal aspects of CMC. Our book can readily serve as the sole or primary textbook in a course in CMC in Personal Relationships. Additionally, our book can be used as one of multiple textbooks in a traditional undergraduate class or honors seminar in CMC, where the instructor desires students to read interesting and well-written literature reviews about how users enact personal relationships via CMC.

To graduate instructors. Our book can assist professors of graduate seminars to cope with the large body of scholarship published on CMC in two ways:

- Our book provides focus (i.e., interpersonal CMC) for a narrowly drawn CMC seminar. It provides an obvious choice of textbook for a course on Interpersonal Aspects of CMC, perhaps, but not necessarily, supplemented by original research reports.
- When an instructor desires to teach a "broad sweep" seminar in CMC that offers a sampling of the theory and topical foci, our book could serve as one of multiple texts.

In sum, our book addresses the needs of both graduate and undergraduate instructors by offering a text with a narrow and popular focus of interest to both students and instructors alike. It contains multiple chapters that review original research representing a diversity of relational foci as well as methodological and theoretical approaches. The chapters, each written specifically for this book, review cutting-edge research that will remain relevant for many years to come. In sum, our book is designed to function as the core text for a specialized graduate seminar in CMC and as a primary text for undergraduate courses focused on CMC in personal relationships.

Selecting the chapters. Whenever editors attempt to gather literature review essays written especially for a volume they are faced with the choice of two traditional processes: editors either (a) conduct research to determine the finite list of scholars who are writing extensively on the topic of their book and then invite review essays from these "heavy hitters" or (b) issue a call for chapter proposals that allows the best proposals to emerge in open competition from junior as well as senior scholars in the field. To gather the strongest possible author list, we did both.

We conducted a detailed content analysis of the Communication Institute for Online Scholarship (CIOS) database ComAbstracts using appropriate key words. ComAbstracts is a searchable, online data base containing the bibliographic entries (as well as typical abstracts and key words) for more than 50,000 articles published in 92 scholarly journals and annuals in the communication discipline. The search yielded a finite list of well-published scholars who examine interpersonal communication via CMC. These nine scholars were contacted individually and invited to prepare a chapter. We are delighted that so many accepted our invitation. Second, we issued an open call for proposals that we distributed widely (newsletters and/or list serves of nine professional associations of communication scholars). We asked for brief proposed literature reviews examining "relational, task, or channel influences on interpersonal CMC."

This book contains essays authored or coauthored by names that will look very familiar to any scholar who reads or writes in this area of research. The remaining essays were selected from among the 29 proposals we received in response to our open call. We received many excellent proposals—far more than we could use. Nonetheless, we experienced little difficulty selecting the essays. We looked for unique, effective, and theoretical treatments of distinct bodies of research and then simply selected the well written proposals that offered us more diversity than others.

Our author list. Our author list includes new, "rising stars" in the scholarly world of CMC as well as most of the best CMC scholars writing about interpersonal communication today. Our author list represents an eclectic mix of published scholars in the communication discipline who hold professorial positions at primarily graduate institutions, and who vary in age, sex, geographic location, and type of affiliated institution (public versus private). In addition, the chapter authors represent diverse career stages from graduate students to tenured full professors. We believe such diversity benefits our edited volume as the authors provide diverse viewpoints that represent the diversity of thinking in the discipline today.

Reviewing the work. Ultimately, it matters little what we set out to do or how we came to our decisions. What will matter, ultimately, is the opinion of the readers and users of this edited volume. We hope that users of this book find not simply one or two but almost all of the chapters very useful in augmenting their understanding of how everyday people communicate with their relational partners via CMC. Here is our synopsis of the chapters that we found so compelling:

In Chapter One, Bryant, Marmo, and Ramirez examine social network sites (SNSs), such as Facebook, Twitter, and MySpace, from a functional perspective. These authors focus on central functions of SNSs, such as how individuals use them to maintain relationships, initiate relationships, seek social information about potential relationship partners, construct individual or multiple identities, manage interpersonal impressions and relationships, and enact metacommunication.

In Chapter Two, Child and Petronio consider the viability of using communication privacy management theory to investigate the privacy-regulating decisions people make in relationships and how they manage communication and information flow within a variety of new media contexts, including social networking sites, microblogs, and other Internet-based technologies.

In Chapter Three, Toma and Hancock offer a general theoretical framework for online self-presentation that examines the role of both psychological antecedents and of the communication medium in shaping self-presentation. They then apply this framework to the distinct context of online dating and provide empirical support for it by reviewing findings from both a large study on this topic conducted by the authors and other relevant studies.

In Chapter Four, Balard-Reisch, Rozzell, Heldman, and Kamerer focus on the development, maintenance, and dissolution of interpersonal relationships within various forms of microchannel communication, such as instant messaging, SMS, and Twitter. These authors explore social construction theory, systems theory, uses and gratifications theory, social exchange theory, social penetration theory, and their corollary theories as potential frameworks to increase our understanding of relational processes among individuals who use microchannel communication media.

In Chapter Five, Sanders and Amason examine the impact of CMC on relationships formed and primarily grounded in face-to-face interaction. The authors explore several issues within this relational context, including impression management, self-disclosure, communication competence, and communication apprehension. Sanders and Amason argue that the nature of CMC provides a unique context that alters our traditional understanding of these concepts and provides theoretical challenges for CMC and interpersonal researchers.

In Chapter Six, Tong and Walther examine current trends in relational maintenance surrounding contemporary CMC technologies such as email, blogs, social network sites, and microblogging. These authors also focus on relationships among intimates and families who cohabitate or are in frequent physical contact, but use CMC to supplement relational maintenance as well as long-distance relationships of various kinds. They conclude by providing recommendations for future research in terms of studying CMC applications and their effects on relational outcomes.

In Chapter Seven, High and Solomon look at social support in CMC relationships. Specifically, these authors discuss how different modes of CMC provide fundamentally different contexts for social support. They begin by providing various definitions of social support and computer-mediated communication, and then they focus on various types of support and social support processes that are frequently observed within several CMC contexts, including online support groups, instant messaging, discussion boards, and virtual worlds.

In Chapter Eight, Wright and Muhtaseb assess interpersonal issues related to the giving and receiving of social support within computer-mediated support groups. Towards that end, he explores the link between social support and health outcomes, relational dilemmas surrounding the provision of social support, and advantages and disadvantages of online support groups/communities. In addition, the chapter discusses several theoretical frameworks used in previous empirical work and offers suggestions for future research.

In Chapter Nine, Kim and Dindia assess the state of the literature on online self-disclosure and suggest future directions for online self-disclosure research. They do this by examining the qualities of online self-disclosure and reviewing empirical findings regarding the relational differences between features of CMC self-disclosure and face-to-face self-disclosure. The chapter concludes with theoretical and methodological suggestions for future computer-mediated self-disclosure research.

In Chapter Ten, Turner and Reinsch focus on the constructs of multicommunication and interpersonal presence. They take an in-depth look at the definitions of both constructs and also link them to existing computer-mediated communication research and theory. The authors point out that the study of multicommunication is in its relative infancy, and they point to several promising future areas of research as well as methodological and theoretical challenges.

In Chapter Eleven, Edley and Houston offer an interdisciplinary perspective focusing on how new communication technologies intersect with family communication. These authors focus on cross-disciplinary literature that connects work-life studies with the adoption and uses of new communication technologies to reveal how family relationships and responsibilities are managed.

In Chapter Twelve, Johnson and Becker investigate the enactment of friendship through computer-mediated communication. The authors begin by looking at traditional features of friendship and how these have been influenced by the advent of CMC technologies. In addition, they examine the concept of flexibility in friendships and relational maintenance process within geographically close and long-distance friendships through various forms of CMC. Finally, they discuss several promising directions for future CMC research on friendship.

In Chapter Thirteen, Maguire and Connaughton bring together theory and research from the distanced relationship and social presence literatures in an effort to understand how partners maintain their distanced relationships. Specifically, the authors argue that distanced relational partners utilize communication technologies that enhance (or diminish) feelings of social presence to maintain and sustain their relationships. These authors focus on a variety of issues related to social presence theory and their implications for long distance relationship and CMC research.

In Chapter Fourteen, Avtgis, Polack, Staggers, Wieczorek explore the multitude of ways provider-patient relationships have been influenced by computer-mediated communication. The authors begin by discussing the history of earlier technologies on provider-patient interaction and then discuss a variety of current issues surrounding the Internet and related communication technologies and their effects on provider-patient relationships. These issues include demographic differences in CMC technology-use among patients, patient and provider expectations surrounding CMC usage, financial and legal issues, and various benefits and limitations of CMC technology for healthcare.

In Chapter Fifteen, Mesch and Frenkel provide a comprehensive review of studies dealing with the effects of information and communication technologies on families with adolescents. The authors frame the review with the context of family systems theory and a life span developmental approach. The authors contend that these technologies are a challenge for the preservation of family boundaries because they provide new sources of information and may compete with parental values. In addition, the authors highlight several promising future areas of research stemming from these perspectives.

In Chapter Sixteen, Hans, Selvidge, Tinker, and Webb review three lines of research dealing with gender performance and interpret the findings via performative theory while focusing on gendered blogging, gender-bending, and cybersex. These authors examine both public and scholarly discourses on CMC and gender performance as well as viewing online performances of gender.

In Chapter Seventeen, Dunbar and Jensen discuss deception within computer-mediated relationships. They begin by defining deception and how it occurs within computer-mediated environments as well as several theoretical frameworks drawn from the deception literature. The authors then move to several relevant issues, including deception detection and CMC, motives behind using CMC for deception vis-à-vis face-to-face relationships, and features of CMC that may facilitate the discovery of deceptive behaviors, and effects of deception in interpersonal relationships.

In Chapter Eighteen, Phillips and Spitzberg delve into the dark side of computer-mediated relationships by focusing on social network site surveillance and obsessive relational intrusion within social networking sites. Moreover, the chapter deals with a variety of theories relevant to understanding these trends and their implications for relationship development, including identity formation, online disclosure and intimacy development, dialectical theory, and privacy management theory. Finally, the authors offer a critique of existing literature in this area and directions for future research.

Finally, in Chapter Nineteen, Schrock and boyd summarize and provide their perspectives on research dealing with two deleterious aspects of computer-mediated relationships: sexual solicitation and cyberbullying. Specifically, they discuss legal aspects of these online activities among young Internet users, characteristics of perpetrators, Internet-initiated online encounters, and characteristics of victims of online sexual solicitation and cyberbullying. In addition, the authors examine online contexts for the problematic activities, effects on victims, and directions for future research.

There are a number of people who are responsible for this project, and we would like to acknowledge them here. First, we want to thank all of the authors who contributed their expertise and experience to this volume. We feel honored that so many well-respected CMC and interpersonal communication scholars were willing to participate in this project. We are very grateful for their contributions. Next, we would like to thank the publication team at Peter Lang for their suggestions and expertise in terms of putting this volume together. In particular, we would like to thank Mary Savigar, Senior Editor, for her assistance and support of this project. Finally, we sincerely hope that this volume will be an important catalyst for CMC stimulating discussion and future research among interpersonal scholars, graduate students, and undergraduate students.

References

ICA (2008). *2008 International Communication Association publication report.* Washington, DC: ICA.

PART I

The Influence of Technology on How Relational Partners Communicate Online

CHAPTER ONE

A Functional Approach to Social Networking Sites

Erin M. Bryant

Jennifer Marmo

Artemio Ramirez, Jr.

The widespread use of social networking websites (SNSs) is one of the most groundbreaking communication trends to emerge in recent years. Since its creation in 2004, sites such as Facebook have become immensely popular among college students. Many SNSs continue to experience exponential growth. Facebook, for example, reached 100 million active users in August 2008 and proceeded to quadruple this membership base to surpass 400 million active users by July 2010 (Facebook.com). In addition to maintaining astronomically high membership rates, SNSs also appear to be part of user's daily schedules. In one study assessing Facebook use, Ellison, Heino, and Gibbs (2006) found that participants reported using the site an average of 10 to 30 minutes each day, with 21% of participants spending more than an hour on the site every day. As a result, high membership and usage rates suggest SNSs hold significant power as a relational and social tool for users.

Coinciding with the growth in membership have been significant advances in research that suggest SNSs may serve important interpersonal and relational functions for users. For instance, although the growing body of literature investigating SNSs is still in its infancy, most suggest one of their central functions is the maintenance of existing off-line relationships. Other research, however, reports SNSs may be used to initiate relationships, seek social information about potential relationship partners, construct individual or multiple identities, manage interpersonal impressions and relationships, and enact metacommunication (via, for example, comment postings, photographs, and relationship status indicators). In addition, less studied functions have also emerged via anecdotal evidence or popular press reports such as how SNSs may serve to aid in relational reconnection (or reconnecting with relational partners for one's past). Although SNSs may fulfill other functions, we propose in this chapter that these are their central functions.

The advantage of taking a functional approach to SNSs, and thus organizing the literature based upon each function, is that it shifts the focus of the chapter from specific SNSs (e.g., Facebook or MySpace only) or contexts (e.g., business/professional SNSs) onto what communicative functions SNSs are used to achieve (for a discussion, see Walther & Ramirez, 2009). This approach borrows from the area of nonverbal communication where it is more clearly articulated through multifunctional "meta-principles" (e.g., multiple functions may be achieved by a single cue; multiple cues may achieve the same function). It also allows for the integration of seemingly disparate lines of research that fulfill the same underlying function. This latter point is particularly important since research based on niche SNSs–such as LinkedIn, Twitter, LiveJournal, and others–may identify functions similar to those of the more popular Facebook and MySpace sites allowing for their synthesis.

The primary purpose of this chapter then is to use these aforementioned functions as guiding principles in discussing SNSs and provide an assessment of these functions as reflected in existing research. A secondary focus of this chapter is to provide a review of literature relevant to each function that may guide future research on SNSs. Much in the same manner that general research functions such as relational communication, social influence, and social support help guide research in several content areas in the field of communication, the functions identified in this chapter may provide interested scholars a new manner of conceptualizing the study of SNSs.

The Primary Relational and Social Functions of SNSs

Relationship Initiation

Although people primarily report using the SNSs to maintain or deepen existing relationships, a secondary function of SNSs is the initiation of new relationships. The term *social networking site* implies that networking will occur to some extent; however, the primacy of the relationship initiation function is largely contingent on the goals of a particular user and the structure and purpose of each specific SNS.

Anonymous SNSs are generally most conductive towards true networking and initiating relationships with previously unknown persons. SNSs such as LiveJournal and Twitter focus on blogging, journaling, and story writing and do not necessarily require users to reveal their offline identity. As a result these SNSs are not structured around "friend" networks, but rather, allow users to subscribe as a "fan" of someone whose writing they enjoy, or "follower" of someone that frequently blogs on a topic of interest. Although users might know many of their contacts on a personal level in anonymous SNSs, these sites provide increased opportunities to broaden social networks by initiating relationships with users that possess common interests.

Similarly, there are a large number of niche SNSs aimed at very specific populations such as pet owners (Catster and Dogster), Christians (MyChurch), music fans (Last.fm), and social activists (Change.org). Interaction on niche SNSs might aid in facilitation of *specialized friendships* (Wellman & Gulia, 1999) by enabling users to form connections based on interests their offline friends may or may not share. Indeed, Baym and Ledbetter (2009) found that shared interests predicted the initiation of friendship on the music SNS Last.fm; however, shared interests did not lead to the development of a more personal relationship. SNS relationships initiated based on shared interests generally remain weak ties; however, partners can develop a more personal focus by utilizing multiple modes of communication.

Some research suggests that people form more cross-sex friendships online than offline (Parks & Roberts, 1998), a finding that has been mirrored in regards to the SNS Last.fm (Baym & Ledbetter, 2009). It might be easier to initiate cross-sex friendships in an anonymous online environment free of the scrutiny of individuals that might judge the relationship or suspect that time spent with a cross-sex friend equates to romance. Similar research is needed to determine whether SNSs structured around offline networks would impose offline relational rules that dissuade the initiation of cross-sex friendship.

As previously mentioned, the primary goal of these SNSs based on offline networks appears to be communicating with and maintaining existing social ties, even if these ties are extremely weak (Ellison, Steinfield, & Lampe, 2007). SNSs might also be used to initiate personal relationships with peripheral members of a user's social circle. As noted by Ellison, Lampe, and Steinfield. (2009), we regularly meet new people that we would like to learn more about yet we often are not willing to exert the effort or engage in the risk necessary to exchange phone numbers and personal information. Searching for these individuals on SNSs and adding them as a "friend" provides an easy way to initiate a weak social tie that ensures continued contact with that person. This within network relationship initiation process might be relatively common given that more than half of Facebook and MySpace users in Raacke and Bonds-Raacke's study (2008) reported using the sites to make new friends.

SNSs might even serve as a relatively risk-free way to initiate romantic relationships. People repeatedly deny using SNSs for overt romantic invitations and random dating searches (i.e., Stern & Taylor, 2007), although, this does not mean SNSs are not used for dating or romantic purposes. Indeed, our currently unpublished focus group data suggests romantic initiation might be common within existing social circles. In our data, Facebook users denied conducting random searches for dating purposes, yet willingly admitted they use SNSs to find out if they are compatible with someone they know socially such as a friend of a friend or a classmate. If a person's SNS profile lists them as single and makes a positive impression, students said they might pursue a romantic relationship by asking the person out on a date. SNSs might therefore serve as a valuable uncertainty-reduction tool for users attempting to determine their compatibility with someone they have already met.

Relational Maintenance

Relational maintenance refers to the active and routine behaviors individuals use to sustain their relationships at a desired state (Canary & Stafford, 1994). Key markers of successful interpersonal relationships such as trust, liking, and commitment have been found to increase when partners engage in more strategic and routine maintenance behaviors (Canary & Stafford 1994; Stafford, Dainton, & Haas, 2000; Weigel & Ballard-Reisch, 1999). Sometimes individuals are unwilling or unable to perform relational maintenance strategies, which can lead to the deterioration of a relationship.

As technology continues to advance new opportunities for relational maintenance are enabled. Early research assumed email and other text-based computer-mediated communication (CMC) were useful for relaying task-related communication, yet could not support the nonverbal cues necessary for relational communication. Walther's (1996) social information processing (SIP) theory articulates the alternative viewpoint that CMC users find ways to communicate rich personal information using the cues at their disposal. SIP has arguably become the guiding theory of CMC research, as numerous studies demonstrate CMC is a rich site of relational communication. In fact, email and instant messaging are primarily used for relational maintenance purposes and use of these media has been linked with increased closeness in relationships (Cummings, Lee, & Kraut, 2006).

The prevalent use of online media to enact relational maintenance led researchers to explore the potential relationship maintenance functions of SNSs. Human and Lane (2008) predicted email would be the primary mode of communication between friends seeking to maintain their relationship and surprisingly found that most participants (35%) checked "other" and wrote in "Facebook" on their survey. Other research has found that SNS use is linked with increased levels of intimacy and liking (Kim & Yun, 2007), closeness and trust (Human & Lane, 2008; Wright, 2004), and willingness to share information (Dwyer, 2007). Pearson (2009) argues that the performative nature of SNSs creates an intimate atmosphere that encourages the flow of information and allows users to feel maintained relationships. Clearly, SNSs are becoming increasingly popular venues for individuals to maintain relationships.

SNSs are embedded in users' daily lives as means to maintain relationships with a variety of people. Indeed, relational maintenance has been cited as the primary function of various SNSs such as MySpace and Facebook (boyd, 2008) and the popular Korean SNS Cyworld (Choi, 2006; Kim & Yun, 2007). Many SNSs such as Facebook are structured around the display of existing offline social networks so it is no surprise people primarily use Facebook to maintain or solidify pre-existing offline relationships and keep in touch with relational partners (Ellison et al., 2007; Lampe, Ellison, & Steinfield, 2006; Joinson, 2008; Hargittai, 2007). SNSs enable users to broadcast updates regarding their lives to a large network of "friends" and also to stream updates about these friends in a single location. SNSs serve as an effective way to stay in the loop regarding social information such as a friend getting engaged, moving, or making other significant life changes. Possessing this information can help two people maintain the minimal level of contact necessary to feel their relationship is being maintained even if they rarely converse in a one-on-one manner.

SNSs serve as powerful relational maintenance tools for many reasons. SNSs provide an easy way for users to locate friends and maintain relationships with the multitude of people they might not see on a regular basis (Dwyer, 2007). Most SNSs provide reminders when a friend has a birthday and broadcast updates when a friend posts news of success or hardship. Knowing these events are occurring enables users to enact relational maintenance strategies such as offering congratulations or wishing friends happy birthday. In fact, SNSs have been noted to reduce the costs associated with relational maintenance (i.e., time, effort, etc.) and therefore enable users to maintain relationships with an extremely large number of people.

Furthermore, most SNSs are asynchronous, meaning users do not have to be on the site at the same time in order to communicate. The asynchronous nature of SNSs can therefore facilitate relational maintenance between users with different schedules and also provide users with increased control over their impressions by allowing extra time to compose and edit messages (O'Sullivan, 2000; Walther & Boyd, 2002). Cost is also cited as a benefit of SNSs and other free online media; however, it may have limited explanatory power given that most CMC users also have access to cell phones, text-messaging devices, and other communication tools (Walther & Ramirez, 2009). It is therefore safe to assume that people use SNSs because they enjoy doing so and are able to efficiently gratify needs (Bryant, 2008).

Our current understanding of relational maintenance on SNSs rests on knowledge accumulated through the secondary findings of numerous studies. Existing research clearly demonstrates that SNSs are functioning to maintain relationships; however, few studies to date have focused entirely on the relational maintenance uses of SNSs. Thus, we agree with Walther and Ramirez's (2009) argument that, "the greatest utility of social networking systems has yet to be explored. These systems provide a dramatically new way to enact *relational maintenance*" (p. 302). Relational communication scholars should devote extensive attention toward unpacking the relational maintenance function of SNSs as the sites should continue to alter the way people maintain relationships into the foreseeable future.

Relational Reconnection

Whereas accumulated research suggests that the primary function of SNSs may be to facilitate the maintenance of existing off-line relationships (boyd, 2008; Ellison et al., 2007; Lampe et al., 2006; Joinson, 2008; Hargittai, 2007), a less studied function is that of how SNSs serve to aid in relational reconnection. Popular press reports of how users employ SNSs in locating long = lost friends and family members are becoming increasingly common (e.g., Porter, 2009). In fact, a recent workplace survey found that an overwhelming majority of respondents (82%) reported that reconnecting with family or friends was their primary motivation for using SNSs (Steelcase, 2008). With respect to the present chapter, relational reconnection refers to the re-establishment of a previously existing relationship that, for one or more reasons, the parties involved lost contact with each other.

SNSs may aid users in fulfilling the relational reconnection function in various ways. As with relationship maintenance, SNSs and the tools they offer users make them ideal for locating and re-establishing contact with relational partners from the past. For instance, users can utilize search tools to locate target profiles. Users may also locate targets indirectly (and possibly accidently) by examining photographs and wall postings on others' profiles within their own personal network to seek contact with targets. Many SNSs such as Facebook even provide users with a list of "people you might know" based on overlapping social networks. In addition, the various communication tools available on most SNSs make it possible to initiate contact fairly efficiently once a target is located and begin testing the possibility of reconnecting.

Unfortunately, academic research has lagged behind anecdotal and popular press reports of relational reconnection. To date, no published academic study documenting relational reconnection exists. However, recently collected data from six college campuses across the United States currently under analysis by one of the co-authors of this chapter provides some preliminary insight into the relational reconnection process. Overall, relational reconnection among college students is quite common. Approximately 75% of participants reported involvement in at least one relational reconnection attempt within the preceding year; almost 60% of the participants reported being recipients of such attempts. Not surprisingly, the relationship targeted most commonly for reconnection, approximately 80%, was that with a friend from the past. Of these friendships, the majority were same-sex. Although not conclusive, these preliminary findings suggest that the relational reconnection function is worthy of further academic study in order to better understand the role of SNSs in facilitating this process.

Identity Experimentation

An individual's identity is complex and multi-faceted; people share similar traits, beliefs, values, characteristics, and interests with different people. People are generally drawn to form social ties with individuals they see as possessing similar traits and identity markers (Feld, 1981). In an online world, however, a person's identity is "mutable and unanchored by the body that is its locus in the real world" (Donath, 1998, p. 29). The corporeal body becomes detached, leaving a person to not fully exist online until he/she writes him/herself into being through "textual performances" (Sundén, 2003); word choice, grammar, spelling, and sentence structure are the text through which we create our online identity. "You are what you type" captures the essence of identity experimentation (Slater, 2002, p. 536).

In fact, SNSs have been argued to offer limitless ability for people to shape, alter, or completely change their identity; however, this sentiment has been challenged.

Early CMC research examined identity experimentation in anonymous online settings such as chat rooms and bulletin boards (Rheingold, 1995; Surratt, 1998; Turkle, 1995). In these environments individuals were found to predominantly engage in role-playing by being someone else, acting out underlying aggressions, using anti-normative behavior, or putting on personae differing from their "real-life" identity (Stone, 1996; Turkle, 1995). The disembodiment and anonymity of these sites was concluded to allow these forms of identity experimentation. Reinventing one's identity can be especially empowering for individuals who feel disadvantaged in face-to-face settings (McKenna, Green, & Gleason, 2002). Identity experimentation in anonymous SNSs also provides an outlet for the expression of "hidden selves," those bubbling under the surface and not shared with most people (Suler, 2002); as well as the exploration of various non-conventional identities we are not meant to be (Rosenmann & Safir, 2006). In sum, anonymous SNSs might provide users a venue to essentially try on different identity markers to experiment with who we are, who we wish to be, and who we might become.

Recently, research has shifted to focus on identity experimentation on anonymous sites such as MySpace, Facebook, and dating sites. There is less space for experimentation on these sites because they are structured around offline networks and identities. The function of identity experimentation is less prevalent in anonymous sites for a few reasons. First, these sites are typically high in offline-online integration (Ellison et al., 2007) because a person's social network will know if they are being untrue to his/her identity (Donath & boyd, 2004). Additionally SNSs are designed to facilitate the exchange of information requiring honesty from users; deviance from social norms where users are expected to present their "real" selves may be punished or ridiculed (Zhao, Grasmuck, & Martin, 2008). Honesty is particularly valued on dating websites where people are attempting to find true love and wish to avoid unpleasant surprises if they decide to pursue a relationship by meeting a person face-to-face (Greene, Derlega, & Mathews, 2006; Ellison et al., 2007). For these reasons, non-anonymous website users engage more actively in impression management, the next function to be discussed, than in identity experimentation.

In sum, identity experimentation is one function of SNSs. Online users are like actors playing a role; they can claim to be whomever they want, whenever they want (Pearson, 2009). Although questions and disagreements are raised regarding manufactured and mediated identities, identity experimentation is popular in anonymous SNSs and likely occurs to a lesser extent in all SNSs.

Impression Formation and Management

Impression formation and management is another critical function of SNSs (boyd & Ellison, 2007) pertaining to the methods people employ to control their own image and make inference regarding other users. As Goffman (1959) observed, individuals strive to influence how others perceive them by minimizing the appearance of characteristics contrary to their idealized self. Impression management occurs in all settings; however, it is becoming an increasingly prominent topic in regards to SNSs and other CMCs.

SNSs are popular venues for young adults to interact via the construction of profiles that "(re)present their public persona (and their networks of connections) to others" (Acquisti & Gross, 2006, p. 2). Impression formation and management occurs in SNSs at both an individual and public level. On an individual level, SNS users attempt to manage their impressions and exert control over their public image by crafting a profile, displaying their likes and dislikes, posting photos, joining groups, and altering the appearance of their profile to fit an idealized notion of their self (boyd & Heer, 2006; Lampe, Ellison, & Steinfield, 2007; Tufekci, 2008). Facebook users generally believe their profiles are accurate and positive representations of their identity (Lampe et al., 2006) demonstrating that non-anonymous SNSs seem to make people more "realistic and honest" (Ellison et al., 2006). Deception and misrepresentation, however, are major concerns for SNS users.

The most common form of deception in SNSs is "stretching the truth" regarding positive attributes to present an attractive and ideal self (Gibbs, Ellison, & Heino, 2006). Indeed, SNS users go to great efforts to create profiles that reveal certain characteristics of their identity, often attempting to exhibit desirable and attractive aspects and downplay less attractive ones (boyd, 2008; Kim & Yun, 2007). Yurchisin, Watchravesringkan, and McCabe (2005) found that people often attempt to hide undesirable features such as being overweight or short. Moreover, many SNSs allow users to limit the viewing of their profile to select individuals enabling them to tailor their identities to particular audiences. SNS users present "highly socially desirable identities" (Zhao et al., 2008, p. 1830) with most engaging in some form of self-enhancement (Gossling, Gaddis, & Vazire, 2007; Ellison et al., 2007). By strategically selecting how and what to convey to certain receivers users are able to enhance their self-image.

SNS users are often judged by the company they keep, meaning impression formation and management also involves inherently social processes (Mazer, Murphy, Simmons, 2007; Walther et al., 2008). Third party perspectives are often considered to be objective and are therefore used as "a signal of the reliability of one's identity claims" (Donath & boyd, 2004, p. 73). For example, Walther and colleagues (2008) found that if a person's friends were attractive the target was more likely to be found physically attractive. Complimentary and pro-social statements by friends enhanced the profile owner's social and task attractiveness, as well as the target's credibility whereas negative statements depicting normatively undesirable traits raised male attractiveness while lowering female attractiveness. Furthermore, number of friends can have a significant impact on social attractiveness (Tong, Van Der Heide, Langwell, & Walther, 2008). Walther, Van Der

Heide, Hamel, and Shulman (2009) found that friends' comments overrode an individual's comments about him/herself, pinpointing the importance of friends to an individual's impression formation and management. Similarly, Zhao et al.'s (2008) results revealed a triangular relationship between desires/interests of user, displayed friends/mates, and the audience. Impressions of a person's identity can be greatly hindered or enhanced by friends and their comments; we are known by the company we keep.

Identity is not an individual characteristic, but rather a "social product, the outcome of a given social environment" (Zhao et al., 2008, p. 1831). Impression management is a necessary component to social interaction as identity evolves and changes to reflect social networks and communities. Additionally, with SNSs being used for legal, employment, academic, athletic, and admission purposes users need to be aware of the impressions they are giving off. For this and other reasons, forming and maintaining impressions appears to be a prevalent function of SNSs. Additional research is needed to examine the extent that SNS impressions carry over into offline interactions by reflecting and shaping users' reputations in the offline world.

Information Seeking

Another important function of SNSs involves information seeking. Ramirez, Walther, Burgoon, and Sunnagrank (2002) conceptualize information seeking as the goal-driven "pursuit of desired information about a target" (p. 217). SNS users possess near instant access to the wealth of information provided by friends and other members of their social networks. Additionally, users can select how to use the information they find, take it out of context, and/or reproduce it without the knowledge of the person that posted it (boyd, 2007). As a result, SNSs essentially function as archives of social information at the disposal of users in pursuit of information about a target. Unfortunately, there is no way for SNS users to know whether their personal information is being sought for noble or nefarious purposes.

Early theoretical approaches, collectively coined as the "cues filtered-out" perspective (Culnan & Markus, 1987), suggested that CMC would hinder interpersonal communication processes such as information seeking. More recent theoretical approaches such as social information processing theory suggest that people will find ways to utilize the contextual cues afforded by online environments to follow the same information-seeking strategies they employ in offline settings (Walther, 1994, 1996).

According to uncertainty-reduction theory (Berger & Calabrese, 1975) people try to relieve uncertainty by seeking information that allows them to assign predictable characteristics to people. To do so, people employ a number of tactics such as active, passive, interactive, and extractive strategies (Ramirez et al., 2002). Active strategies involve asking a third party for information about a person of interest. Passive strategies include observing a person's actions and making inferences to collect information about them. Interactive strategies are utilized by directly interacting with the person of interest and engaging in reciprocal self-disclosure. Perhaps unique to online environments, extractive strategies involve using search engines like Google to indirectly gather multiple sources of stored information available about a person of interest. Extractive strategies differ from passive strategies because the person of interest often has no control over the information and/or does not know it is being collected and used for purposes they did not originally intend.

The multiple forms of contextual cues provided by different SNSs might enable all of the above information-seeking strategies. Sanders (2008) examined how active, passive, and interactive strategies function to reduce uncertainty on Facebook and found that interactive strategies most effectively reduced uncertainty, passive strategies were only mildly effective as reducing uncertainty, and active strategies were unimportant in reducing uncertainty on Facebook. Additionally, intense Facebook users reported more confidence in their ability to reduce uncertainty with information collected on Facebook, signaling that intense users might view Facebook as a valuable source of social information. SNSs might also enable a mixture of passive and active strategies to the extent that users gather information that third parties have posted on a person of interest's profile, yet do not directly ask that third party for information.

SNSs are created and maintained by the millions of users who provide information and interact on the sites. This information serves as potential data to be collected by users and used for various purposes. SNS users who are skeptical of other members' identity claims might strategically seek information to confirm or discredit claims made by other users or new relational partners (Walther et al., 2009). SNSs can also be used to seek information concerning what a user's existing friends are doing as well as with whom these friends associate. Specific profile elements such as a person's contact information, current status, schedule, or hobbies might also be sought. Users might even seek information such as the interests or relationship status of a romantic interest.

As SNSs increase in popularity they will likely continue to serve as a primary source of information about users that create profiles. Continued research is needed to explore how the information-seeking experience of SNSs is similar to and distinct from other information-seeking tools such as search engines, email, text messaging, chat rooms, and instant messenger. Given that SNSs are often only one component of larger relational dynamics, research should also aim to understand how SNSs operate in conjunction with other information seeking tools. Research by Lampe and colleagues (2006) demonstrates that Facebook users typically use the site for social searching (finding out more information about current contacts) rather than social browsing (randomly searching for new contacts). Future research should apply the concept of searching versus browsing within the context of existing networks. How often do SNS users engage in goal-driven information-seeking and how often do they engage in passive browsing such as following newsfeed updates. Such an understanding will help provide a better understanding of the ways SNS users obtain information.

Metacommunication

Metacommunication is assumed to play an important role in all forms of relationships yet it is currently understudied in many contexts, including SNSs. The term *metacommunication* can be loosely defined as communication about communication (DeVito, 2001). More specifically, metacommunication addresses that communication is a complex process involving numerous situational and contextualization factors that determine the availability and appropriateness of specific cues. Metacommunication also refers to the unarticulated meanings transmitted through communication that reinforce, contradict, or distract the surface meaning of a message (Young, 1978). Message attributions are made based on the sender's chosen mode of communication, sensitivity to the setting, and use of all available cues. The numerous cues enabled by SNSs could facilitate several forms of metacommunication.

In many cases, metacommunication functions as actual talk or communication about a previous message. On SNSs, metacommunication manifests in the form of comments made in regard to other users' wall posts, status updates, and notes. Facebook even installed a "Like" feature that allows users to communicate whether they approve of the comments and status updates made by friends. Many SNS posts provoke additional comments from multiple users, essentially forming a conversational thread akin to those seen in blogs and newsgroups. This form of interaction can therefore be examined as a form of metacommunication; however, research has yet to examine this potentially valuable line of research.

SNSs might serve a particularly important role in romantic-relationship metacommunication. Romantic partners who engage in more metacommunication report higher relational satisfaction than partners who engage in less metacommunication. Mann (2003) found that Facebook users display their romantic-relationship status by indicating whether they are "single," "in a relationship," "in an open relationship," or even "in a complicated relationship." Samp and Palevitz (2008) point out that changing one's Facebook relationship status has become an important public ritual akin to the traditional bonding and terminating stages described by Knapp and Vangelisti (2005). It is not uncommon to see a couple's relationship status change numerous times over the course of a single night, indicating that SNS users are aware of the impact they can make by publicizing their relational woes.

SNS users might also monitor the "relationship status" of their partner to ensure their relationship is viewed as official. Existing research suggests metacommunication is prevalent during relational turning points such as the beginning or ending of a relationship (Baxter & Bullis, 1986). It therefore makes sense that SNS users stress the importance of relationship status indicators. A person's relationship status on SNS profiles should be updated to reflect offline communication such as agreeing to begin or end a monogamous relationship; however, the manipulation of SNS relationship status can also provoke offline metacommunication if one partner's status does not coincide with the other partner's vision of the relationship. A person might feel uncomfortable if the person they went on one date with suddenly declared them as a romantic partner on SNS. Similarly, a person might feel hurt or confused if the person they are dating does not declare their relationship via SNS. In many cases, couples change their status to "in a relationship" or "single" at nearly the same time, suggesting that some form of offline metacommunication occurred in the form of a state-of-relationship talk.

SNSs also provide an underexplored site of metacommunication regarding the multiple attributions that can be made based on any given message. Indeed, SNS users learn to read a great deal of meaning into seemingly innocuous behaviors with the unfortunate side effect that inaccurate message attributions can be made. Facebook users in Stern and Taylor's (2007) study reported using the site to monitor the fidelity of their current romantic partner yet noted that doing so can cause problems when comments or photos are taken out of context. Knowing this, some SNS users might purposely manipulate their profile to make romantic partners and friends jealous by publicly communicating with someone their relational partner dislikes, removing that person from their "top friends" list, or posting photos that will arouse jealousy. SNSs can even be used to enact revenge on a relational partner (romantic or friendship) by posting pictures and comments that display the user has moved on, or even by sharing information that will destroy the reputation of their former partner. In this way, SNS profiles can take on a life of their own and indirectly communicate messages that may or may not have been intended by the user.

There is much room for metacommunication research in the field of relational research at large and in the specific context of SNSs. The prevalence of SNSs as a site of relational communication, coupled with the potential ambiguity of messages in CMC contexts, creates a need to unpack the connections of SNSs and metacommunication. Indeed, we believe that the metacommunication functions of SNSs could provide a fruitful line of research for relational communication scholars.

Conclusion

SNSs are playing an increasingly important role in the communication patterns of young adults and are even beginning to gain reputation as a viable networking tool for adults. It is important to continue questioning the ways that SNSs are impacting personal relationships; however, it is equally important to note that SNSs are only one development in the ongoing evolution of communication technology (Bargh & McKenna, 2004). As such, this chapter aimed to shift the focus of SNS research away from describing individual sites by reorganizing SNS research using a functional approach.

This chapter discussed many of the primary functions of SNSs: relationship initiation, relational maintenance, relational reconnection, identity experimentation, impression formation and management, information seeking, and metacommunication. This chapter focused on describing the relational and pro-social functions of SNSs; however, SNSs are clearly not utilized entirely for pro-social relational purposes. For example, research has shown that people use SNSs as a complementary source of political information (Kushin & Yamamoto, 2009). In fact, a 2007 study by the Pew Center found that 27% of young adults sought information regarding the 2008 presidential election from SNSs (Kohut, 2008). The political information sought from SNSs is likely of a more social nature than other political sources because content is created, supplied, and endorsed by members of their social network. Similarly, people commonly report using SNSs for diversionary and entertainment purposes; SNSs relieve boredom, kill time, and are a procrastination tool for students wishing to postpone work (Bryant, 2008). Diversion and entertainment uses are not commonly grouped as a relational function; however, the relational communication and socialization opportunities provided by SNSs might actually drive this function. SNSs might be yet another venue where people virtually "hang-out" with their friends, even if those friends are not physically present.

There are undoubtedly many anti-social functions of SNSs as well. For example, surveillance and "Facebook stalking" can be used to keep track of friends' actions and maintain relationships, yet true stalking also occurs on SNSs. Indeed, the openness that many users display might attract cyber-stalkers who can use SNSs to harass, threaten, or bully other users. Privacy is also an issue as SNS profiles are being used by admission officers, employers, and even legal officials as a way to judge the character of applicants. Furthermore, many of the SNS social functions rely on users' willingness to disclose information; however, privacy and safety concerns might limit the information users are willing to provide. SNSs are increasingly being infiltrated by hackers and so-called "phishing scams" (Jagatic, Johnson, Jakobsson, & Menczer, 2007) that manipulate user's information to commit identity theft (Gross & Acquisti, 2005).

As SNSs and other mediated forms of communication increase in popularity the line between the offline and online world will become difficult to distinguish. The terms online and offline relationships are becoming increasingly entwined (boyd & Ellison, 2007) as people engage in mixed mode relationships (Walther & Parks, 2002) that are formed and maintained using multiple communication venues. Imagining an unmediated relationship might become difficult as offline relationships increasingly matriculate to SNSs and use the sites as a mundane form of interaction (Beers, 2008). As a result, there is a need to understand how relational schemas and social scripts are adjusting to incorporate the growing presence of online technology such as SNS (Samp & Palevitz, 2008). Future research needs to examine how SNSs are incorporated into the initiation, maintenance, and dissolution of the relationship trajectory as well as how these relationship stages manifest themselves in SNS interaction.

SNSs are also becoming increasingly mobile with popular sites such as Facebook and MySpace accessible via mobile devices, and new MSNs (mobile social network systems) like Twitter and Dodgeball enabling the broadcasting of microblogs and text messages within networks (Humphreys, 2007). Understanding SNSs in a tumultuous technological landscape requires embracing an approach that maintains focus on functionality and common theoretically driven research.

The functional approach described in this chapter asserts that SNSs are essentially another interpersonal communication tool being used to fulfill various relational functions such as initiating and maintaining relationships, managing and forming impressions, and seeking information about relational partners. Highlighting the centrality of relational functions on SNSs should allow for the continued study of SNSs both in isolation and in the larger context of interpersonal communication.

References

Acquisti, A., & Gross, R. (2006). Imagined Communities: Awareness, Information Sharing, and Privacy on the Facebook. *Proceedings of the 6th Workshop on Privacy Enhancing Technologies*, Cambridge, UK.

Bargh, J. A., & McKenna, K. Y. (2004). The Internet and social life. *Annual Review of Psychology, 55*, 573–590.

Baxter, L., & Bullis, C. (1986). Turning points in developing romantic relationships. *Human Communication Research, 12*, 469–493.

Baym, N. K., & Ledbetter, A. (2009). Tunes that bind? *Information, Communication, and Society, 12*, 408–427.

Beers, D. (2008). Social network(ing) sites... revisiting the story so far: A response to danah boyd & Nicole Ellison. *Journal of Computer-Mediated Communication, 13*, 516-529.

Berger, C., & Calabrese, R. (1975). Some explorations in initial interaction and beyond: Toward a developmental theory of interpersonal communication. *Human Communication Research, 1*, 99–112.

boyd, d.m. (2007). Social network sites: Public, private, or what? *The Knowledge Tree*. Retrieved May 13, 2010, from http://kt.flexiblelearning.net.au/tkt2007/?page_id=28

boyd, d. (2008). Why youth (heart) social network sites: The role of networked publics in teenage social life. In David Buckingham (Ed.), *Youth, identity, and digital media* (pp. 119–142). Cambridge, MA: MIT.

boyd, d., & Ellison, N. B. (2007). Social network sites: Definition, history, and scholarship. *Journal of Computer-Mediated Communication, 13*, 210–230.

boyd, d., & Heer, J. (2006). Profiles as conversation: Networked identity performance on Friendster. *Proceedings of Thirty-Ninth Hawaii International Conference on System Sciences (HICSS–39), Persistent Conversation Track*. Los Alamitos, CA: IEEE.

Bryant, E. (2008, November). *Uses and gratifications of Facebook*. Paper presented at the human communication and technology division of the National Communication Association 94th Annual Convention, San Diego, CA.

Canary, D., & Stafford, L. (1994). Maintaining relationships through strategic and routine interaction. In D. J. Canary & L. Stafford (Eds.), *Communication and relational maintenance* (pp. 3–22). San Diego, CA: Academic.

Choi, J. H. (2006). Living in *Cyworld*: Contextualising cy-ties in South Korea. In A. Bruns & J. Jacobs (Eds.), *Use of blogs (digital formations)* (pp. 173–186). New York: Peter Lang.

Culnan, M. J, & Markus, M. L. (1987). Information technologies. In F. M. Jablin, L. L. Putnam, K. H. Roberts, & L. W. Porter (Eds.), *Handbook of organizational communication: An interdisciplinary perspective* (pp. 420–443). Newbury Park, CA: Sage.

Cummings, J. M., Lee, J. B., & Kraut, R. E. (2006). Communication technology and friendship during the transition from high school to college. In R. E. Kraut, M. Brynin, S. Kiesler (Eds.), *Computers, phones, and the Internet: Domesticating information technology* (pp. 265–278). New York: Oxford University Press.

DeVito, J. (2001). *The Interpersonal Communication Book* (9th ed.). New York: Addison-Wesley Longman.

Donath J. (1998). Identity and deception in the virtual community. In P. Kollock and M. Smith (Eds.), *Communities in Cyberspace* (pp. 29–59). London: Routledge.

Donath, J., & boyd, d. (2004). Public displays of connection. *BT Technology Journal, 22* (4), 71-82.

Dwyer, C. (2007). *Digital relationships in the 'MySpace' generation: Results from a qualitative study.* Paper presented at the 40th Hawaii International Conference on System Sciences (HICSS), Waikoloa, HI.

Ellison, N. B., Heino, R., & Gibbs, J. (2006). Managing impressions online: Self-presentation processes in the online dating environment. *Journal of Computer-Mediated Communication, 11*(2), article 2. Retrieved July 11, 2010 from http://jcmc.indiana.edu/vol11/issue2/ellison.html

Ellison, N. B., Lampe, C., & Steinfield, C. (2009). Social Network Sites and Society: Current Trends and Future Possibilities. *Interactions Magazine, 16* (1), 6–9.

Ellison, N. B., Steinfield, C. & Lampe, C. (2007). "The benefits of Facebook 'friends:' Social capital and college students' use of online social network sites." *Journal of Computer-Mediated Communication, 12*, 1143–1168.

Facebook.com. *About Facebook: Press Room.* Retrieved July 11, 2010, from http://www.facebook.com/press/info.php?statistics

Feld, S. L. (1981). The focused organisation of social ties. *American Journal of Sociology, 86*, 1015-1035.

Gibbs, J. L., Ellison, N. B., & Heino, R. D. (2006). Self-presentation in online personals: The role of anticipated future interaction, self-disclosure, and perceived success in Internet dating. *Communication Research, 33*, 152–177.

Goffman, E. (1959). *The presentation of self in everyday life.* New York: Doubleday Anchor.

Gossling, S. D., Gaddis, S., & Vazire, S. (2007). *Personality impressions based on Facebook profiles.* Paper presented at the proceedings of the Thirteenth Americas Conference on Information Systems, Keystone, CO.

Greene, K., Derlega, V. L., & Mathews, A. (2006). Self-disclosure in personal relationships. In A. Vangelisti & D. Perlman (Eds.), *Cambridge handbook of personal relationships* (pp. 409–427). Cambridge, UK: Cambridge University Press.

Gross, R., & Acquisti, A. (2005, November). *Information revelation and privacy in online social networks.* Paper presented at WPES'05 (pp. 71–80) Alexandria, VA.

Hargittai, E. (2007). Whose space? Differences among users and non-users of social network sites. *Journal of Computer-Mediated Communication, 13*, article 14. Retrieved July 11, 2010 from http://jcmc.indiana.edu/vol13/issue1/hargittai.html

Human, R., & Lane, D. R. (2008, November). *Virtually friends in Cyberspace: Explaining the migration from FtF to CMC relationships with Electronic Functional Propinquity Theory.* Paper presented at the National Communication Association 94th Annual Convention, San Diego, CA.

Humphreys, L. (2007). Mobile social networks and social practice: A case study of Dodgeball. *Journal of Computer-Mediated Communication, 13*, article 17. Retrieved July 11, 2010 from http://jcmc.indiana.edu/vol13/issue1/humphreys.html

Jagatic, T., Johnson, N., Jakobsson, M., & Menczer, F. (2007). Social phishing. *Communications of the ACM, 5 (10)*, 94–100.

Joinson, A. N. (2008). *Looking at, looking up or keeping up with people? Motives and use of Facebook.* In proceedings of the SIGCHI 2008, 1027–1036.

Kim, K. & Yun, H. (2007). Crying for me, crying for us: Relational dialectics in a Korean social network site. *Journal of Computer-Mediated Communication, 13*, article 15. Retrieved July 11, 2010 from http: onlinelibrary.wiley.com/journal/10.111/(ISSN)1083-6101

Knapp, M. L., & Vangelisti, A. L. (2005). *Interpersonal communication and human relationships* (5th ed.). Boston: Allyn & Bacon.

Kohut, A. (2008, January 11). Social networking and online videos take off: Internet's broader role in campaign 2008. The Pew Research Center for the People and the Press. Retrieved July 11, 2010 from http://www.pewinternet.org/pdfs/Pew_MediaSources_jan08.pdf

Kushin, M. J. & Yamamoto, M. (2009, February). *Searching for media complementarity: Use of social network sites and other online media for campaign information among young adults.* Paper presented at the annual convention of the Western States Communication Association, Phoenix, AZ.

Lampe, C., Ellison, N., & Steinfield, C. (2006). A face(book) in the crowd: social searching vs. social browsing. *Proceedings of CSCW-2006* (pp. 167–170). New York: ACM Press.

Lampe, C., Ellison, N.,& Steinfield, C. (2007). A *familiar Face(book): profile elements as signals in an online social network.* In proceedings of the SIGCHI Conference on Human Factors in Computing Systems (pp. 435–444). New York: ACM Press.

Mann, T. M. (2003). Relationship between metacommunication among romantic partners and their level of relationship satisfaction. *UW-L Journal of Undergraduate Research, 6*, 1–8.

Mazer, J. P., Murphy, R. E., & Simonds, C. J. (2007). I'll See You on "Facebook": The Effects of Computer-Mediated Teacher Self-Disclosure on Student Motivation, Affective Learning, and Classroom Climate. *Communication Education, 56*, 1–17.

McKenna, K. Y. A., Green, A. S., & Gleason, M. E. J. (2002). Relationship formation on the Internet: What's the big attraction? *Journal of Social Issues, 58*, 9–31.

O'Sullivan, B. O. (2000). What you don't know won't hurt me: Impression management functions of communication channels in relationships. *Human Communication Research, 26*, 403–421.

Parks, M. R., & Roberts, L. D. (1998). "Making MOOsic": The development of personal relationships on line and a comparison to their offline counterparts. *Journal of Social and Personal Relationships, 15*, 517–537.

Pearson, E. (2009). All the World Wide Web's a stage: The performance of identity in online social networks. *First Monday, 14*(3). Retrieved July 11, 2010 from http://firstmonday.org/htbin/cgiwrap/bin/ojs/index.php/fm/article/view/2162/2127

Porter, W. (2009, June 5). Old friends using social networking to reconnect. *Denver Post*. Retrieved from www.capecodonline.com/apps/pbcs.dll/article?AID=/20090605/LIFE/ 90604022.

Raacke, J., & Bonds-Raacke, J. (2008). MySpace and Facebook: Applying the uses and gratifications theory to exploring friend-networking sites. *CyberPsychology and Behavior, 11*, 169–174.

Ramirez, A. Jr., Walther, J. B., Burgoon, J. K., & Sunnafrank, M., (2002). Information-seeking strategies, uncertainty, and computer-mediated communication: Toward a conceptual model. *Human Communication Research, 28*, 213–228.

Rheingold, H. (1995). *The virtual community: Finding connection in a computerized world.* London: Secker & Warburg.

Rosenmann, A., & Safir, M. P. (2006). Forced online: Push factors of Internet sexuality: A preliminary study of online paraphilic empowerment. *Journal of Homosexuality, 51*, 71–92.

Samp, J. A., & Palevitz, C. E. (2008, November). *Dating and romantic relationships: Taking tradition into the future with a computer.* Paper presented at the National Communication Association 94th Annual Convention, San Diego, CA.

Sanders, W. S. (2008, November). *Uncertainty reduction and information-seeking strategies on Facebook.* Paper presented at the National Communication Association 94th Annual Convention, San Diego, CA.

Slater, D. (2002). Social relationships and identity online and offline. In L. A. L. S. Livingstone (Ed.), *Handbook of new media: Social shaping and consequences of ICTs* (pp. 533–546). Thousand Oaks, CA: Sage.

Stafford, L., Dainton, M., & Haas, S. (2000). Measuring routine and strategic relational maintenance: Scale revision, sex versus gender roles, and the prediction of relational characteristic. *Communication Monographs, 67*, 306–323.

Steelcase (2008, October 22). Steelcase study shows businesses not harnessing the power of social networking. Retrieved May 13, 2009 from http://www.steelcase.com/na/study_shows_businesses_not_har_News.aspx?f=36628

Stern, L. A., & Taylor, K. (2007). Social networking on Facebook. *Journal of the Communication, Speech & Theatre Association of North Dakota, 20*, 9–20.

Stone, A. A. (1996). *The war of desire and technology at the close of the mechanical age.* Cambridge, MA: MIT Press.

Suler, J. R. (2002). Identity management in cyberspace. *Journal of Applied Psychoanalytic Studies, 4*, 455–459.

Sundén, J. (2003). *Material Virtualities: Approaching Online Textual Embodiment.* New York: Peter Lang.

Surratt, C. G. (1998). *Netlife: Internet citizens and their communities.* New York: Nova Science.

Tong, S. T., Van Der Heide, B., Langwell, L., & Walther, J. B. (2008). Too much of a good thing? The relationship between number of friends and interpersonal impressions on Facebook. *Journal of Computer-Mediated Communication, 13*, 531–549.

Tufekci, Z. (2008). Can you see me now? Audience and disclosure regulation in online social network sites. *Bulletin of Science, Technology and Society, 28*, 20–36.

Turkle, S. (1995). *Life on the screen: Identity in the age of the Internet.* New York: Simon & Schuster.

Walther, J. B. (1994). Anticipated ongoing interaction versus channel effects on relational communication in computer-mediated interaction. *Human Communication Research, 20*, 473–501.

Walther, J. B. (1996). Computer-mediated communication: Impersonal, interpersonal, and hyperpersonal interaction. *Communication Research, 23*, 3–43.

Walther, J. B., & Boyd, S. (2002). Attraction to computer-mediated social support. In C. A. Lin & D. Atkin (Eds.), *Communication technology arid society: Audience adoption and uses* (pp. 53–188). Cresskill, NJ: Hampton.

Walther, J. B., & Parks, M. R. (2002). Cues filtered out, cues filtered in: Computer-mediated communication and relationships. In M. L. Knapp & J. A. Daly (Eds.), *Handbook of interpersonal communication* (3rd ed.) (pp. 529–563). Thousand Oaks, CA: Sage.

Walther, J. B., & Ramirez, Jr., A. (2009). New technologies and new direction in online relating. In S. Smith, & S. Wilson (Eds.), *New directions in interpersonal communication* (pp. 287–307). Thousand Oaks, CA: Sage.

Walther, J. B., Van Der Heide, B. Hamel, L., & Shulman, H. (2009). Self-generated versus other-generated statements and impressions in computer-mediated communication: A test of warranting theory using Facebook. *Communication Research, 36*, 229–253.

Walther, J. B., Van Der Heide, B., Kim, S., Westerman, D., Tong, S. T., & Langwell, L. (2008). The role of friends' appearance and behavior on evaluations of individuals on Facebook: Are we known by the company we keep? *Human Communication Research, 34*, 28–49.

Weigel, D., & Ballard-Reisch, D. (1999). All marriages are not maintained equally: marital type, marital quality, and the use of maintenance behaviors. *Personal Relationships, 6*, 291–304.

Wellman, B., & Gulia, M. (1999) Virtual communities as communities: Net surfers don't ride alone. In M. Smith & P. Kollock (Eds.), *Communities in cyberspace* (167–194). New York: Routledge.

Wright, K. B. (2004). On-line relational maintenance strategies and perceptions of partners within exclusively Internet-based and primarily Internet-based relationships. *Communication Studies, 55*, 239–253.

Young, M. C. (1978). Four problems relating to awareness of metacommunication in business communication. *Journal of Business Communication, 16*, 39-47.

Yurchisin, J., Watchravesringkan, K., & McCabe, D. B. (2005). An exploration of identity re-creation in the context of Internet dating. *Social Behavior and Personality, 33*, 735–750.

Zhao, S., Grasmuck, S., & Martin, J. (2008). Identity construction on Facebook: Digital empowerment in anchored relationships. *Computers in Human Behavior, 24*, 1816–1836.

CHAPTER 2

Unpacking the Paradoxes of Privacy in CMC Relationships: The Challenges of Blogging and Relational Communication on the Internet

Jeffrey T. Child

Sandra Petronio

In a recent posting, a parent uploaded and tagged a picture of their 18 year old son as a little boy dressed in a tutu. Once tagged, the photo became accessible to the son's entire Facebook friend network with a corresponding newsfeed of the action. The mortified son quickly responded to the posting by untagging and disassociating the image from his identity. He also made it clear to his parents that there needed to be a mutually agreed upon posting criterion with his parents so this kind of privacy invasion did not occur again. As this example illustrates, privacy management is a challenging enterprise with social media such as Facebook. While the creation of personal-journal diary blogs, social networking, and Twitter sites provide users multiple opportunities to engage in computer-mediated communication (CMC), privacy, communication, and technology are interwoven in critical ways.

boyd and Ellison (2008) define a social network site (SNS) through three criteria (1) construction of a profile in a system that can be bounded or restricted if desired, (2) inclusion of others with whom they share some type of connection, and finally (3) viewership and surfing capabilities among the list of contacts if desired. Concerning this definition of a SNS, Beer (2008) contends that it potentially hides differences within the applications and notes that "we should be moving toward *more* differentiated classifications of the new online cultures not away from them" (emphasis original, p. 519). A critical aspect of these networks is to uncover the online cultural values concerning privacy. An alternative to highlighting the similarities and differences in these SNSs is to consider variations in disclosure and privacy management practices.

There are many variations in the way people, both young and old, tend to manage protection of their private information when using social network sites in general. For some, privacy boundaries are very closed (creating thick, impermeable boundary walls), restricting access. For others, their privacy boundaries are very open (allowing high permeability), granting significant access. There are also those who slide between these two extremes depending on their needs, adjusting access as necessary (Petronio, 2002). Young adults have characteristically engaged in more privacy protection behaviors than older adults who use SNSs (Madden, Fox, Smith, & Vitak, 2007). However, older adults have learned from the younger generation and are now engaging in comparable levels of concern about SNS profile access (Lenhart, 2009). For example, as of 2007, 60% of adults allowed anyone to access their SNS, while only 40% of teenagers allowed total open access of their SNS profile to others (Madden, Fox, Smith, & Vitak, 2007). In 2009, only 36% of adults permitted no restrictions on access to their SNS profile (Lenhart, 2009).

Access issues have grown in importance for all SNS users (Lenhart & Madden, 2005). However, reports about the use of privacy features on social networking sites, specifically among Facebook.com users, tend to contradict these findings (Facebook, 2009). According to Facebook company research, few users appear to customize their privacy settings to make their sites more secure. Instead, by leaving the original open settings in place, they overlook the fact that they have not been protecting the information they have put on their sites. In fact, the company estimates that less than 20% of individuals utilize these options or change the defaults, which are set to more in network openness than privacy protection (Facebook, 2009; Stone, 2009a). While the Facebook research rests on actual privacy changes made by users, self-report research contradicts these findings suggesting that people believe they are increasing their level of privacy protection on the social networking sites as companies adapt their privacy policies to allow more in-network sharing of information (Christofides, Musie, & Desmarais, 2009). As Facebook increases in popularity among all Internet users, it is likely that more individuals will experience unexpected privacy violation and intrusions by parents, employers, and unknown others. As a consequence, users are apt to change the way they protect their privacy and do more to guard private information in ways that move beyond simply adapting disclosure practices (Allen, 2009; Gavin, 2009; Schonfeld, 2008; Stone, 2009a, 2009b; Stross, 2007).

Twitter is a unique social media, insomuch as 90% of users allow their microblog Twitter page to be completely public while other SNSs have a higher proportion of individuals who restrict or render their privacy boundary around this information as entirely impermeable (Graham, 2008). Yet, simply focusing on the form of social media does not give enough information to determine the calculus people use to make decisions about the way they regulate their privacy in these circumstances. There are underlying issues that help explain the nature of privacy management and the structure that is used to grant or deny access. Communication Privacy Management (CPM) theory (Petronio, 2002) is a useful framework from which variations in CMC disclosure and privacy practices on SNSs can be conceptualized and explored (Child, Pearson, & Petronio, 2009).

Much of the research from a CPM perspective explores privacy within face-to-face interpersonal relationships, with increasing research spanning more diverse contexts such as CMC interactions (Metzger, 2007). Walther (2009) contends that more research in CMC should test assumptions tied to competing theories and examine more fully the mechanisms outlined within our theories. In a recent study, Child et al. (2009) illustrate the feasibility and capacity of applying CPM theory to CMC interactions occurring on SNSs (primarily focusing the diary-based blogging format of an SNS) and develop a theory-based blogging privacy management measure. This chapter considers the viability of using CPM theory to investigate the privacy-regulating decisions people make and understand the way they manage information flow.

Communication Privacy Management

CPM is an evidence-based theory about how people manage private information, both theirs and others' who have granted access to their information (Petronio, 2002). In addition, CPM gives apparatus to understand, not only when privacy is managed in a coordinated, effective way, but also when and how mistakes are made with privacy management (Petronio, 2002). As an evidence-based theory, CPM asserts that individuals have both access and privacy needs forming a dialectical tension that drives choices for privacy management. CPM theory incorporates metaphorical privacy boundaries to illustrate individual versus collective information ownership (Petronio, 1991, 2002). Relational or personal needs are met by giving access or revealing private information, thereby creating a collective privacy boundary with others. On the other hand, concealing information from others, thereby retaining a personal privacy boundary, works to protect an individual's privacy.

CPM stipulates five principles about the privacy management that give a route to better understand both the times when access to the information is granted and when access is denied (Petronio, 2002). The *first principle* states that individuals equate private information with personal ownership. That is, from a behavioral standpoint, people feel they own their private information in the same way that they own other possessions (Child et al., 2009). For example, when individuals disclose or share information on an SNS, they continue to retain their ownership rights over the information. The *second principle* predicts that because people believe they own their information, they also believe that they have the right to control the flow of the information to others. Accordingly, even though individuals may contribute private information to an SNS, they still believe that they retain rights and responsibilities to regulate how much of that information is subsequently shared with others.

Principle three predicts that people develop and use privacy rules to control the flow of information to others (Durham, 2008; Petronio, 2002). For example, individuals who have higher or lower self-monitoring skills develop different rules governing CMC interactions and privacy management practices (Bello, 2005; Child & Agyeman-Budu, 2010; Flynn, Reagans, Amanatullah, & Ames, 2006). The development of privacy rules is predicated on criteria such as cultural expectations. The use of privacy rules can be gender specific. These rules are driven by motivations and frequently take into account a risk-benefit ratio. Finally, privacy rule development and implementation are often impacted by critical incidents or situations that can serve as a catalyst to change existing rules (Petronio, 2002; Petronio & Durham, 2008).

Principle four predicts that once private information is disclosed or others are granted access, the information moves from individual ownership to collective ownership. Collective boundaries imply a joint responsibility and obligation by the original owner and co-owners together to regulate the flow of this information in a mutually agreed upon fashion (Petronio, 2006; Petronio & Gaff, in press; Petronio, Jones, & Morr, 2003; Petronio & Reierson, 2009). This coordination of agreed upon privacy rules may stem from negotiations that the original owner and co-owners enact or through socialization where co-owners learn accepted privacy rules, like children in families (Petronio, 1994). CPM stipulates that typically, people coordinate three different types of privacy rules to manage a collectively held privacy boundary. Thus, the original owner and co-owners coordinate the management of information through the use of *privacy boundary permeability rules*, *privacy boundary ownership rules*, and *privacy boundary linkage rules* (Child et al., 2009; Petronio, 2002; Petronio & Reierson, 2009). These different types of rules are coordinated to control the extent to which third-party dissemination of information may occur from information disclosed within the collective boundary.

Child et al. (2009) describe how each of the three distinct types of privacy rules employed to manage collective boundaries apply to CMC and SNS usage and develop a measure to assess the privacy rules people employ. The measure taps into the extent to which individuals consider the three types of collective boundary management rules when interacting on diary-based SNS sites.[1] From this instrument, it is possible to assess boundary permeability rules that identify which individuals are more public or private about the depth and breadth of their disclosures on SNSs. In addition, the measure evaluates boundary ownership rules assessing the extent to which individuals minimize or expand on others' capabilities to disseminate information within the SNS collective boundary further. In other words, individuals who are concerned about others having access to their private information may utilize more coded language or stipulate restrictions on how the information is to be safeguarded. Finally, boundary linkage rules isolate the characteristics of individuals (e.g., sharing common interests, attraction, potential friendship development) that contribute to an individual making decisions to engage in either more or less privacy management on an SNS (Child et al., 2009).

The *fifth principle* concerns the prediction that if owners and co-owners do not coordinate the privacy rules to regulate information flow, disruption will occur and boundary turbulence will result. When this type of disruption happens, the outcome exposes implicit or taken-for-granted expectations that have been violated. CPM also predicts that this boundary turbulence requires the owners and co-owners to recalibrate and readjust privacy management practices because it becomes clear that they are not functioning adequately or as intended. When boundary turbulence occurs, individuals discover that information they have moved into a collective boundary is not appropriately being managed by the individuals within the collective. Thus, boundary turbulence occurs when violations, disruptions, or unintended consequences occur as a result of privacy management practices (Petronio, 2002). Original owners of the information often expect that co-owners, composing the collective privacy boundaries, will know and follow the privacy rules they use for management.

As a response to boundary turbulence, individuals revisit, readjust, or renegotiate privacy expectations with other members of the collective boundary where the disruption occurred. Through experiencing boundary turbulence, individuals learn about the adequacy of current privacy management practices and dynamically adjust and readjust privacy rules over time to meet evolving privacy needs and expectations (Petronio, 2002). Boundary turbulence may be experienced by bloggers who assume that others will not communicate information they disclose to other co-workers or adults. If a breakdown occurs and the transgression is discovered, the blogger needs to assess privacy rules regulating disclosure he or she has used that has lead to uninvited co-workers having access. Repairs to this privacy breakdown necessitate adapting the level of boundary permeability, ownership, and linkage rules to mend the integrity of the privacy boundary and prevent further breeches from occurring (Child et al., 2009).

CPM and SNS Disclosure Practices

There are many clues to privacy management strategies people use online but they have not been connected within a meaningful framework to see the larger picture. CPM theory allows a rich context to predict and examine CMC disclosure practices on SNS profile pages (Allen, Coopman, Hart, & Walker, 2007; Child et al., 2009; Cochran, Tatikonda, & Magid, 2007; Mazer, Murphy, & Simonds, 2007; Metzger, 2007; Petronio & Durham, 2008; Tyma, 2007). Previous research demonstrates that incorporating others either without any restrictions or allowing only certain categories of people to access an SNS profile and postings is the most common way people utilize an SNS to manage disclosures (Graham, 2008; Lenhart, 2009; Madden et al., 2007). Given this pattern, the creation of an SNS for most users functionally establishes a collective privacy boundary between the user and those who access the site according to CPM predictions (Child et al., 2009; Petronio, 2002). Consequently, allowing, inviting, and encouraging others to share the SNS profile space, fundamentally gives permission to become a co-owner of the posted information. A collective boundary is established through granting access permission, for any given social network site. The collective privacy boundary is further enhanced by disclosure contributions that site visitors make adding their information to the information already posted. Interestingly, because CMC interactions on SNS spaces are written, they offer a unique opportunity to see the sequencing of how collective privacy boundaries are formed and managed according to the theoretical propositions of CPM.

Influence of Decision Criteria in Privacy-Disclosure Choices

Applying the theoretical frame of CPM opens up many avenues of inquiry not considered previously. From this theoretical base, one of the important CPM predictions concerns the basis for choice-making that influences online users to either disclose or remain private. In CPM terms, the decision criteria drive the kinds of privacy rules that people apply to communicative situations (Durham, 2008; Petronio, 2002). From existing research, it is clear that there are a number of decision criteria leading to the development and implementation of privacy rules that function in the background (Petronio, 2002). For example, whether someone has a higher or lower level of self-consciousness, it impacts choices for privacy regulations and by implication, the privacy rules are influenced by the criteria that are used as a result (Child et al., 2009).

Individuals with higher levels of self-consciousness spend more time in general considering their own internal thoughts (private self-consciousness) as well as how their own thoughts might be interpreted by others (public self-consciousness) before acting or making decisions (Buss, 1980; Child et al., 2009; Fenigstein, Scheier, & Buss, 1975). In both cases, bloggers with higher public and private self-consciousness levels employed more of a public orientation towards the management of blogging disclosure and privacy management practices, disclosing more and seeking a wider array of individuals to provide feedback about their thoughts than individuals with less self-consciousness (Child et al., 2009). Miura and Yamashita (2007) found that higher private self-consciousness also ultimately strengthened an individual's overall blogging intentions. Thus, bloggers' internal personality dispositions are aligned with SNS privacy management disclosure practices and blogging persistence (Child & Agyeman-Budu, 2010; Child et al., 2009; Guadagno, Okdie, & Eno, 2008; Miura & Yamashita, 2007).

In addition, research shows that self-monitoring and concern for appropriateness (CFA) dispositions also function as base decision criteria influencing the privacy rules that are used. Child and Agyeman-Budu (2010) explored self-monitoring and concern for appropriateness dispositions (CFA) as types of motivational influences on blogging privacy management practices within CPM theory. Higher self-monitors were more likely than lower self-monitors to enact a more private orientation in all of their blogging privacy management practices, where more coded language was used and less permeability of private information and fewer linkages occurred. While high self-monitors were more cautious about enacting overly public CMC disclosure and privacy management practices on their blogs, they ultimately blog more frequently than do low self-monitors. Individuals who were more concerned about enacting socially appropriate behaviors in their personal relationships employed more of a public orientation towards their blogging boundary permeability rules (Child & Agyeman-Budu, 2010). As such, high-CFA individuals are increasingly likely to utilize a blog as a forum to disclose more information about a variety of topics; consequently, there is a tendency for high-CFA individuals to blog more frequently than do low-CFA individuals.

Curiously, the findings about self-monitoring and CFA dispositions are consistent with previous research supporting that higher levels in both of these personality dispositions are related to greater sensitivity to message misinterpretation and using communication to proactively and carefully manage impressions (Bello, 2005; Flynn et al., 2006; Shaffer & Pegalis, 1998; Tardy & Hosman, 1982). Thus, self-monitoring skills and CFA dispositions are associated with the amount of privacy control bloggers exercise over online disclosure practices, a prediction of CPM theory (Child & Agyeman-Budu, 2010).

From this body of research, an initial set of conditions emerge that serve as online decision criteria that drive the way that privacy rules are developed ultimately guiding choices about CMC interactions. Thus, decisions about privacy rules to use are predicated on the degree to which people are private or public in their orientation to self-consciousness, whether they engage in high or low self-monitoring, and what their public-private orientation is to behavioral appropriateness when interacting online.

CPM argues that another possible basis for decisions leading to the establishment of privacy rules concerns gender (Petronio, 2002; Petronio & Martin, & Littlefield, 1984; Petronio & Martin, 1986). Because men and women tend to have a different set of needs where privacy is concerned, they use different privacy rules to regulate the disclosures that they make to others. Child (2007) explored connections between bloggers' orientation towards privacy management on blogs and gender as an individual difference criterion. From this study, women were more concerned with blogging privacy management and as such used more coded language on their blogs, blogged in ways that limit public information ownership, and were more cautious than men about who was allowed to link to their blog. Men enacted more of a public orientation in the blogging privacy management rules guiding their CMC interactions overall on blogs. These findings are also supported by recent research among Facebook profiles and privacy management practices (Lewis, Kaufman, & Christakis, 2008).

Influence of Family Privacy Orientations

CPM research demonstrates that families have a significant role in socializing the children to learn the kinds of privacy rules to which the family ascribes (Morr Serewicz & Canary, 2008; Petronio, 2002). Further, families also attempt to socialize new members, such as spouses of children, teaching them the family privacy orientations that are held by the members as a whole (Morr, 2002; Morr Serewicz & Canary, 2008; Morr Serewicz, Dickson, Morrison, & Poole, 2007; Petronio, 2002). Family privacy orientation refers to rules that have been developed and endorsed overtime by a family of origin (Petronio, 2002; Petronio, in press). The privacy orientation represents a value structure of the family and is a whole family perspective concerning how they define privacy. Family privacy orientations can range from very open to completely closed (Petronio, in press). Within these orientations, families manage two types of privacy boundaries. First, there are internal privacy cells, where private information is held and controlled by only certain members. These cells shift and change depending on the disclosure and privacy needs of the particular members with the privacy chamber. Second, families also have an external privacy boundary where the whole family ascribes to a rule about what can and cannot be disclosed to outsiders as well as the general level of access to information outsiders are given.

Typically, family privacy orientations serve as a guideline for choices about dissemination of family-private information. However, as children grow into adults, they often develop their own set of standards about privacy regulation regarding information they own that fit their needs, thereby moving away from closely following their family of origin's privacy orientation (Hawk, Keijsers, Hale, & Meeus, 2009; Petronio, 1994, 2002). This process of deindividuation that adolescents and young adults experience represents their claim to their own personal privacy boundary with rules that they develop and control apart from the family (Petronio, 2002; Youniss & Smollar, 1985). Clearly, the family orientation to privacy is an influencing factor, but it is unclear how the privacy rules, ascribed to by the family as a whole, impact choices that might be seen as more individualistic for adolescents and young adults when it comes to managing their own web interactions.

With the concept of family privacy orientation in mind, Child (2007) examined whether parental socialization about privacy orientations was related to the way that bloggers managed their privacy on their blogs. Interestingly, the findings suggest that families advocating a more open orientation did not result in the young adults applying the family's values to their choices about information access in their blogging rules. Furthermore, when the family privacy orientation advocated being more closed about information to outsiders, the young adult blogging choices regarding privacy management practices also did not match the family's orientation (Child, 2007; Morr Serewicz & Canary, 2008). Therefore, family privacy orientations do not seem to carry over to the decision-making processes occurring on blogs by young adults.

Because living in their parents' household, for young adults, may be an influential factor in the degree to which they ascribe to their family privacy orientation, Child (2007) also assessed the impact that being under the same roof might have on privacy choices. The findings indicate that blogging privacy management practices for individuals who lived with their parents were no different from individuals who did not live with their parents. The influence of being under the same roof as the parents does not appear to change decisions about privacy management for these young adults where blogging management is concerned. However, this study found another clue to unpacking this riddle because the majority of participants (94%) were certain their parents did not read or know anything about their blogging disclosure practices. These finding may be tapping into choices and conditions of privacy rules that are seen as more pertinent to management of personal privacy boundaries for these young adults rather than applying to the whole family. This may be especially true for blogging, given that the SNS culture has largely been developed and reinforced by young adults for more peer versus parental interaction (Pempek, Yermolayeva, & Calvert, 2009).

Influence of Context

Another predictive factor that CPM advocates is the influence of context. Many times context serves as a catalyst for changing personal or collectively held privacy rules, because there is a need to reach a particular goal (Westerman, Van Der Heide, Klein, & Walther, 2008). When people want to use online banking, for example, they trade a certain amount of privacy to attain the ease of managing their money (Petronio, 2002). Metzger (2007) examined CMC from a CPM perspective within e-commerce online settings, demonstrating that CPM theory propositions extend to understanding CMC interactions between organizations and individuals beyond the more commonly relationally-oriented applications of the theory exploring CMC interactions. In e-commerce settings, individuals often withheld private information or falsified more sensitive information (such as social security numbers or credit card information) when interacting with organizations, such as banks. Organizations have a common practice of soliciting a substantial amount of privacy information in exchange for promises of free promotional products. Withholding or falsifying private information is a common way to enact privacy protection rules. The strategy is similar to the way bloggers, or social network site users, may utilize more coded language or allow less permeability to protect private information they choose to retain in their individual boundary versus allowing it to reside within the collective boundary (Child, 2007; Child & Agyeman-Budu, 2010; Child et al., 2009). As such, individuals who utilize CMC develop appropriate ways to manage the inherent tensions with the public/private dialectic.

Facebook is another popular context where privacy is managed, a second type of SNS. Facebook allows substantial opportunity for CMC and variations in privacy management practices through the upgraded feature of status updates that took place in 2006 (Thompson, 2008). For instances, through such options as photo tagging/commenting, and open-ended wall discussions, individuals can regulate their private information more effectively than in the past. Some of these features are similar to the CMC that takes place on a personal-journal type blog or SNS. As a result, Facebook users have a central profile page where they can upload pictures, provide status updates, and post wall comments that are archived on the profile. Because Facebook offers several unique opportunities for social networking, the landscape created offers the ability to gain insights into distinctive aspects of privacy management, particularly from a CPM perspective (Child et al., 2009; Petronio, 2002; Stross, 2009; Thompson, 2008). In particular, a central tenet of CPM is privacy control. For Facebook, control plays a fundamental role in this context and therefore offers ways of understanding contextual constraints and latitudes. Christofides et al. (2009) explore individual's disclosure patterns in comparison to face-to-face interactions and information control needs on Facebook.

The findings of the Christofides et al.'s study bring into question the perceptions and rule differences people have for face-to-face interactions as opposed to those they have on Facebook. Consequently, this research underscores that individuals tend to be more likely to reveal information on Facebook than in their face-to-face relationships. Yet, a strong individual predictor of Facebook disclosure practices tends to be how people disclose relationally when they are face-to-face. In other words, their privacy rules are set on the same wave length, using the same criteria for both Facebook and face-to-face interactions. However, the amount of information they tell is mediated by the medium that is used. Possibly, in face-to-face interactions the discloser receives immediate feedback and adjusts the amount depending on the reactions of the receiver, whereas, in the Facebook interactions, the feedback is in writing (losing the non-verbal messages) and lag behind an already constructed message (Child et al., 2009; Petronio, 2000, 2002). This research also found that the need for popularity has some impact on the Facebook disclosure practices.

Privacy control needs on Facebook appear to be discernibly different than in face-to-face interactions (Christofides et al., 2009). Accordingly, individuals with lower overall propensity to disclosure in face-to-face interactions tend to have higher information control needs on Facebook. In CPM terms, this finding indicates that there appears to be a consistent privacy rule across both communicative situations regarding more emphasis on privacy protection than open disclosure. Thus, higher control needs are manifested in regulating privacy boundaries by controlling the flow of information to others, regardless of conversational context. This research also suggests that when individuals have lower levels of trust for the use of the medium, the target of the information, or the unknown others who might gain access and higher levels of self-esteem they feel greater concern for the ability to control their information on Facebook (Christofides et al., 2009). Because people believe they own their information and have the right to control the information, the ability to retain jurisdiction over personal information is paramount to feeling that it still belongs to the person (Petronio, 2002). As a result, any time control, even perceived control, is compromised; the turbulence that erupts causes significant consequences for the information owner and those sharing the information.

Mechanisms of privacy control are also seen in the study by Lewis et al. (2008). Because social groups influence each other on multiple levels, it stands to reason that the same types of issues would be found with SNSs and particularly with Facebook use. This study finds that individuals who have more friends who use private profiles are more likely to maintain their own Facebook profiles in a similar way. While it makes sense that friends would influence each other, interestingly, roommates also have an impact on the way they influence each other to ascribe to specific ways to regulate their privacy online. In this manner, friends and roommates were influenced to restricted access to their information when their friends used high-control privacy setting. The social use and general cultural adaptations of privacy norms exert a strong influence on privacy management practices as theorized and suggested by CPM theory (Petronio, 2002).

Camouflage as a Privacy Protection

Where privacy protection is concerned, there are many strategies that people use when they want to implement this privacy rule. Their boundaries can be various levels of thickness, letting in some information, no information, or a lot of certain information (Quin & Scott, 2007). Through the use of such strategies as coded language, bloggers can restrict access to others or camouflage content as ways of protecting their privacy (Child et al., 2009). While the use of coded or ambiguous language provides a way for bloggers to limit co-ownership of private information on a blog, such privacy protection strategies have been connected to lower-quality interactions in other types of CMC.

In particular, Henderson and Gilding (2004) interviewed individuals about their chatroom interactions. They found that when engaging in synchronous chat sessions, individuals who used pseudonyms and overly ambiguous or less revealing language often developed less trust and rapport with their corresponding CMC chat partners. However, the study also found that given the limited cues available with CMC chatroom interactions, engaging in deeper levels of disclosure was possible because individuals did not have to worry about someone looking at them, making eye contact, or feeling embarrassed. The users also had the opportunity to select each word and have more control over impression management. Given the unique differences of chat-based interactions from SNS interactions (i.e., use of images and a permanent profile with a wide range of information) these findings may not be related to CMC processes on SNSs. However, it is important to examine a wide array of relational outcomes associated with variations in CMC privacy management practices on SNSs.

Communication privacy management theory (Petronio, 2002) provides a rich and integrative framework to explore disclosure and interaction processes occurring through social media (Child et al., 2009). As more individuals are drawn to interaction through SNSs, discovering as well as testing the ways that individual privacy rules develop deserves attention in future research. Examining the decision criteria behind the rule development and usage to learn how issues such as motivations, gender, context, and cultural factors influence privacy management practices ultimately provides deeper understanding to the way people regulate their privacy (Gavin, 2009; Lenhart, 2009; Lenhart & Fox, 2009; Madden et al., 2007). In addition, exploring the way individuals manage collective privacy boundaries for relationships established and maintained through social media and computer-mediated communication is also feasible because CPM theory gives the tools to ask meaningful questions and interpret information in consequential ways.

Future Research

Investigations into this new cultural phenomenon are just beginning to piece together insights into the "hows" and "whys" of privacy management. Fortunately, there is a long history of examining the management of private information found in the research and theoretical development of communication privacy management theory (Petronio, 1991, 1994, 2002, 2006, in press). This chapter offers applications of CPM that illustrate the utility of the theory and research underscores how researchers can benefit from using this body of information. Yet, much work needs to be done to gain a clear picture of how privacy management is enacted and why choices are made by different populations within varying contexts.

Several areas are suggested for future research capitalizing on the CPM concept of boundary turbulence. By examining instances where privacy management expectations are not met, it is possible to isolate some of the fundamental assumptions people make about the medium and identify why privacy management might be compromised and result in privacy breakdowns. Among the many possibilities, CPM theory promises to produce productive results related to two current themes surrounding SNSs, disclosure, and privacy management practices. These include (1) boundary crossings: navigating professional, personal, and familial privacy boundaries; (2) an interface between identity and privacy management.

"Boundary Crossing:"Navigating Professional, Personal, and Familial Privacy Boundaries

One of the most obvious issues emerging from the impact of social network site use is the challenge of drawing boundary lines that denote where relationships begin and end. Essentially, these are privacy boundaries that mark ownership of information. When there is a transgression or "boundary crossing," however unintended, the person feeling aggrieved makes clear that there has been a breach in some way (Petronio, 2002). In the discussion of "fuzzy boundaries" that occur when there is a disruption in the way privacy boundaries can be effectively managed, CPM argues that this state is often caused by ambiguities in who has rights to access the private information (Petronio, 2002). Clearly, these uncertainties underpin the challenges to "boundary crossing" situations when someone is attempting to manage both personal and professional boundaries, but has not had the opportunity to negotiate mutually agreed upon privacy rules for how or if such "crossing" should take place. "Boundary crossing" dilemmas are seen in a variety of circumstances and need further explication with the help of CPM theory.

For example, many parents are now joining Facebook and other SNSs like MySpace in an attempt to reach out to their adult children as well as reestablish their own friendship circles through social media (Lenhart, 2009; Madden et al., 2007; Schonfeld, 2008). Given that more parents are learning to use Facebook, they see both the advantage of access to their children's Facebook page and have learned to appreciate personal rewards themselves from owning a Facebook page. Nevertheless, parents also see that access to their child's page allows a certain degree of surveillance that they never had before to keep tabs on their children's activities (Fletcher, 2009). Increased parental involvement on Facebook and SNSs in general often results in children, especially adolescents and young adults, having to consider the potential ramifications of parental friend requests and devise responses for the increasing reality of parents asking for permission to be included as members of this collective boundary, a relatively new development for young adult children (Child, 2007; Child et al., 2009). Clearly, there are boundary management issues that call for new ways of establishing parameters for how much parents know about their children's activities. Likewise, young adult access to their parents' Facebook or SNSs mean that they have access to the parent's lives in ways not possible before multi-generational use of SNSs and social media. Obviously, these circumstances are ripe for conflicts over the management of inter-family privacy boundary that likely challenge parent/child relationships (Hawk et al., 2009).

Employing a CPM framework to examine the boundary navigation that parental Facebook requests prompt allows exploration of how existing family factors may impact a child's decision to accept, modify content, or change rules regulating disclosure before accepting, ignoring, or outright rejecting parental friend requests on SNSs. Subsequent disclosure practices may also be altered by young adults in the way they manage their collective SNS boundary, knowing that their parents will have access to their general SNS disclosure practices. Likewise, exploring how the parents cope with the same kinds of requests from their children is potentially a viable and productive area of future research. These boundary negotiations may represent a new way of understanding how parents and children keep or yield access, changing permeability rules for privacy in SNS interactions. Making these changes may represent new kinds of decisions that parents and children enter into regarding private information going across boundary lines in both directions. The process of these negotiations likely holds insights into expectations that both parents and children have about the other when it comes to privacy issues.

In particular, exploring how young adults interpret parental friend requests as either a type of privacy invasion behavior or not has implications for parent/child relational quality assessments. Petronio (1994) found that when young adults perceive that their parents invade their privacy, the invasive behaviors create more openness in the parent/child collective boundary often at the expense of overall relational quality. The application of CPM to parent/child face-to-face interaction and CMC allows addressing the impact of the privacy management practices among the different generations mutually drawn to SNS utilization.

"Boundary crossing" also prevails in many other kinds of situations, particularly when personal and professional privacy boundaries collide or intersect. For example, research shows that accountability and professionalism can be at risk for pharmacy students when they post personal information on their Facebook and compromise professional judgment (Cain, Scott, & Akers, 2009). Physicians may also find that if their patients try to "friend" them, the request crosses the borders of their private lives making them feel uncomfortable or realize that their professionalism is compromised having a Facebook site (Guseh, Brendel, & Brendel, 2009; Thompson, Dawson, Ferdig, Black, Boyer, Coutts, & Black, 2008). As a consequence, negotiating a professional relationship is likely a challenge for physicians or other medical providers when they maintain a personal Facebook that exposes aspects of their lives they wish to remain private.

Likewise, "boundary crossing" occurs for educators and students (Carter, Foulger & Ewbank, 2008; Dippold, 2009; Greenhow & Robelia, 2009). Students google their teachers and find their Facebook site or other "private" online social networking sites. Even if the students are not given permission, the "public" information may be more than the teachers want their students to know about them. The same kind of "boundary crossing" may occur for students who wish to enter a particular college or apply for a position in the business world (Kluemper & Rosen, 2009). In the same way these "boundary crossings" happen so do they in organizations of all sorts (e.g., Allen, Walker, Coopman, & Hart, 2007; Petronio, 2002; Shadur, Kienzle, & Rodwell, 1999). Such diverse boundary crossing applications deserve greater attention in future research.

Interface of Identity and Privacy Management

Obviously, much work is needed to better understand the nature of identity within the world of privacy in online sites. Because the basic assumption of identity speaks of sustaining a sense of autonomy in today's world, it is clear that the ways identity and privacy management are interrelated matter for people as they traverse Facebook and other SNSs. For many, identity regulation is analogous with sustaining security of their person when they are using Internet communication. Many businesses are spending large amounts of money to create encryption programs that will protect their customers (e.g., Zhang & Imai, 2009). The scope is far-reaching and includes such issues as security of medical research data of tissue samples or copyright protection for digital information, for instance (Manion, Robbins, Weems, & Crowley, 2009).

Identity management also takes into account interpersonal relationship issues where privacy and SNSs are concerned. Substantial research in interpersonal communication has explored how interpersonal motives impact relational outcomes and interpersonal processes (Barbato, Graham, & Perse, 2003; Graham, 1993; Rubin, Perse, & Barbato, 1988). Because CPM identifies "privacy motives" as an important criterion used to make judgments about privacy rule development and adjustments (Child et al., 2009; Petronio, 2002), this existing body of work is useful in examining variations in individual and collective boundary management processes. The CPM framework surrounding interpersonal privacy motives determines how diverse CMC interaction goals and needs are associated with an individual's current disclosure and privacy management practices on SNSs (Schmidt, 2007).

In particular, the way people balance social capital (an identity management strategy) they gain from having a network of friends and telling them about their trips, choices, decisions, or feelings with the risks of disclosing those things on their social networking sites. CPM predicts that when one person makes a larger contribution of information than others linked as co-owners of the now, collective boundary, doing so often means that the differential information contribution increases the likelihood of the co-owners having more power than the original discloser (Petronio, 2002; Petronio & Kovach, 1997). In this way, the web-disclosure has increased risk while trying to increase social capital by not only having people to tell (the traditional definition of social capital) but by telling personal information often designed to impress, entice, and appeal to others (Ellison, Steinfield, & Lampe, 2007; Valenzuela, Park, & Kee, 2009).

Clearly, the interface of privacy and identity management also speaks of trust as a prerequisite of starting the Internet conversations in the first place (Petronio, 2002). Although trust is a critical factor when people are considering online shopping or whether to reveal something personal on their site, the nature of trust and its place within the privacy calculus is not always understood. Through the conceptual apparatus of CPM theory, a better grasp of how trust works is offered and provides predictable ways people make decisions about trust needs (Petronio, 2002).

Learning more about the evolution of disclosure and privacy management practices of diary-based and networking-based SNSs across time and factoring in trust issues helps to illustrate how identity and privacy are collectively managed. Further, we advocate research that moves beyond cross-sectional survey designs to ultimately clarify how the development of unique privacy trajectories can occur within different kinds of Internet needs. One privacy trajectory might track how online consumer health information changes the patient's willingness to disclose symptoms to the doctor because the symptoms no longer seem ambiguous or embarrassing to the patient. Another possible privacy trajectory could be exploring how young adults adjust SNS privacy management practices in light of encountering significant life changes such as moving to college, starting a new career, or developing and maintaining a significant committed romantic relationship. Research illuminating the moments and contexts that such changes take place is needed. Thus, significant events have the potential to impact privacy trajectories and can be identified and studied through soliciting descriptions about how people make decisions to reveal more or less information than is typical in light of important turning point events that mark significant changes in patterns of behaviors.

Finally, capturing the relationship between identity and privacy management requires more integrative research securing an understanding of the complexity found in the available types of SNSs that may be used by individuals. For example, many young adults have all three types of SNSs (diary-based blogs, Facebook social network sites, and Twitter sites). Little is known about how individuals make decisions about what information to post on which collective SNS boundary and how or if they consider issues of identity management when they make these choices.

Individuals may choose to differentiate their personal and professional networks through the three types of SNSs and interact with diverse others in nuanced ways that allows more or less protection of privacy and personal identity. Greater understanding about how individuals make privacy management and CMC decisions that take into account diverse CMC interactions in multiple venues would provide more integrative understanding of how social media work in concert. Thompson (2008) highlights how young adults easily move among the diverse SNSs in ways that result in almost constant awareness and documentation of others' thoughts, opinions, routines, habits, behaviors, and locations. This phenomenon occurs so often that some individuals are responding by creating businesses, spaces, and events where privacy expectations are explicitly reinforced with greater privacy expectations, thereby allowing the opportunity to exhale (Puente, 2009; Salkin, 2009).

Conclusion

Because social networking is new in the world and the communication system used has to borrow from existing features of social interaction, we are scratching an unexplored surface. Not only are our social relationships changing because we have access to this form of interaction with others, so too is our sense of autonomy and therefore privacy in ways we cannot fully comprehend at the moment. This chapter offers a functional beacon to begin the process of understanding the way privacy management functions within this larger mediated communicative system. Exploring privacy regulation in the SNS context pushes many of the assumptions we have made theoretically. The yield of understanding is very promising, but challenges basic hypotheses and beliefs about the way people communicate with each other. We are witnessing evolution in the making and must stand ready with an arsenal of tools to keep pace with the changes we are experiencing because of this new way to interact, always balancing both connectedness and autonomy in our socially driven world.

Note

1 Although the blogging privacy management scale (Child et al., 2009) was developed based on interactions on diary-based blogs, general modifications made to the scale language from blog to social network website, Facebook, or Twitter allows use of the items to explore either general privacy management practices or more specific practices tied to the other types of SNSs, beyond diary-based blogs. Further refinement and adaptation of the scale in general to a variety of SNS domains is currently under development.

References

Allen, M. W. (2009, June 21). A warning about tweeting vacation plans. *The Plain Dealer*, p. F2.

Allen, M. W., Coopman, S. J., Hart, J. L., & Walker, K. L., (2007). Workplace surveillance and managing privacy boundaries. *Management Communication Quarterly, 21*, 172–200.

Barbato, C. A., Graham, E. E., & Perse, E. M. (2003). Communicating in the family: An examination of the relationship of family communication climate and interpersonal communication motives. *Journal of Family Communication, 3*, 123–148.

Beer, D. (2008). Social network(ing) sites...revisiting the story so far: A response to danah boyd & Nichole Ellison. *Journal of Computer-Mediated Communication, 13*, 516–529.

Bello, R. (2005). Situational formality, personality, and avoidance-avoidance conflict as causes of interpersonal equivocation. *Southern Communication Journal, 70*, 285–300.

boyd, d. m., & Ellison, N. B. (2008). Social network sites: Definition, history, and scholarship. *Journal of Computer-Mediated Communication, 13*, 210–230.

Buss, A. H. (1980). *Self-consciousness and social anxiety.* San Francisco: Freeman.

Cain, J. Scott, D. T., & Akers, P. (2009). Pharmacy students' Facebook activity and opinions regarding accountability and e-professionalism. *American Journal of Pharmaceutical Education, 73*, 1–6.

Carter, H. L., Foulger, T. S., & Ewbank, A. D. (2008). Have you googled your teacher lately? Teachers' use of social networking sites. *Phi Delta Kappan, 89*, 681–685.

Child, J. T. (2007). *The development and test of a measure of young adult blogging behaviors, communication, and privacy management.* (Unpublished doctoral dissertation), North Dakota State University, Fargo, ND.

Child, J. T., & Agyeman-Budu, E. A. (2010). Blogging privacy rule development: The impact of self-monitoring skills, concern for appropriateness, and blogging frequency. *Computers in Human Behavior, 26*, 957–963.

Child, J. T., Pearson, J. C., & Petronio, S. (2009). Blogging, communication, and privacy management: Development of the blogging privacy management measure. *Journal of the American Society for Information Science and Technology, 60*, 2079–2094.

Christofides, E., Muise, A., & Desmarais, S. (2009). Information disclosure and control on Facebook: Are they two sides of the same coin or two different processes? *CyberPsychology & Behavior, 12*, 341–344.

Cochran, P. L., Tatikonda, M. V., & Magin, J. M. (2007). Radio frequency identification and the ethics of privacy. *Organizational Dynamics, 36*, 217–229.

Durham, W. T. (2008). The rules-based process of revealing/concealing the family planning decisions of voluntarily child-free couples: A communication privacy management perspective. *Communication Studies, 59*, 132–147.

Dippold, (2009). Peer feedback through blogs: Student and teaching perceptions in an advanced German class. *ReCall, 21*, 18–36.

Ellison, N. B., Steinfield, C., & Lampe, C. (2007). The benefits of Facebook "friends": Social capital and college students' use of online social network sites. *Journal of Computer-Mediated*

Communication, 12, article 1. Retrieved February 22, 2010, from http://jcmc.indiana.edu/vol13/issue1/hargittai.html

Facebook (2009). Press room. Palo Alto, CA: Facebook. Retrieved November 10, 2009, from http://www.facebook.com/press/info.php?statistics

Fenigstein, A., Scheier, M. F., & Buss, A. H. (1975). Public and private self-consciousness: Assessment and theory. *Journal of Consulting and Clinical Psychology, 43,* 522–527.

Fletcher D. (2009, July 8). Oh crap! My parents joined Facebook. *Time.* Retrieved November 10, 2009, from http://www.time.com/time/business/article/0,8599,1909187,00.html

Flynn, F. J., Reagans, R. E., Amanatullah, E. T., & Ames, D. R. (2006). Helping one's way to the top: Self-monitors achieve status by helping others and knowing who helps whom. *Journal of Personality and Social Psychology, 91,* 1123–1137.

Gavin, J. (2009, July 2). *Russia has world's most engaged social networking audience.* Retrieved November 10, 2009, from http://www.comscore.com/Press_Events/Press_Releases/2009

Graham, E. E. (1993). The interpersonal communication motives model. *Communication Quarterly, 41,* 172–186.

Graham, J. (2008, July 21). Twitter took off from simple to 'tweet' success; surprisingly hot social-network service keeps pals in touch and puts companies on their toes. *USA Today.* Retrieved October 5, 2009 from LexisNexis Academic database.

Greenhow, C., & Robelia, B. (2009). Old communication, new literacies: Social network sites as social learning resources. *Journal of Computer-Mediated Communication, 14,* 1130–1161.

Guadagno, R. E., Okdie, B. M., & Eno, C. A. (2008). Who blogs? Personality predictors of blogging. *Computers in Human Behavior, 24,* 1993–2004.

Guseh, J. S., Brendel, R. W., & Brendel, D. H (2009). Medical professionalism in the age of online social networking. *Journal of Medical Ethics, 35,* 584–586.

Hawk, S. T., Keijsers, L., Hale, W. W., & Meeus, W. (2009). Mind your own business! Longitudinal relationship between perceived privacy invasions and adolescent-parent conflict. *Journal of Family Psychology, 23,* 511–520.

Henderson, S., & Gilding, M. (2004). 'I've never clicked this much with anyone in my life': Trust and hyperpersonal communication in online friendships. *New Media & Society, 6,* 487–506.

Kluemper, D. H., & Rosen, P. A. (2009). Future employment selection methods: Evaluating social networking web sites. *Journal of Managerial Psychology, 24,* 567–580.

Lenhart, A. (2009). *Adults and social network websites.* Retrieved October 1, 2009, from the Pew Internet and American Life Project Website: http://www.pewinternet.org

Lenhart, A., & Fox, S. (2009). *Twitter and status updating.* Retrieved October 1, 2009, from the Pew Internet and American Life Project Web site: http://www.pewinternet.org

Lenhart, A., & Madden, M. (2005). *Teen content creators and consumers.* Retrieved October 1, 2009. from the Pew Internet and American Life Project Website: http://www.pewinternet.org

Lewis, K., Kaufman, J., & Christakis, N. (2008). The taste for privacy: An analysis of college student privacy settings in an online social network. *Journal of Computer-Mediated Communication, 14,* 79–160.

Madden, M., Fox, S., Smith, A., & Vitak, J. (2007). *Digital footprints.* Retrieved October 1, 2009, from the Pew Internet and American Life Project Website: http://www.pewinternet.org

Manion, F. J., Robbins, R. J., Weems, W. A., & Crowley, R. S. (2009). Security and privacy requirements for a multi-institutional cancer research data grid: An interview-based study. *Medical Informatics and Decision Making, 9,* 1–40.

Mazer, J., Murphy, R., & Simonds, C. (2007). I'll see you on "Facebook": The effects of computer-mediated teacher self-disclosure on student motivation, affective learning, and classroom climate. *Communication Education, 56,* 1–17.

Metzger, M. J. (2007). Communication privacy management in electronic commerce. *Journal of Computer-Mediated Communication, 12,* 335–361.

Morr, M. C. (2002). *Private disclosure in a family membership transition: In-laws' disclosures to newlyweds.* Unpublished doctoral dissertation, Arizona State University, Tempe, AZ.

Morr Serewicz, M. C., & Canary, D. J. (2008). Assessments of disclosure from the in-laws: Links among disclosure topics, family privacy orientations, and relational quality. *Journal of Social and Personal Relationships, 25,* 333–357.

Morr Serewicz, M. C., Dickson, F. C., Morrison, J. H., & Poole, L. L. (2007). Family privacy orientation, relational maintenance, and family satisfaction in young adults' family relationships. *Journal of Family Communication, 7,* 123–142.

Miura, A., & Yamashita, K (2007). Psychological and social influences on blog writing: An online survey of blog authors in Japan. *Journal of Computer-Mediated Communication, 12,* 1452–1471.

Pempek, T. A., Yermolayeva, Y. A., & Calvert, S. L. (2009). College students' social networking experiences on Facebook. *Journal of Applied Developmental Psychology, 30,* 227–238.

Petronio, S. (1991). Communication boundary management: A theoretical model of managing disclosure of private information between marital couples. *Communication Theory, 1,* 311–335.

Petronio, S. (1994). Privacy binds in family interactions: The case of parental privacy invasion. In W. R. Cupah & B. H. Spitzberg (Eds.), *The dark side of interpersonal communication* (pp. 241–257). Hillsdale, NJ: Lawrence Erlbaum.

Petronio, S. (2000). The ramifications of a reluctant confidant. In A.C. Richards & T. Schumrum (Eds.), *Invitations to dialogue: The legacy of Sidney M. Jourard* (pp. 113–150). Dubuque, IA: Kendall Hunt.

Petronio, S. (2002). *Boundaries of privacy: Dialectics of disclosure.* New York: State University of New York Press.

Petronio, S. (2006). Communication privacy management theory: Understanding families. In D. O. Braithwaite & L. A. Baxter (Eds.), *Engaging theories in family communication: Multiple perspectives* (pp. 35–49). Thousand Oaks, CA: Sage.

Petronio, S. (in press). Communication privacy management theory: What do we know about family privacy regulation? *The Journal of Family Theory and Review.*

Petronio, S., & Durham, W. T. (2008). Communication privacy management. In L. A. Baxter & D. O. Braithwaite (Eds.), *Engaging theories in interpersonal communication: Multiple perspectives* (pp. 309–322). Thousand Oaks, CA: Sage.

Petronio, S., & Gaff, C. (in press). Managing privacy ownership and disclosure. In C. Gaff & C. Bylund (Eds.), *Talking about Genetics.* New York: Oxford University Press.

Petronio, S., Jones, S., & Morr, M. C. (2003). Family privacy dilemmas: Managing communication boundaries within family groups. In L. R. Frey (Ed.), *Group communication in context: Studies of bona fide groups* (pp. 23–55). Mahwah, NJ: Lawrence Erlbaum.

Petronio, S., & Kovach, S. (1997). Managing privacy boundaries: Health providers' perceptions of resident care in Scottish nursing homes. *Journal of Applied Communication Research, 25,* 115–131.

Petronio, S., & Martin, J. (1986). Ramifications of revealing private information: A gender gap. *Journal of Clinical Psychology, 42,* 499–506.

Petronio, S., Martin, J., & Littlefield, R. L. (1984). Prerequisite conditions for self-disclosure: A gender issue. *Communication Monographs, 51,* 268–273.

Petronio, S., & Reierson, J. (2009). Regulating the privacy of confidentiality: Grasping the complexities through communication privacy management theory. In T. D. Afifi & W. A. Afifi (Eds.), *Uncertainty, information management, and disclosure decisions: Theories and applications* (pp. 365–383). New York: Routledge.

Puente, M. (2009, April 15). Relationships in a twist over Twitter; glued to your gadget? You may be losing human link. *USA Today*. Retrieved September 17, 2009, from LexisNexis Academic database.

Quin, H., & Scott, C. R. (2007). Anonymity and self-disclosure on weblogs. *Journal of Computer-Mediated Communication, 12*, 1428–1451.

Rubin, R. B., Perse, E. M., & Barbato, C. A. (1988). Conceptualization and measurement of interpersonal communication motives. *Human Communication Research, 14*, 602–628.

Salkin, A. (2009, August 9). Party on, but not tweets. *The New York Times*. Retrieved September 17, 2009, from LexisNexis Academic database.

Schmidt, J. (2007). Blogging practices: An analytical framework. *Journal of Computer-Mediated Communication, 12*, 1409–1427.

Schonfeld, E. (2008, December 31). *Top social media sites of 2008 (Facebook still rising)*. Retrieved September 17, 2009, from the TechCrunch website: http://www.techcrunch.com/2008/

Shadur, M. A., Kienzle, R., & Rodwell, J. J. (1999). The relationships between organizational climate and employee perceptions of involvement. *Group & Organization Management, 24*, 479–503.

Shaffer, D. R., & Pegalis, L. J. (1998). Gender and situational context moderate the relationship between self-monitoring and induction of self-disclosure. *Journal of Personality, 66*, 215–234.

Stone, B. (2009a, March 29). Is Facebook growing up too fast? *The New York Times*. Retrieved September 17, 2009, from LexisNexis Academic database.

Stone, B. (2009b, September 25). Twitter appears to raise $100 million, valuing it at $1 billion. *The New York Times*. Retrieved October 5, 2009, from LexisNexis Academic database.

Stross, R. (2007, December 30). How to lose your job on your own time. *The New York Times*. Retrieved September 17, 2009, from LexisNexis Academic database.

Stross, R. (2009, March 8). When everyone's a friend, is anything private? *The New York Times*. Retrieved September 29, 2009, from LexisNexis Academic database.

Tardy, C. H., & Hosman, L. A. (1982). Self-monitoring and self-disclosure flexibility: A research note. *Western Journal of Speech Communication, 46*, 92–97.

Thompson, C. (2008, September 7). I'm so digitally close to you. *The New York Times*. Retrieved September 17, 2009, from LexisNexis Academic database.

Thompson, L. A., Dawson, K., Ferdig, F., Boyer, J., Coutts, J., & Black, N. (2008). The intersection of online social networking with medical professionalism. *Journal of General Internal Medicine, 23*, 954–957.

Tyma, A. (2007). Rules of interchange: Privacy in online social communities—A rhetorical critique of MySpace.com. *Journal of the Communication Speech and Theater Association of North Dakota, 20*, 31–39.

Valenzuela, S., Park, N., & Kee, K. F. (2009). Is there social capital in a social network site?: Facebook use and college students' life satisfaction, trust, and participation. *Journal of Computer-Mediated Communication, 14*, 875–901.

Walther, J. B. (2009). Theories, boundaries, and all of the above. *Journal of Computer-Mediated Communication, 14*, 748–752.

Westerman, D., Van Der Heide, B., Klein, K. A., & Walther, J. B. (2008). How do people really seek information about others?: Information seeking across Internet and traditional communication channels. *Journal of Computer-Mediated Communication, 13*, 751–767.

Youniss, J., & Smollar, J. (1985). *Adolescent relationship with mothers, fathers, and friends*. Chicago: University of Chicago Press.

Zhang, R., & Imai, H. (2009). Strong anonymous signatures. In M. Yung, P. Liu, & D. Lin (Eds.), *Information security and cryptology* (pp. 60–71). Heidelberg, Germany: Springer.

CHAPTER THREE

A New Twist on Love's Labor: Self-Presentation in Online Dating Profiles

Catalina L. Toma

Jeffrey T. Hancock

Perhaps nowhere are first impressions as important as in romantic encounters. Romance can thrive if first impressions are positive, or may not even take off if they are negative. An important question, then, is what kind of information people rely on to form these first impressions. In traditional forms of dating, such as being introduced by a mutual friend or simply sharing a glance across the room, first impressions are typically based on the other's physical appearance, dress, and conversational style. This limited amount of information allows people to gauge romantic "chemistry," but it tends to lack breadth (i.e., information about exact age, occupation, family) and depth (i.e., information about personality and core beliefs).

With the growing popularity of social network websites, such as online dating, impression formation in romantic contexts has witnessed a significant change. Online dating involves constructing detailed profiles describing the self and then browsing others' profiles with a view to finding a good "match." In contrast with face-to-face dating, online profiles reveal information that has both breadth (e.g., age, height, weight, education, occupation, income) and depth (e.g., personality traits, political beliefs, religious beliefs). While this repository of information may be beneficial when scrutinizing others' profiles, it presents a significant challenge when constructing one's own. How do online daters construct their self-presentation, given the pressure to reveal a wealth of personal information and the importance of making a good first impression? How do they strategically select which information to disclose and which to circumvent? Most importantly, how honest are their self-presentations?

The purpose of this chapter is to examine the process of self-presentation in online dating profiles, from both theoretical and empirical lenses. We first offer a general theoretical framework for online self-presentation that encompasses both psychological factors and the role of the communication medium. Then we apply this framework to the distinct context of online dating and provide empirical support for it by reviewing findings from a large study we conducted on this topic (Hancock & Toma, 2009; Toma & Hancock, 2010; Toma, Hancock, & Ellison, 2008) as well as from other relevant studies (Ellison, Heino, & Gibbs, 2006; Gibbs, Ellison, & Heino, 2006; Whitty, 2008).

Online Self-Presentation: A Theoretical Framework

We begin by introducing a general framework for online self-presentation. While our main focus in this chapter is on self-presentation in online dating profiles, we intend this framework to be applicable more generally to self-presentations in online environments. As briefly mentioned earlier, online self-presentation differs from face-to-face self-presentation in form (i.e., the type of information that gets included, the manner in which the self-presentation is constructed, the audience to whom it is available), but it can be similar in function (i.e., it serves the same fundamental purpose of finding love, connecting with friends, or impressing employers) (see also Walther, 2007). This theoretical framework draws on the quintessential psychological factors that guide self-presentation and offers a detailed discussion on how features of the communication medium are expected to interact with these factors.

Following Leary and Kowalski's (1990) two-component model of self-presentation, we view self-presentation as consisting of (1) motivational processes, or the degree to which self-presenters are motivated to control how others see them; and (2) construction processes, or the actual implementation of a desired impression. Below we elaborate on both processes, with an emphasis on how the computer-mediated environment can alter them.

Motivational Processes

When do people care about how others see them? Generally speaking, people care about others' impressions when those impressions are relevant to the fulfillment of their goals (Goffman, 1959; Schlenker, 2002). Simply put, if others' opinions matter in achieving certain goals, one will invest time and effort into influencing those opinions. For instance, if the goal is to attract a desirable mate, it is vital that potential mates perceive one as attractive. Similarly, if the goal is to land a job, it is important to impress the interviewer with one's knowledge and competence. Let it be noted that these goals are fundamental to human nature and should not be affected by the online environment.

A corollary of this principle is that the motivation to control others' impressions should increase as the importance of the goals increases (Beck, 1983; Leary & Kowalski, 1990). For instance, if one perceives finding a marriage partner as more important than connecting with friends, one will be more motivated to carefully control one's online dating self-presentation than one's self-presentation on friendship-related websites, such as Facebook.

A second corollary of this principle is that the anticipation of future interaction with others should increase the motivation to control how others see the self (Walther & Parks, 2002). This is the case because people who are a more stable presence in one's life are more likely to affect the fulfillment of one's goals. The online environment facilitates encounters where there is (1) an anticipation of future *face-to-face* interaction, such as online dating; (2) an anticipation of future *online* (but not face-to-face) interaction, such as many discussion boards or blogs; and (3) no anticipation of future interaction, such as anonymous chat. Generally speaking, the anticipation of future interaction, whether face-to-face or mediated, should result in greater motivation to control impressions than when there is no anticipation of future interaction. The anticipation of face-to-face interaction should result in greater motivation to impression-manage than the anticipation of mediated interaction when the self-presenter's goals need to be accomplished in the face-to-face environment, such as dating.

A third corollary is that publicness, or the degree to which one's behavior is visible to others, should increase motivation to manage impressions. Public behaviors are more likely to be relevant to the accomplishment of goals than private behaviors, and hence people should be more motivated to control them (Leary & Kowalski, 1990). The online environment alters the publicness of one's behavior in several important ways. At one end of the spectrum, it can render online behaviors completely anonymous, in a way that is impossible in face-to-face settings (see McKenna & Bargh, 2000; Turkle, 1995). In this case, the motivation to impression-manage should be low. At the other end of the spectrum, the online environment can offer a considerable degree of publicness to people who may never have such arenas for public behavior in face-to-face environments. For instance, personal websites and blogs are visible to audiences of millions, and users are in a position to broadcast their opinions and thoughts to the public at large. The online world can also make self-presentational acts more permanent by preserving a record of the self-presentation that is available for long periods of time (see Hancock, Thom-Santelli, & Ritchie, 2004). By contrast, face-to-face self-presentations tend to be fleeting—only available to observers who are present when the behavior occurs. When online behaviors are permanent and have a large audience, we expect self-presenters to be highly motivated to control how others see them.

Construction Processes

Depending on how much self-presenters are motivated to control how others see them, they will invest time and effort into constructing a desired image. But how exactly do they go about constructing this image? According to Leary and Kowalski's (1990) framework, self-presenters first need to decide how they want to come across to their audience, and then to implement this desired image by engaging in various self-presentational strategies.

Deciding on a desired impression. The first step in image management occurs a priori to the actual self-presentational behaviors: people decide how exactly they *want* to be perceived by their audience. Two factors determine the construction of this desired image: how self-presenters see themselves and what they perceive the values of their audience to be. In other words, self-presenters select out of their repertoire of self-images the ones they think will best mesh with the values of their audience. Let us first examine the role of the self-concept, or how people perceive themselves, in the construction of a desired image.

The self-concept includes several dimensions (Higgins, 1987): 1) the *actual self*, which consists of characteristics one currently possesses; 2) *the ideal self*, which consists of characteristics one would like to and could possess in the future, but does not possess currently; and 3) *the ought self*, which consists of characteristics one thinks one should possess given social norms and expectations from others. Self-presenters can choose characteristics from all these three dimensions when constructing their desired image, and research has shown that they tend to strike a balance between presenting themselves candidly (i.e., selecting characteristics of the actual self) or presenting themselves at their best potential (i.e., selecting characteristics of their ideal self) (Schlenker, 2002). However, the more motivated self-presenters are to impress their audience, the more they tend to select aspects of their ideal self (Leary & Kowalski, 1990; Schlenker, 2002). It is also noteworthy that people are often proud of aspects of their actual self, and when the motivation to impress the audience is high, they are likely to make a deliberate effort to display these aspects.

A further and critical consideration in choosing which self-aspects to present is the audience. As strategic self-presenters, people consider the values of the audience and then tailor their images to those specific preferences. For instance, people constructing self-presentations for job searches will emphasize job-related skills and competence, whereas those looking for romantic partners will play out to the preferences of these potential partners. These characteristics often correspond to the ought self, or how people think they should be in order to be successful in a certain social arena. A particularly important of such characteristics is honesty, as social norms dictate that people are who they appear to be (Goffman, 1959). Self-presenters then face pressures to present themselves in a flattering manner, while not deviating from the truth substantially, or in ways that cannot be justified.

Implementing the desired impression. Once self-presenters have decided precisely how they want to come across to their audience, they need to implement this desired self-presentation—a process which is profoundly affected by the communication medium: self-presentations conveyed in face-to-face settings differ substantially from those conveyed over the telephone or over email.

Generally speaking, the communication medium affects self-presentation in terms of 1) content, or what gets presented; and 2) delivery, or the manner in which the self-presentation is conveyed to the audience. Let us first consider the content of the self-presentation. As mentioned earlier, the key characteristic of face-to-face self-presentation is that it is embodied, meaning that the physical self is directly displayed. By contrast, online self-presentations are disembodied, with self-presenters interacting with others in the absence of the physical self. As a result, information about self-presenters' physical appearance is usually transmitted via photographs or textual self-descriptors (i.e., participants describing their appearance verbally). Some of these descriptors are objective and straightforward, such as mentioning height, weight, age, eye color, or hair color. However, other descriptors, such as photographs and subjective assessment of one's appearance (e.g., "average build" or "curvy") are malleable and open to interpretation. In fact, an argument can be made that photographs can never portray reality the way we see it with our own eyes, because they are bi-dimensional and generated by technologies that differ from the way our eyes operate (Hancock & Toma, 2009). At their best, photographs can offer viewers an impression of verisimilitude, or a good enough approximation of how the person portrayed in the photograph looks in real life. Photographs are also highly susceptible to manipulation: they can be staged, digitally altered, or simply old (thus depicting a younger version of the self).

More generally, textual statements describing any aspects of the self can be easily altered in order to present a more favorable version of the self. For instance, it is easy to claim a higher salary or a more prestigious occupation without having to prove the veracity of those claims immediately. For this reason, the textual and photographic elements of online self-presentations can be conceptualized as *conventional signals* (Donath, 2007), or statements that are not costly to produce and hence are easy to fake.

An additional issue to consider is the role of technological parameters in dictating the content of online self-presentation. While many online environments do not instruct users on what information to disclose (e.g., discussion boards, personal websites), the majority of profile-based Websites request users to answer a set of pre-determined questions (e.g., height, weight, occupation, income). This may place pressure on users to manage the self-presentation of very personal pieces of information that are not normally disclosed in face-to-face environments (i.e., exact age, weight, income).

Let us now turn our attention to the manner in which self-presentation is conveyed to the audience. According to Walther's (1996) hyperpersonal model of impression formation in computer-mediated environments, online self-presenters may have increased opportunities to control and carefully manage information flow compared to their face-to-face counterparts. This occurs because many online environments offer users a set of affordances that are absent in face-to-face communication:

- *Editability*, or the opportunity to revise one's self-presentation after it has been posted. By contrast, face-to-face self-presenters cannot "take back" an unthoughtful remark or a bad hair day.
- *Asynchronicity*, or the time lag between composing a self-presentation and making it available to others for scrutiny. This allows self-presenters to take as much time as they need to prepare their desired self-presentation. Asynchronicity is an important technological affordance for all profile-based websites, but it may not be available in more interactive environments, such as instant messenger.
- *Re-allocation of cognitive resources*, or the ability to focus solely on composing the self-presentation, without distractions or interferences. Unlike face-to-face settings, where self-presenters need to be mindful of their environment *while* delivering their self-presentation,

online users can compose their self-presentations in the privacy of their home, where they do not have to attend to any distracters.

Together, these affordances allow online communicators to engage in *selective self-presentation* (Walther, 2007), a more controlled and optimized version of face-to-face self-presentation. Additionally, the conventional nature of online self-descriptors (both textual and visual) makes it possible for online self-presenters to truly put their best foot forward.

To summarize, this theoretical framework of online self-presentation postulates that online self-presenters are often highly motivated to control how other people see them, because of the publicness and permanence of their self-presentational acts, and because of their ability to achieve important interpersonal goals online, such as finding love, connecting with friends, or seeking jobs. Additionally, highly motivated self-presenters have the opportunity to create flattering self-presentations, which cater to their audiences and draw upon the best aspects of their actual self, and also aspects of their ideal self. Let us now apply this theoretical framework to self-presentation in online dating profiles and summarize current empirical research that provides support for it. To facilitate the presentation of these results, we begin with an overview of the large study we conducted on self-presentation in online dating profiles.

Study Overview

To assess the self-presentational strategies of online daters, we selected a sample of 80 online daters (40 men and 40 women) from the New York City metropolitan area, who subscribed to one of four online dating services: Match.com, Yahoo Personals, American Singles, and Webdate. These services were selected because they were widely popular, catered to mainstream as opposed to niche audiences, and requested subscribers to compose detailed online self-presentations.

Participants were invited to a study of "self-presentation in online dating profiles" through advertisements in a local newspaper, the Village Voice, and on a popular online forum, craigslist.com. Interested online daters were invited to the study if they were heterosexual and over 18 years of age. We also attempted to match participants' age as closely as possible to the age of a national sample of online daters (Fiore, 2004), in order to increase the generalizability of results.

Participants were invited for a research appointment at the New School University in Manhattan. Prior to their arrival, their online dating profile was archived and printed out. The research procedure comprised several steps. First, participants were asked to go through their profiles and rate the accuracy of their statements on each profile element (e.g., age, height, education, occupation, activities, photographs). Participants were also asked to rate the acceptability of lying on each profile element (e.g., "How acceptable is it to lie about age?"). Second, participants' exact deviations from the truth were measured on three profile elements: height, weight, and age. Participants' height was measured using a standard measuring tape; their weight was measured using a standard scale; and their age was recorded from their driver's licenses. Third, several photographs were taken of each participant: a head shot, a full-body shot, and a photograph in which participants were asked to replicate the pose of their main profile photograph. Fourth, participants filled out a questionnaire about their online dating experiences and self-presentational tactics. Finally, participants were interviewed about their profile self-presentation. All participants were compensated $30 each for their time and effort.

After the completion of the study, participants' lab photographs were shown to a group of judges in order to derive (1) measures of the accuracy of profile photographs and (2) measures of daters' overall physical attractiveness. Specifically, to obtain measures of photographic accuracy, participants' lab photograph was shown side-by-side with their profile photograph to a group of judges, who were undergraduate students at Cornell University. Judges were told that the lab photograph represents daters' everyday appearance, and they were instructed to rate the accuracy of the profile photograph compared to the lab photograph. Another group of judges was shown all three photographs that were taken during daters' research appointment (i.e., headshot, full-body shot, and replica of profile photograph) and were asked to rate each dater's physical attractiveness based on them.

We now apply the theoretical framework of online self-presentation to online dating profiles and provide support for it with the findings from this study, as well as several other relevant studies.

Self-Presentation in Online Dating Profiles

Impression motivation. Consider first online daters' motivation to control their profile self-presentation. The above framework postulates that (1) the basic function of self-presentation is to aid in the fulfillment of personal goals, and (2) self-presenters are more likely to devote time and effort to the creation of their self-presentation the more they value these goals. It goes without saying that the goal of online dating is to establish personal relationships, which many people believe to be the single most important source of personal happiness (Kelley, 1982). Given the importance of establishing romantic relationships, we expect online daters to devote a significant amount of thought to the creation of flattering profiles that can help them succeed in finding a desirable partner.

While all daters are expected to care about their self-presentation because of the importance assigned to romantic relationships in general, an interesting question is whether daters' specific relational goals (i.e., whether they are looking for serious or casual relationships) impact their motivation to control their self-presentation. It can be argued that finding a long-term relationship partner is a more important goal for most people than finding casual dates. Additionally, establishing long-term relationships requires more frequent future interactions than casual relationships. Hence, we expect daters looking for serious relationships to be more motivated to carefully manage their self-presentation. Consistent with this prediction, the participants in our study who were looking for serious relationships wrote more about themselves in the "about me" section of their profile and reported more of a tendency to alter the profile than those looking for casual relationships. Also, serious-minded daters posted photographs that independent judges deemed more realistic. Given the subjectivity of photographs and the difficulty of selecting photographs that present a realistic view of the self, one possible conclusion is that daters motivated to find serious relationships put more thought into the presentation of their physical appearance. Similarly, Gibbs, Ellison, and Heino (2006) found that online daters seeking long-term relationships engaged in higher levels of self-disclosure in their profiles, disclosed more personal information, and reported making more conscious and intentional disclosures.

Another factor that should influence online daters' motivation to control their self-presentation is publicness. All online dating profiles are, by definition, available for others to scrutinize and this audience is potentially very large. In fact, the number of potential mates who could scrutinize one's personal information in online dating is much larger than in more traditional forms of dating, such as meeting others in a bar or at a party. We then expect this high degree of publicness to increase online daters' motivation to control their self-presentational behaviors.

One interesting aspect of publicness in online dating is that the audience is not only large, but also undifferentiated—that is, in many online dating services, it is impossible to know who exactly views one's profile. A distinct possibility is that the profile might be viewed not just by potential mates, but also by people from one's own social circle. As strategic self-presenters who wish to be perceived positively, online daters should be mindful of this possibility and engage in self-presentational acts that take into account this particular audience. Consistent with this hypothesis, we found that online daters whose friends and acquaintances were aware of their online dating profiles posted photographs that were more accurate. Additionally, participants who posted profile photographs and hence made themselves visually identifiable to anyone who accessed their profile, were generally more accurate about their profile statements, and in particular about their relationship status. In other words, online daters were mindful of their audience and engaged in self-presentational acts meant to maintain their credibility in front of the people who knew them well.

Finally, online daters should be motivated to control their profile self-presentation because the profile is a permanent record of their claims, that can remain posted on the online dating site for an unlimited amount of time. This contrasts with face-to-face self-presentations, which are transitory and leave no tangible "residue." Indeed, the online daters in our sample reported that they often save their dates' profiles with a view to comparing them with information that transpires later on in the relationship.

To summarize, the motivation to engage in impression management should be elevated for online daters, because (1) the purpose of online dating profiles is to facilitate romantic relationships, which is a valued goal for most people; (2) statements about the self on online dating profiles are available to a large and undifferentiated audience, which includes both potential mates and friends and acquaintances; and (3) the profile stands as a permanent record of self-presentational claims, that can be scrutinized for a long period of time. Having established that the motivation to control one's image is high, what kind of self-presentations do online daters construct?

Impression construction. Recall that image construction processes involve two steps: figuring out what the "desired" image is, and then implementing it. The desired image is based on the self-presenters' self-concept (i.e., how they see themselves), but also incorporates the self-presenters' perceptions of what their target audience values.

Generally speaking, online daters' desired image should be flattering and positive, such that it attracts potential mates, but also realistic, such that it makes it possible to develop and sustain relationships. Indeed, deception can have catastrophic consequences for relationship development because (1) it makes it hard to gauge whether somebody who appears attractive online will also appear attractive in person; and (2) it undermines trust (Whitty & Joinson, 2008). Consistent with these predictions, the participants in our study reported that encountering deception in others' profiles was generally unacceptable, with deception about important relationship parameters, such as relationship status or having children, considered completely unacceptable.

Given the norms against deception in online dating profiles, we expect online daters to draw upon their actual self, or the characteristics they currently possess, when constructing their self-presentation. Indeed, online daters reported that it is important to present their actual self in their profiles because presenting oneself honestly is necessary for relationship development (Whitty, 2008). Similarly, the online daters interviewed by Ellison and colleagues (2006) reported they felt it was necessary to articulate versions of themselves that were grounded in reality, such that future meetings would not be unpleasant or surprising. When asked to report the accuracy of their profiles, the online daters in our sample reported a high degree of accuracy for all profile statements: on a scale from 1 (completely inaccurate) to 5 (completely accurate) all profile elements were rated higher than 4.

It is important to note that presenting one's actual self is not necessarily a simple, uncalculated exercise—in fact, it takes conscious effort to select and display the aspects of one's actual self that one is proud of (see Leary & Kowalski, 1990). Consistent with this, we found that online daters who were judged as physically attractive by a group of independent judges posted more photographs of themselves than their less attractive counterparts, presumably in an effort to showcase their physical attractiveness.

While online daters face pressure to portray themselves accurately in their profiles, they also need to present a version of the self that is attractive and worthy of pursuit. This may lead them to present aspects of their ideal self—or characteristics that they would like to possess, but do not currently possess. One way to introduce elements of the ideal self without undermining one's credibility is to select aspects of the self that are either attainable in the near future or justifiable in some other way. For instance, the online daters interviewed by Ellison and her colleagues (2006) reported presenting themselves as thinner online because (1) being thinner was part of their ideal self-conceptualization; and (2) they thought they could lose weight before meeting other daters in person, thus eliminating the deception. In fact, presenting a thinner persona in the online profile served as a motivation to lose weight. Similarly, when presenting their "activities" (e.g., hiking, skiing), some online daters reported selecting activities in which they engage sporadically, but which they would like to pursue if they had more time. This presentation of the ideal, but not actual, self is justifiable by the ambiguity of the online profile, which doesn't specify whether the activities are part of daters' current or past routine, or whether the activities are practiced frequently or not.

Another key criterion in constructing online dating self-presentations is catering to the preferences and values of the audience. What exactly do daters look for in potential partners? It is a widely accepted notion that, in addition to idiosyncratic preferences, people generally look for two characteristics in potential partners: physical attractiveness and social status. According to evolutionary psychology, both these characteristics enhance reproductive fitness (i.e., the ability to pass on our genes to the next generation), which is why we have evolved to favor them. Specifically, youth and physical attractiveness serve as an honest indicator of people's health, good genes, and overall mate quality (e.g., Barber, 1995; Buss & Schmitt, 1993; Daly & Wilson, 1995; Gangestad & Thornhill, 1997; Symons, 1979), whereas high social status serves as an indicator of ability to provide and protect (Feingold, 1992; Sprecher, 1989; Trivers, 1985).

While physical attractiveness and social status are generally favored, research shows that men tend to prefer physical attractiveness and youth in potential mates as an indicator of fertility, whereas women tend to prefer social status in potential mates as an indicator of ability to provide and protect (Lance, 1998; Woll & Cozby, 1987). A robust body of research provides support for these claims. For example, when composing newspaper personals, women emphasized their physical attractiveness and body shape (Ahuvia & Adelman, 1992; Hirschman, 1987; Jagger, 2001) and men spent more time seeking information about women's youth and physical appearance (Lynn & Bolig, 1985); richer men tend to pursue more physically attractive women, while women are more attracted to men with higher status occupations (Hitsch, Hortacsu, & Ariely, 2004); attractive people are considered more desirable dating partners, are more popular with the opposite sex, and are able to attract more desirable partners (Gangestad & Scheyd, 2005; Riggio, Widaman, Tucker, & Salinas, 1991; Singh, 2004). We conclude that online daters' "desired" impression should include youth and physical attractiveness (particularly for women) and high social status (particularly for men).

As strategic self-presenters, we expect online daters to seek to incorporate these desired images into their self-presentation. But does the medium of communication—in this case, the online dating profile—allow them to do so? As discussed earlier, the online environment constitutes an ideal venue for putting one's best foot forward, because it provides communicators with a great degree of control over their self-presentation, much more so than the face-to-face environment (Walther, 2007). Indeed, online profiles are composed under conditions of (1) asynchronicity, meaning that online daters have as much time as they wish to compose their self-presentations; (2) editability, meaning that daters have the opportunity to revise them until they are fully satisfied with them; and (3) reallocation of cognitive resources, meaning that online daters can dedicate their undivided attention to the creation of flattering profiles (much unlike face-to-face daters, who must attend to the conversation and to environmental distracters while trying to come across as desirable to potential mates). Further, the elements of online dating presentation are either purely textual (i.e., online daters verbally describe themselves) or visual (i.e., online daters post photographs to describe their physical appearance) and as such they are very malleable and subject to control. Recall that these elements can be construed as conventional signals (Donath, 2007), because they are quite inexpensive to produce—daters only need to type out their desired self-presentations, or simply upload older photographs that present a more attractive version of the self—and hence very easy to fake. How do online daters take advantage of these affordances to present their desired image of attractiveness but also honesty?

We have already discussed that, generally speaking, online daters reported a high degree of honesty in all of their profile statements. Let us now take a closer look at the presentation of their physical appearance through (1) verbal self-descriptors of height, weight, and age; and (2) photographs. We focus on physical appearance because it is one of the most important criteria of what daters look for in potential partners (Whitty, 2008) and because it is prominently featured by the online dating profile.

Recall that physical attractiveness is a highly prized asset in the dating world, particularly for women, and that the online environment enables daters to present an optimized version of themselves. As a result, a full 81% of the daters in our sample misrepresented their height, weight, or age. Although frequent, these misrepresentations were small in magnitude: on average, height deviations were about 0.77 inches, with actual height ranging from 3 inches taller to 1.75 inches shorter than profile statements; weight deviations were about 9 lbs, ranging from 35 lbs heavier to 20 lbs lighter than reported in profile; and age deviations were about half a year, with real age ranging from 3 years younger to 9 years older than what was reported in the profile. It can be argued that while these deviations presented a slightly more flattering version of the self, they were small enough not to be detected in face-to-face meetings. Interestingly, the element that was most frequently misrepresented was weight, with about two thirds of the participants presenting inaccurate weight measurements. Compared to height and age, lying about one's weight is more justifiable because (1) one can claim ignorance about one's precise weight, but not about one's age or height; and (2) weight can be adjusted to match profile statements, by either losing or gaining weight, whereas this adjustment is impossible for height or age. Lies about height and weight were also tailored to the preferences of the opposite gender: men lied more about their height, given that women prefer taller men (as an indicator of high status), and women lied more about their weight, given that men prefer thinner women. Importantly, the magnitude of these lies was tailored to online daters' actual *need* to self-enhance: less attractive daters lied more about these indicators than attractive daters, presumably in an effort to boost their perceived attractiveness. Together, these data suggest that online daters lied strategically in the verbal description of their physical appearance: they deviated from the truth just enough to present a slightly more favorable version of the self without appearing dishonest, they catered to the preferences of their audience, and they compensated for specific shortcomings, such as lack of attractiveness.

The same strategic approach to self-presentation was observed in the depiction of the physical self through photographs. Because physical attractiveness is a more valued characteristic of women than of men, women posted more photographs than men (an average of about four compared to an average of about two), presumably in an attempt to display their physical self, and they also engaged in more photographic self-enhancement. Specifically, women's profile photographs were rated as more attractive than their everyday photographs (taken in the lab), while this was not the case for men. Women's photographs were also perceived as less accurate than men's. A group of coders identified the specific discrepancies between daters' everyday photographs and the photographs they had posted on their profiles. These discrepancies referred to physical characteristics, such as age, hair, skin, or photographic processes, such as retouching or hiring a professional photographer. Results show that (1) women's photographs tended to be retouched or professionally taken more so than men's, (2) women's photographs contained more discrepancies related to physical characteristics, such as age, hairstyle, and skin, than men's photographs; and (3) women's photographs contained, on average, more discrepancies than men's. Also noteworthy is that women posted older photographs (about 17 months) than men (about 6 months), thus displaying younger and potentially more attractive versions of themselves. Given that photography is a subjective and elastic medium that can be easily manipulated, we conclude that women strategically manipulated the presentation of their physical self through photographs such that they cater to men's preferences towards youth and physical attractiveness.

Another important finding regarding the accuracy of photographs is that less attractive daters engaged in more photographic self-enhancement than their more attractive counterparts, with unattractive women posting the least accurate photographs. Again, this underscores the strategic aspect of self-presentation, where online daters take into account their own strengths and weaknesses and use deception as a resource to create more flattering self-presentations that are tailored to the preferences of their audiences.

Conclusion

This chapter introduces a theoretical framework for online self-presentation that considers both the motivation to engage in image management and the actual construction of this image, given the affordances and limitations of the communication medium. We then apply this theoretical framework to self-presentation in online dating profiles and provide empirical support for it by reviewing several studies on this topic.

Online dating self-presentation is different from face-to-face self-presentation in that it requires the disclosure of a wealth of very private information, but it also gives self-presenters a broad arsenal of tools to control these disclosures. Highly motivated to create favorable impressions, online daters appear to handle this situation by taking advantage of the affordances of the online world to put their best foot forward. Results show that online daters are highly strategic in their self-presentational choices: the profile presents a version of daters' actual self that is slightly improved through small and strategically placed deceptions. These deceptions cater to the specific preferences of potential mates, such as men's preference for thin women and women's preference for tall men, and also are meant to redress daters' shortcomings, such as reduced physical attractiveness.

It can be argued that this kind of highly controlled self-presentation is only possible in computer-mediated environments. For instance, traditional daters only have limited options for enhancing their attractiveness: wearing flattering clothes, having their makeup and hair professionally done, or wearing a nice perfume. They also have to be highly spontaneous in order to make the best impressions during conversation. In the online environment, however, daters have a wide array of tools for boosting their attractiveness: selecting the most flattering photographs out of large repositories of photographs accumulated over the years, hiring professional photographers or retouching their photographs, laboring for days or weeks over their self-descriptions, or even asking for friends' help to compose the most flattering profile.

Because of these increased opportunities for selective self-presentation, the online environment may raise questions about users' behaviors that couldn't be examined in face-to-face environments (see also Walther, Gay, & Hancock, 2005). For instance, when given such liberty to take license with the truth, how do online self-presenters manage their images? We find that online daters do not lie discriminately, simply because they can, but rather that they use deceptions strategically in order to accomplish their face-to-face goals. This is consistent with self-presentational tactics used in face-to-face environments (see Leary & Kowalski, 1990), where people also use deception cautiously and make an effort to present themselves both positively and accurately. We conclude that, even though the online environment is barely a few decades old, users' behavior in it can be predicted in systematic ways by relational goals and preferences that have been hardwired through millennia of evolution.

References

Ahuvia, A. C., & Adelman, M. B. (1992). Formal intermediaries in the marriage market: A typology and review. *Journal of Marriage and the Family, 54,* 452–463.

Barber, N. (1995). The evolutionary psychology of physical attractiveness: Sexual selection and human morphology. *Ethology and Sociobiology, 16,* 395–424.

Beck, R. C. (1983). *Motivation: Theories and principles* (2nd ed.). Englewood Cliffs, NJ: Prentice Hall.

Buss, D. M., & Schmitt, D. P. (1993). Social strategies theory: An evolutionary perspective on human mating. *Psychological Review, 100,* 204–232.

Daly, M., & Wilson, M. (1995). Discriminative parental solicitude and the relevance of evolutionary models to the analysis of motivational systems. In M. S. Gassaniga (Ed.), *The cognitive neurosciences,* (pp. 1269–1286). Cambridge, MA: MIT Press.

Donath, J. (2007). Signals in social supernets. *Journal of Computer-Mediated Communication, 13,* 231–351.

Ellison, N., Heino, R., & Gibbs, J. (2006). Managing impressions online: Self-presentation processes in the online dating environment. *Journal of Computer-Mediated Communication, 11,* article 2. http://jcmc.indiana.edu/vol11/issue2/ellison.html

Feingold, A. (1992). Gender differences in mate selection preferences: A test of the parental investment model. *Psychological Bulletin, 112,* 125–139.

Fiore, A. T. (2004). *Romantic Regressions: An Analysis of Behavior in Online Dating Systems.* (Unpublished master's thesis). Massachusetts Institute of Technology, Cambridge, MA.

Gangestad, S. W., & Scheyd, G. J. (2005). The evolution of human physical attractiveness. *Annual Review of Anthropology, 34,* 523–548.

Gangestad, S. W., & Thornhill, R. (1997). Human sexual selection and developmental stability. In A. Simpson & D. T. Kenrick (Eds.), *Evolutionary social psychology* (pp. 169–195). Mahwah, NJ: Lawrence Erlbaum.

Gibbs, J. L., Ellison, N. B., & Heino, R. D. (2006). Self-presentation in online personals: The role of anticipated future interaction, self-disclosure, and perceived success in Internet dating. *Communication Research, 33,* 1-26.

Goffman, E. (1959). *The Presentation of Self in Everyday Life.* New York: Anchor

Hancock, J., Thom-Santelli, J., & Ritchie, T. (2004). Deception and design: The impact of communication technology on lying behavior. In E. Dykstra-Erickson & M. Tscheligi (Eds.), *Proceedings of the 2004 Conference on Human Factors in Computing Systems* (pp. 129–134). New York: ACM.

Hancock, J. T., & Toma, C. L. (2009). Putting your best face forward: The accuracy of online dating photographs. *Journal of Communication, 59,* 367–386.

Higgins, E. T. (1987). Self-discrepancy theory. *Psychological Review, 94,* 1120–1134.

Hirschman, E. C. (1987). People as products: Analysis of a complex marketing exchange. *Journal of Marketing, 51,* 98–108.

Hitsch, G. J., Hortacsu, A., & Ariely, D. (2004). *What makes you click: An empirical analysis of online dating* (Working Paper). Retrieved from http://papers.ssrn.com/sol3/Papers.cfm?abstract_id=895442

Jagger, E. (2001). Marketing Molly and Melville: Dating in a postmodern, consumer society. *Sociology, 35,* 39–57.

Kelley, H. H. (1982). *Personal relationships: Their structure and processes.* Hillsdale, NJ: Lawrence Erlbaum.

Lance, L. (1998). Gender differences in heterosexual dating: A content analysis of personal ads. *Journal of Men's Studies, 6,* 297–305.

Leary, M. R., & Kowalski, R. M. (1990). Impression management: A literature review and two-component model. *Psychological Bulletin, 107,* 34–47.

Lynn, M., & Bolig, R. (1985). Personal advertisements: Sources of data about relationships. *Journal of Social and Personal Relationships, 2,* 377–383.

McKenna, K. Y. A., & Bargh, J. (2000). Plan 9 from Cyberspace: The implications of the Internet for personality and social psychology. *Personality and Social Psychology Review, 4.* 57–75.

Riggio, R. E., Widaman, K. F., Tucker, J. S., & Salinas, C. (1991). Beauty is more than skin deep: Components of attractiveness. *Basic and Applied Social Psychology, 12,* 423–469.

Schlenker, B. R. (2002). Self-presentation. In M. R. Leary & J. P. Tangney (Eds.), *Handbook of self and identity* (pp. 492–518). New York: Guilford.

Singh, D. (2004). Mating strategies of young women: Role of physical attractiveness. *The Journal of Sex Research, 41,* 43–54.

Sprecher, S. (1989). The importance to males and females of physical attractiveness, earning potential, and expressiveness in initial attraction. *Sex Roles, 21,* 591–607.

Symons, D. (1979). *The evolution of human sexuality.* New York: Oxford University Press.

Toma, C. L., & Hancock, J. T. (2010). Looks and lies: Self-presentation in online dating profiles. *Communication Research, 37,* 335–531.

Toma, C. L., Hancock, J. T., & Ellison, N. B. (2008). Separating fact from fiction: An examination of deceptive self-presentation in online dating profiles. *Personality and Social Psychology Bulletin, 34,* 1023–1036.

Trivers, R. (1985). *Social evolution.* Menlo Park, CA: Benjamin Cummings.

Turkle, S. (1995). *Life on the Screen: Identity in the Age of the Internet.* New York: Simon & Schuster.

Walther, J. B. (1996). Computer-mediated communication: Impersonal, interpersonal, and hyperpersonal interaction. *Communication Research, 23,* 3–44.

Walther, J. B. (2007). Selective self-presentation in computer-mediated communication: Hyperpersonal dimensions of technology, language, and cognition. *Computers in Human Behavior, 23,* 2538–2557.

Walther, J. B., Gay, G., & Hancock, J. T. (2005). How do communication and technology researchers study the Internet? *Journal of Communication, 55,* 632–657.

Walther, J. B., & Parks, M. R. (2002). Cues filtered out, cues filtered in: Computer-mediated communication and relationships. In M. L. Knapp & J. A. Daly (Eds.), *Handbook of Interpersonal Communication* (3rd ed., pp. 529–563). Thousand Oaks, CA: Sage.

Whitty, M. T. (2008). Revealing the "real" me, searching for the "actual" you: Presentations of self on an internet dating site. *Computers in Human Behavior, 24,* 1707–1723.

Whitty, M. T. & Joinson, A. N. (2008). *Truth, lies and trust on the internet.* London: Psychology Press, Routledge.

Woll, S., & Cozby, P. C. (1987). Videodating and other alternatives to traditional methods of relationship initiation. *Advances in Personal Relationships, 1,* 69–108.

CHAPTER FOUR

Microchannels and CMC: Short Paths to Developing, Maintaining, and Dissolving Relationships

Deborah Ballard-Reisch

Bobby Rozzell

Lou Heldman

David Kamerer

Abstract

The purpose of this chapter is to examine what foundation and emerging communication theories might offer an understanding of how individuals develop, maintain and dissolve interpersonal relationships using microchannel media. This chapter will examine six communication theories that have fruitfully been applied to analysis of the development of personal relationships: social construction theory (Berger & Luckmann, 1966), systems theory (Watzlawick, Beavin, & Jackson, 1967), social exchange theory (Thibault & Kelley, 1959), social penetration theory (Altman & Taylor, 1973), strong and weak ties theory (Granovetter, 1973), bridging and bonding theory (Putnam, 2000). In addition the potential contributions of four theories, either those not traditionally applied to relationship development, or newly emerging theories will be explored: uses and gratifications theory (Katz et al., 1974), social identity and deindividuation theory (SIDE) (Lea & Spears, 1992), social information processing theory (SIP) (Walther, 1992) and hyperpersonal theory (Walther, 1996). The objective of this chapter is to identify the unique contributions each theory might make to research on how individuals use microchannel media to facilitate relationship processes.

Since the advent of the Internet, communication researchers on one end of the spectrum have decried the threat of computer-mediated communication to meaningful social relationships (Bos, Olson, Gergle, Olson, & Wright, 2002; Herring, 1999; Joinson, 2001; Miller, 2008). Miller (2008) sees the threat increasing with microchannel communication options like Twitter that lead to the loss of important exchanges and the domination of what he terms phatic communication, or communication that has "purely social (networking) and not informational or dialogic intents" (p. 387). At the other end are communication scholars who view computer-mediated communication as a tool for relationship development. These scholars have attempted to identify the underlying processes that allow for the development and maintenance of interpersonal relationships through computer-mediated communication (Tidwell & Walther, 2002; Walther, 1992, 1996; Wright, 2004). These issues are of increasing importance as more computer-mediated communication takes place in short-form environments.

Three popular forms of microchannel communication have emerged and become part of mass culture: instant messaging, short message service and, more recently, microblogging. Instant messaging, as popularized by AOL's Instant Messenger (IM) client software, allows real-time text-based communication with selected "buddies" over the Internet. Pew (2004) found that 53 million Americans, or 42% of Internet users, used instant messaging in 2004. The channel is most popular with adults 18–27, with 35% of heavy IM users coming from this group. Instant messaging networks are intimate and contained; two-thirds in the study regularly communicated with five people or fewer; only 9% IM'd more than 10 people. Growth on instant messaging networks has been contained in part by client incompatibility; products from the three leaders in client software—AOL's AIM, Microsoft's Windows Live Messenger, and Yahoo Messenger—don't directly communicate with one another. Even so, worldwide audiences for IM are large: a study conducted by AOL in 2006 found that 12 billion instant messages are sent daily, with a worldwide user base is 300 million people, 80 million in the United States. In the United States, 17% of all adults send and receive instant messages on mobile devices (Pew, 2008).

Short message service, or SMS, informally called text messaging, is a way to send short (up to 160 character) text messages over mobile telephone networks. While typically one-to-one, the SMS protocol can also be used to broadcast messages, vote in contests, or donate to a cause. SMS is very popular in Europe, Asia, and worldwide, in part because for many consumers SMS messages are cheaper than phone calls. According to Nielsen Mobile (2008) even in the United States, the average consumer sends more SMS messages (357 a month) than makes phone calls (204 a month) on their mobile phones. For all age groups from 12-44 years, text messages per month surpass phone calls. SMS use nearly tripled in the U.S. from 2007 to 2008, to one trillion messages (CTIA, 2009).

In 2009, microblogging became part of popular culture, with Twitter featured on the *Oprah Winfrey Show* and the cover of *Time* magazine. Entering the year, 11% of U.S. online adults were using Twitter or a similar service, according to Pew (2009). Microblogging shares modal qualities of instant messaging, but the messages are shared publicly online. Microblogs limit user posts to a set number of characters, typically fewer than 200. Services that conform to this definition include Jaiku (www.jaiku.com), Plurk (www.plurk.com), and Twitter (www.Twitter.com). Twitter, launched in 2006, had 94,000 users by April, 2007 (Java, Song, Finin, & Tseng,

2007) and is currently estimated to have more than seven million users. According to the web information company Alexa.com (2010), Twitter is the ninth most popular website in the United States. Twitter has become a worldwide, multi-lingual social network that limits its users to 140 characters per post (Huberman, Romero, & Wu, 2009). Communication through Twitter ranges from phatic communication to information sharing, from updates and URLs to news reporting by professional and citizen journalists (Krishnamurthy, Gill, & Arlitt, 2008), to interpersonally and interactionally based messages.

A description of Twitter helps illustrate the concept of microblogging. The service requires low bandwidth, is predominantly text-based, and has a system-imposed short length of 140 characters. The barrier to entry is very low. There is no charge to join, no software to install and anyone with an Internet connection can become a member in a few minutes and begin following other members immediately. Twitter is a predominantly open-ended forum where the activity is primarily social, but a significant portion is also commercial or news-based.

Updates are called Tweets and users are called Twitterers or Tweeters. Interaction among users is limited. One can only see messages from people one chooses to follow. Following someone requires clicking the appropriate icon unless users have locked access, which means they must approve those who request to follow them. In essence, Twitter is a self-moderating communication-based community. The number of people one can follow has recently been modified from unlimited to 2,000, unless one has more than 2,000 people following them.

Twitter, like Facebook and other social media, is both one-to-one, and one-to-many communication. Except for direct messages exchanged between two participants, and accounts with locked access, all Twitter communication takes place in the open, and may be observed by followers of both Tweeters. Twitter participants may or may not have an offline relationship with those they are followed by and/or following. Twitter updates may also be sent to or from handheld devices or cell phones via SMS.

Although Twitter provides for a virtual relationship, Twitter users do not necessarily limit themselves to virtual communication. The phenomenon of Tweetups has also spread worldwide. Tweetups are often spontaneous or minimally planned opportunities to meet other Twitterers at an identified location at a particular point in time. Invitations to Tweetups are informal and spread throughout follower networks so that anyone available and interested can attend.

Microblogging is also an integral part of the world's largest social network, Facebook, which according to ComScore had more than 484 million active accounts as of March, 2010 (Techcrunch, 2010). The primary way in which people communicate with others on Facebook, the status update, is limited to 160 characters. While Facebook features a mixture of photos, quizzes, trivia, games, and other activities, the status update is analogous to an Update on Twitter.

A growing body of literature applies communication theory and methodology to the development of various types of computer-mediated relationships. Examples include Chan and Cheng's (2004) use of relationship development models (Duck, 1985; Knapp, 1984); Wright's (2004) exploration of online relationship maintenance approaches, (Canary, Stafford, Hause, and Wallace, 1993); Tidwell and Walther's (2002) use of uncertainty-reduction theory (Berger & Calabrese, 1975); and Ruggiero's (2000) application of uses and gratification theory. However, the limited research that exists on communication and microchannels has largely consisted of conference papers and online distributions of analyses focused on networks of followers and attempts to provide broad characterizations of users and their messages (Krishnamurthy, Gill, & Arlitt, 2008; Miller, 2008; Mischaud, 2007). What is absent from the literature, so far, is an emphasis on understanding the development, maintenance, and dissolution of interpersonal relationships through the use of communication microchannels.

The purpose of this chapter is to assess the potential contributions of six traditional relationship development theories and four promising theories to understand the processes through which individuals develop, maintain, and dissolve interpersonal relationships with the use of microchannel media. Specifically, we will review social construction theory, systems theory, social exchange theory, social penetration theory, strong ties/weak ties theory, and the theory of bridging and bonding. Additionally, theories not traditionally applied to relationship development: uses and gratifications theory, social identity and deindividuation theory (SIDE), social information processing theory (SIP), and hyperpersonal theory to identify the unique contributions each can make to an understanding of the relationship between microchannels and relationship development.

Traditional Relationship Development Theories

Social Construction Theory

Berger and Luckmann (1966) in their landmark work *The Social Construction of Reality* (p. 1) advanced the concept that "reality is socially constructed." "Through socialization, interaction and language, individuals, within the context of social institutions ... collectively constitute the realities within which they live (Stamp, 2004, p. 9). Social construction theory, according to Shi-xu (2005, p. 82), "argues that all knowledge, whether as internal consciousness or the external world or both, is constructed in and situated through human, cultural, and historical interaction." Knowledge and reality construction from this perspective are subjective and emergent.

Craig (1999) expanded on Berger and Luckmann (1966) positing that "Communication ... produces and reproduces shared meanings" (Craig, 1999, p. 125) and that it is through interaction, specifically the use of symbols, that the process of social construction occurs. Gergen (1985) articulated four premises of a social construction perspective:

1. Rather than objectively, human beings know the world subjectively, through experience which is influenced largely by language.
2. Through social interaction, categories of language and meaning emerge within a specific context at a particular point in time.
3. Communication conventions determine how meanings are understood at a particular point in time.
4. Patterns of communication lead to the social construction of reality. (pp. 266–269)

Interaction is critical to the construction of reality on a variety of levels including: "personal identity, the meaning of individual behavior, the formation of social structures, and the determination of value" (Deetz, 1992, p. 81). Interaction is "a fundamental mode of explanation" (Craig, 1999, p. 126) and the process by which individuals account for the world and their experiences in it (Gergen, 1985).

As individuals, our identities are socially constructed (Littlejohn, 2002) through the process of social interaction. Additionally, we provide symbolic indicators of our realities through the accounts we offer (Harré, 1979). Finally, through coordinated action and interaction, human beings co-create a shared reality within which they act (Deetz & Mumby, 1990). Meanings attached to events lead to rules of behavior that guide future action.

The reality constructed by an individual, group, community, or culture is an ongoing co-creation constructed through communication. Thus social construction theory allows researchers to examine the construction of meaning at both micro levels (within individual accounts) as well as macro levels (within a culture, community, or group). Fruitful avenues for unpacking the social dimensions of meaning-making in interpersonal relationships include an analysis of myths, metaphors, themes, narratives, and rituals (Sabourin, 2006). The analysis of CMC from a social constructionist perspective could allow for the identification of social constructions reflected in symbols, efforts to assert rules or actual emergent rules that evolve to govern behavior of particular groups, accounts of individual users about their experiences, and perceptions of relationship development, maintenance, and dissolution using these media.

Systems Theory

Systems theory in communication has its foundations in both Weiner's (1948) study of cybernetics or the regulation, control, and feedback processes systems use, and von Bertalanffy's (1950-1972) general systems theory which emphasized the dynamic, interdependent nature of living organisms. The foundation of the use of systems theory in the study of personal relationships is grounded in Watzlawick, Beavin, and Jackson's *Pragmatics of Human Communication* (1967) and anthropologist Gregory Bateson's *Steps to an Ecology of Mind* (1972) which emphasized the nature of communication processes and patterns.

Cushman and Cahn (1985) define a system as "a set of components which influence each other and which constitute a whole or unity" (p. 10). They further posit that "communication in general, and interpersonal communication in particular, can best be understood by describing the systems in which communication takes place" (p. 10). Systems theorists focus on the interdependent and interconnected nature of the parts of a system and the integration of those parts into a larger whole. As Dainton (2004) concluded, "any time that a group of people has repeated interaction with each other, they represent a system" (p. 51). It is through communication that systems, and in this case, interpersonal relationships, are created and sustained (Dainton, 2004; Monge, 1973). This dynamic is summarized by Baxter and Braithwaite (2008): "Messages in combination generate patterns of communication, which in turn generate the larger evolving patterns of relationship" (p. 339). Systems theory further recognizes that systems are embedded within hierarchies such that smaller subsystems and larger suprasystems exist in a mutually influencing dynamic. Systems are composed of four characteristics: objects (parts, elements, variables), attributes (qualities or properties of the system or objects), internal relationships among objects, and an environment (context) (Littlejohn, 1992, p. 41). "Systems approaches center on the mutual influence between system members, as well as between subsystems, systems, and suprasystems" (Dainton, 2004, p. 51). There are seven fundamental elements of systems theory as it applies to communication in human relationships.

1. *Nonsummativity* (Fisher, 1978)—the whole is greater than the sum of its parts—something unique occurs when the elements of the system are brought together that is beyond the capacity of the individual component parts.
2. *Interdependence* (Rapoport, 1968)—the behavior/communication of one system member is influenced by and influences other system members.
3. *Hierarchy* (Ashby, 1964)—systems are embedded within and contain other systems.
4. *Interaction with the environment* (von Bertalanffy, 1968)—contexts exert influence on systems.
5. *Homeostasis* (Ashby, 1964)—systems work to maintain balance to promote stability in reaction to environmental changes.
6. *Morphogenesis* (Hall & Fagen, 1968)—systems continuously change and adapt.
7. *Equifinality* (von Bertalanffy, 1968)—the same ends can be reached by different means.

In short, systems are composed of interdependent parts that interact within a larger environment, adapt and change in response to internal and external influences, and work to achieve balance and stability. Systems are more than their component parts as a dynamic synergy is created when the parts interact and through adaptations, systems can find a variety of ways to achieve desired ends.

Specifically with respect to communication in interpersonal relationships, Watzlawick, Beavin, and Jackson (1967) articulated five axioms:

1. *The impossibility of not communicating*—all behavior, even a lack of behavior can be interpreted; therefore "one cannot not communicate" (p. 51).

2. *The content and relationship levels of communication*—Every communication has a content and a relationship aspect. The relationship aspect creates the context within which the content is to be interpreted (p. 54).
3. *Punctuation of the sequence*—interactants experience communication as a series of beginnings and endings (punctuations) and view their communication as impacted by the communication of others. The nature of a "relationship is contingent upon the punctuation of the communicational sequences" (p. 59).
4. *Digital and analogic communication*—"human beings communicate both digitally and analogically" (p. 67). The meanings of digital communication are contextual and subjective. In analogic communication symbols have objective and generalized meaning.

Symmetrical and complementary interaction—interactions are always based on either equality or difference (p. 70). When communicants interact in similar ways (sarcasm leads to sarcasm; praise leads to praise) they are communicating symmetrically. When they behave in different ways, they are behaving complementarily (a question leads to an answer; an assertion leads to agreement).

As Dainton (2004) summarizes, "systems theories recognize the complexities of interaction. They focus on the patterns of relationships that develop between people who interact" (p. 57).

Research into microchannel communication could profitably focus on both the structural characteristics that enhance or detract from relationship development as well as the patterns of behavior, rules, and norms that develop in microchannel communication. Similarly, limitations on channel capacity to promote digital and relationship meanings as well as punctuation dynamics could be fruitfully analyzed.

Social Exchange Theory

Drawing from common sense, small-group research, economic theory, behavioral psychology, and even animal research, sociologist George Homans (1958) described social exchange theory as "one of the oldest theories of social behavior" (p. 597). He argued that social exchange theory is grounded in the assumption that people engage one another socially within a context of rewards and costs. Thus, social exchange theory offers an economic model for understanding the choices individuals make as they develop, maintain, and dissolve interpersonal relationships (Thibault & Kelley, 1959). Cropanzano and Mitchell (2005) traced SET back to the 1920s and 1930s and the work of anthropologist Malinowski and concluded that "social exchange involves a series of interactions that generate obligations" and that under the right conditions "have the potential to generate high-quality relationships" (pp. 874–875).

Social exchange theory advances that individuals weigh benefits and costs in light of expectations of the relationship, termed the comparison level. The comparison level is a threshold above which a relationship will be seen as attractive. Additionally, individuals weigh outcomes of current relationships against those of available alternatives, the comparison level of alternatives. The comparison level of alternatives is the best outcome anticipated in other available relationships (Kelley

& Thibault, 1978; Thibault & Kelley, 1952). Social exchange theory posits a dynamic process of reward and cost assessments and comparisons with alternative relationships that influence an individual's decisions to engage in, maintain, or dissolve interpersonal relationships. At base, social exchange theory argues that individuals will attempt to maximize rewards and minimize costs in interpersonal relationships (Kelley, 1979).

Key concepts of social exchange theory include reciprocity, fairness, and negotiated rules, with information, approval, respect, power, group gain, and personal satisfaction among the rewards in successful transactions (Eisenberger et al., 2001, p. 42). Emerson (1981) described reciprocity in exchange-based relationships "Benefits exchanged through social processes are contingent upon benefits provided 'in exchange'" (p. 32). Based on their work in organizational settings, Eisenberger et al. (1961) viewed fairness, or support, as an exchange in which individuals are treated fairly and give loyalty and productivity in return. Cropanzano and Mitchell (2005) noted that negotiated rules "tend to be more explicit and quid pro quo than reciprocal exchanges. In addition, the duties and obligations exchanged are fairly detailed and understood" (p. 878). Foa and Foa (1974) identified two types of outcomes in social exchange: economic (such as money, goods, and services) and socioemotional (such as love and status).

Ongoing relationships are built on trust and mutual interest that grow or diminish over the course of time. Molm and her colleagues described "high levels of trust, mutual regard and feelings of commitment" as necessary for a reciprocal exchange. The payoff was in social capital for relationships that are "reciprocal, trusting and positive" (2007, p. 200).

Social exchange theory provides a framework to examine communication microchannel users' assessments of the relative costs and rewards of involvement with the medium within the context of interpersonal relationship development, maintenance, and dissolution. Specifically, it would be instructive to assess the impact of microchannel communication on existing relationships: how microchannels enhance or detract from existing relationships; to what extent microchannel users develop relationships that either remain virtual or transition to face-to-face, and users' perceptions of the benefits and costs of microchannel communication in creating new interpersonal relationships. Further, users' assessment of microchannels versus other technology as relationship development tools would be informative.

Social Penetration Theory

In social penetration theory, psychologists Irwin Altman and Dalmas Taylor (1973) advance that intimacy in relationships develops over time as partners progressively disclose more and more personal information to one another. Key elements of the social exchange process are 1) *breadth*—the number of topics on which relationship partners share information, and 2) *depth*—the intimacy of the information shared in topic areas. Growing relationships are marked by both the breadth and the depth of information shared. Further, maintaining relationships relies on sharing continued breadth and depth (Altman & Taylor, 1973). The dissolution of relationships comes when information is withheld or is perceived as untrustworthy (Altman & Taylor, 1973). They identified a four-stage process of relationship development and marked the characteristics of intimate relationships that begin to dissolve.

1. *Orientation stage*—initial meeting when basic information is exchanged. If this time is pleasurable then there is movement to the next stage.
2. *Exploratory affective stage*—increased exchange of information. Wide breadth of information but not very much depth. Time of discovering common likes and dislikes and deciding if interactants like each other enough to continue further.
3. *Affective stage*—trust has developed to the depth of a close friend or romantic partner. Exchanges will be of more depth and may include negative information about oneself, trusting that the partner will accept both positive and negative characteristics of the discloser.
4. *Stable stage*—a place of open self-disclosure where no topic is out-of-bounds and there are no secrets. This knowledge allows partners to predict the other's responses due to knowledge of their innermost character.
5. *Depenetration*—a decrease in intimacy that follows a reverse pattern of less and less breadth, depth, and frequency of personal information exchange.

Various approaches to social penetration theory and the development of close relationships emerged based on this developmental perspective including Knapp's (1974) stages of coming together and coming apart and Duck's (1985) personal relationships theory.

From a social penetration theory perspective, relationships are built and deepened as relationship partners reveal the most intimate elements of their personalities to one another. This mutual revelation of self is typically gradual and reciprocal (Altman & Taylor, 1973). Individuals tend to reveal information over time (Werner & Haggard, 1985) as they become more and more vulnerable to one another. Each relationship partner must make on-going decisions to continuously self disclose and move towards increased intimacy, maintain the current level of intimacy, or return to more superficial levels of communication.

Miller and Sunnafrank (1982) conceptualized three kinds of information disclosed in growing relationships moving from most impersonal to most interpersonal.

1. *Cultural information*—typically viewed as public, largely superficial, and easily shared with new acquaintances. Cultural information is the easiest to access and reveals the least about an individual's unique characteristics.
2. *Sociological information*—information derived from the social groups and roles to which a person belongs. While more revealing of the unique characteristics of the individual, sociological information is still largely general and impersonal in nature.
3. *Psychological information*—is the most personally revealing type of information. Psychological information consists of the unique characteristics of a person, the individual feelings and the attitudes they hold. This is the kind of information that allows for specific and intimate knowledge of a person.

Miller (2008) is concerned with the potentially negative impact of microchannel communication on relationship development. Microchannels like Twitter provide the data to explore this issue by preserving a complete record of messages between communicants, including limited visual cues (avatar, word spacing, use of capital letters, emoticons, and/or sending photos). Additionally, microchannel communication users can be interviewed regarding their perceptions of the role this medium plays in their relationships.

Strong and Weak Ties Theory

Granovetter (1973) suggests that the dyadic connections in social networks are not all the same. He proposed four relationship variables that could be measured to determine relationship strength: time, emotional intensity, intimacy, and reciprocal service. The greater the intensity of these variables the stronger the tie between the two actors. Strong ties (such as those between family members or close friends) and weak ties (closer to, but not exactly like, acquaintances) serve people in different ways.

Strong ties usually indicate that the actors share many connections with each other and their social networks greatly overlap. These strong relationships tend to be supportive and deeply meaningful but also insular. Often they are shared by people who are similar and have great trust in one another. Because of this Granovetter suggests that strong ties breed cohesion locally but also bring about fragmentation overall. The more tightly knit a group is the less likely they are to interact with or trust outsiders.

Weak ties serve people in different ways. They allow for connections between actors who may have little in common. They do not demand the investment of time and self that a strong tie does so they can be formed quickly and in larger numbers. Information exchanges can reach a much larger number of people and travel greater social distances. Granovetter suggests that weak ties are, "indispensible to individuals' opportunities and to their integration into communities" (1973, p. 1378). Studies on the significance of strong and weak ties have examined their importance in diffusion of innovations (Rogers, 1979), job searches (Lin, Ensel, & Vaughn 1981), social networks (Friedkin, 1980), and CMC (Petroczi, Neusz, & Bazso, 2006).

From the beginning Granovetter believed this approach was not only important for understanding the dynamics of large scale interpersonal networks but also that, "the personal experience of individuals is closely bound up with larger-scale aspects of social structure, well beyond the purview or control of particular individuals" (Granovetter, 1973, p. 1377). Research from this perspective could emphasize the differing functions that both strong and weak ties play in computer mediated relationships as well as the unique characteristics of each.

Bridging and Bonding

Putnam (2000) coined the terms *bridging* and *bonding* in parallel to Granovetter's weak ties and strong ties, to describe two different types of social relationships. Putnam viewed both bridging and bonding relationships as crucial to the healthy functioning of individuals and society. Bridging relationships, which are similar to weak ties, are the kinds of relationships that occur when individuals make connections across social networks. They are inclusive, often tentative and lack depth. They open up opportunities for the sharing of information and/or new resources. Strength of connection is not the key but the diversity and scale of bridging connections enable people to discover information, learn more about those outside their close group, and become more open to outsiders.

Bonding relationships tend to be exclusive, reinforce group identity and maintain membership boundaries. They are the kind of relationship that provides emotional and physical support and are the foundation of family and friendship networks, Putnam calls bonding relationships the "superglue" of personal relationships while bridging relationships function as "WD-40" (2000, p 23). As with strong ties/ weak ties, research from this perspective could emphasize the unique characteristics of bridging and bonding relationships as well as the different functions each play in peoples' lives.

Promising Theories for Understanding Relationship Development in CMC

Four theories have special potential for unpacking the relationship development processes mediated by online communication.

Uses and Gratifications Theory

Uses and gratifications (U&G) theory lies in contrast to the dominant media paradigm of the 20th century, the media effects model, which viewed audiences as passive recipients of media messages (for example, see Lasswell, 1927). U&G is based on the premise that individuals consciously choose and use media to gratify needs. Katz, Blumler, and Gurevitch (1974) outlined a seven-step model of U&G which included "(1) the social and the psychological origins of (2) needs which generate (3) expectations of (4) the mass media or other sources which lead to (5) differential exposure (or engaging in other activities), resulting in (6) need gratification and (7) other consequences, perhaps mostly unintended ones" (p. 20). While elements of U&G have been part of the communications theory lexicon since the 1940s or earlier (see Lazarsfeld & Stanton, 1944), the theory was articulated and advanced by Elihu Katz and his colleagues, notably in *The Uses of Mass Communication* (1974), who identified the following assumptions:

1. Individuals, as audience members, actively engage in goal-seeking behavior.
2. Individuals actively make choices as they use media. Individuals satisfy needs through a variety of mechanisms; media is but one.
3. Audience members are able to self-report the gratifications behind their media choices.

4. Researchers should refrain from making value-judgments about media and message choices (pp. 15–17).

The uses and gratifications approach assumes active, purposeful engagement in media by audience members (Katz, Blumler, & Gurevitch, 1974). Further, different individuals may seek different kinds of gratifications from the same media channels, and can understand and articulate the gratifications they seek.

Gratifications identified across research fall into the categories of personal gratifications and social gratifications. Personal gratifications include surveillance or information-seeking (Blumler & Katz, 1974; McQuail, 1983; McQuail, Blumler, & Brown, 1972), entertainment/tension release/diversion (Blumler & Katz, 1974; Katz, Gurevitch, & Haas, 1973; McQuail, 1983; McQuail, Blumler, & Brown, 1972; Peters, Almekinders, van Buren, Snippers, & Wessles, 2003), and personal identity/personal integration (McQuail et al., 1972; McQuail, 1983; Katz et al., 1973). Peters et. al., (2003) also found immediate access and time efficiency as personal gratifications while Katz, Gurevitch and Haas (1973) identified cognitive and affective gratifications. McQuail, Blumler, and Brown (1972), McQuail (1983), Katz, Gurevitch, and Haas (1973), and Peters et al., (2003) identified social gratifications associated with social integration and the development of personal relationships. Blumler and Katz (1974) identified cultural transmission.

In addition to straightforward analysis of interpersonal relationship development, maintenance and dissolution, uses and gratifications theory offers a number of additional avenues for research on Twitter use as it impacts these dynamics. For example, research could focus on the impact of channel characteristics like interactivity, demassification, and asynchroniety (Ruggiero, 2000) on relationships. Fredin and David (1998) identified three components of interactivity in hypermedia that might offer relevant insights: 1) that hypermedia demands activity or communication simply stops; 2) that individuals choose from a seemingly infinite set of responses; and 3) that individuals' choices seem to be dependent on prior choices. Brand (1987) expanded on the potential implications of interactivity and identified five corollaries:

1. *Interruptibility*–the ability of the user to pace the communication
2. *Granularity*–the breaking of the message units into user-navigable pieces
3. *Graceful degradation*–the ability of the text to accommodate "wrong" requests from the user
4. *Limited look-ahead*–the inability of a user to determine the outcome of a "conversation" in advance
5. *No default* –the absence of a pre-programmed linear path

Each of these corollaries frames a duty and/or an opportunity on behalf of the audience member as she or he communicates; each frames a component of being an active audience member, one of the implicit assumptions of U&G theory. While studies using the U&G theory have traditionally focused on intrapersonal themes, such as surveillance, escape, or desire to pass the time, there is no reason the paradigm cannot be expanded to study relationship-based gratifications as well (see McQuail et al., 1972). Finally, Palmgreen and Rayburn (1982) melded expectancy value theory with uses and gratifications theory to tease apart gratifications sought from gratifications obtained. This approach may be useful in unpacking reasons why some users integrate a tool like Twitter into relationship development, maintenance, and dissolution activities while others do not, or abandon this media use all together.

The following three theories relate to the lack of information cues available in CMC exchanges in contrast to face-to-face (FtF) encounters and their potential impact on the development of CMC relationships.

Social Identity and Deindividuation (SIDE) Model

The SIDE model (Lea & Spears, 1992) recognizes the absence of nonverbal cues in computer-mediated communication. This lack causes users to form impressions that are based on the social groups and categories of communicators rather than any interpersonal cues interactants may share.

Further, it has been assumed that the anonymity (either full anonymity or the partial anonymity offered by not being physically present even if easily identified) afforded CMC participants leads to uninhibited and aggressive behavior such as flaming (internet slang for angry and / or demeaning messages) and trolling (internet slang for a poster's intentional behavior that disrupts online discussion groups) (Donath, 1999). Field research in the CMC environment paints a much more complex, community-influenced, picture (Postmes, Spears, & Lea, 1998). The situational norms of a group, along with the lack of contextual and nonverbal cues, can powerfully influence the behavior of online communicators. The social identity model of deindividuation effects (SIDE) posits that what is unique to the online experience can reinforce conformity to online group norms. Flaming often occurs in CMC situations because of cues that encourage the behavior either intentionally or unintentionally, as when participants give the flamer an inordinate amount of attention (Donath, 1999). Because individuating cues, such as physical context and nonverbal communication, are absent in text-based CMC, the cues that do occur take on greater value and partners often over-attribute meaning to the cues they are given. Due to this lessening of individuation, the communication context changes from one of dealing with idiosyncratic individuals to one of building a shared social identity. This shift from a personal identity to a social identity, in certain CMC contexts, can be a powerful enforcer of the norms of a group that would hold no power at all in a face-to-face setting.

The use of the SIDE model would allow for assessment of the dynamics through which CMC users negotiate relationships and develop relational norms absent common cues in face-to-face interaction.

Social Information Processing (SIP) Model

CMC's scarcity of clues, both non-verbal and contextual (for example, physical setting, spatial relationship, and dress), led many to assume it was an impersonal medium that promoted equal participation and was task-oriented but poorly suited to developing consensus (Walther, 1992). In contrast to these assumptions field studies and anecdotal evidence found that participants in various forms of CMC reported socio-emotional content and consensus building which refuted the assumptions.

In response to this seeming contradiction Walther (1992) proposed the Social Information Processing (SIP) model. He suggested that what CMC seems to lack is not inherent to the medium but a product of time, or more precisely, a product of message quantity. The cues that are lacking at the beginning of a computer-mediated relationship can be communicated but it requires supplying them message by message. If the participants adapt the cues available to them to relationship management then the more messages they exchange the more relational and contextual cues they are able to share.

The SIP model (Walther, 1992) suggests that CMC users adjust to the lack of non-verbal cues by making use of the cues that might be available including typographic (Walther & D'Addario, 2001), chronemic (Walther & Tidwell, 1995; Kalman, et al., 2006), language and content. Misspellings, bad grammar, and over use of exclamation points can lead to a strong negative judgment of the sender. Likewise kind, positive statements can lead to strong positive feelings towards the sender beyond what they would receive in a face-to-face setting (Walther, 1996). SIP offers CMC researchers potential insight into what cues lead to what types of interpretations and how those interpretations impact relationship development, maintenance and dissolution.

Hyperpersonal Model

The hyperpersonal model argues that due to the absence of non-verbal cues that are readily available in FtF exchanges CMC users engage in selective self-presentation and partner idealization. The resulting exchanges can be more intimate than those of FtF encounters and lead to a different relationship than one based on FtF interaction. (Walther, 1997; Walther, Slovacek, & Tidwell, 2001). The hyperpersonal model of CMC (Walther, 1996) builds upon the SIP and SIDE models. The hyperpersonal model is built upon three unique characteristics of online, especially text-oriented, relationships.

1. Because of the lack of cues in CMC, as predicted by the SIDE model, an idealized perception of partners is often the norm. The lack of cues enables a greater control over first impressions and the opportunity to avoid physical/social judgments. Participants are also able to devote more cognitive resources to the communication process at hand and have less concern about their physical self presentations.

2. Because of the asynchronous nature of CMC (meaning that one does not need to immediately respond to another's message and can delay response, the amount of delay depending on the type of CMC), participants have time, should they choose, to carefully plan their responses and construct their presentations. Asynchrony also allows communicators to overcome the temporal limits of conflicting or restrictive schedules.
3. CMC feedback loops, as they are also composed of restricted cues, can lead to stronger positive (or negative) feelings because they will include less information that disconfirms previous perceptions and will be reinforced by the above mentioned characteristics of asynchronous and limited cue CMC.

One may wonder why CMC participants have such a positive perception of hyperpersonal online relationships (Henderson & Gilding, 2004) when they are characterized by selective self-presentation, idealization, and a lack of information. Walther suggests that these types of relationships can be "profoundly rewarding" and "more desirable than we can often manage FtF" (1996, p. 28).

Thus, the hyperpersonal model would allow CMC researchers to assess impacts of idealized perceptions of communication partners, the value of asynchronious communication, and the positive assessment of CMC relationships by participants to unpack the contributors to these dynamics in the development, maintenance and dissolution of interpersonal relationships.

Conclusion

As interpersonal relationships have developed online components, CMC plays a more significant role in their development, maintenance, and dissolution. While a growing body of research exists analyzing the impact of CMC on relationships, technological developments have outpaced the research. The newest emerging technology, the use of microchannels of communication, is largely unstudied. While traditional relationship development theories will not likely address all the dynamics of communication in these relationships, they do offer useful tools for initial inquiry as more specialized, context specific theories emerge. Additionally, CMC requires a reassessment of the value of various media, communication and relationship theories for unpacking the unique relationship dynamics facilitated by and through these media.

Microchannel communication modalities are too new to have generated significant research on relationship formation, development, and dissolution, but foundation and emerging communication theories can provide useful lenses for studying these relationships. The authors, all of whom are active users of microchannel communication, have spent more than two years observing a group of participants in a mid sized U.S. city. Among the four of us, we regularly follow more than 100 local microchannel users at least several times a week.

At the end of a Media Transformation course, in which Wichita State University communication students were required to join and become active in Twitter, a 22-year-old female senior wrote:

> *Mostly I want to thank you for introducing me to Twitter. If you would have told me in December that my life was going to drastically change just because of a social media site I would have thought you were crazy. But now it's a reality. I have learned so many things, gained MANY friendships and explored my opportunities in Wichita–personally and professionally.*

As Cropanzano and Mitchell noted in reviewing research by others in a management science context, "relationship development is not a matter of single-stimulus response .... The goal achieved at one step (successfully grasping the next rung) provides the foundation for an even higher climb" (2005, p. 890). The value of microchannels in advancing this developmental process in relationships has yet to be adequately studied in terms of its potential contributions to the ways in which relationships develop, are maintained, and are dissolved.

References

Alexa., J. (2010) Retrieved June 10, 2010, from http://www.alexa.com/siteinfo/twitter.com#

Altman, I., & Taylor, D. A. (1973) *Social penetration*. New York: Holt.

Ashby, R.W. (1964). *Introduction to cybernetics*. London: Routledge.

Bateson, G. (2000). *Steps to an ecology of mind*. Chicago: University of Chicago Press.

Baxter, L.A., & Braithwaite, D. O. (2008). *Engaging theories in interpersonal communication*. Thousand Oaks, CA: Sage.

Berger, C. R., & Calabrese, R. J. (1975). Some explorations in initial interaction and beyond: Toward a developmental theory of interpersonal communication. *Human Communication Research, 1*(2), 14.

Berger, P. L., & Luckmann, T. (1966). *The social construction of reality: A treatise in the sociology of knowledge*. Garden City, NY: Anchor.

Blumler, J., & Katz, E. (1974). *The uses of mass communications*. Beverly Hills, CA: Sage.

Bos, N., Olson, J., Gergle, D., Olson, G., & Wright, Z. (2002). *Effects of four computer-mediated communications channels on trust development*. Paper presented at the SIGCHI conference on Human factors in computing systems: Changing our world, changing ourselves.

Brand, S. (1987). *The Media Lab: inventing the future at MIT*. New York: Penguin.

Canary, D., Stafford, L., Hause, K., & Wallace, L. (1993). An inductive analysis of relational maintenance strategies: A comparison among young lovers, relatives, friends, and others. *Communication Research Reports, 10*, 10.

Chan, D. K.-S., & Cheng, G. H.-L. (2004). A comparison of offline and online friendship qualities at different stages of relationship development. *Journal of Social and Personal Relationships, 21*(3), 16.

Cropanzano, R., & Mitchell, M. (2005). Social exchange theory: An interdisciplinary review. *Journal of Management, 31* (December 6), 874-900.

Craig, R. T. (1999). Communication theory as a field. *Communication Theory, 9*, 119-161.

CTIA (2009). *Semi-annual wireless industry survey*. Retrieved September 14, 2009 from http://ctia.org/media/ industry_info/index.cfm/AID/10316

Cushman, D.P., & Cahn, D.D. Jr., (1985). *Communication in interpersonal relationships*. Albany, State University of New York: SUNY Press.

Dainton, M. (2004). Explaining theories of interpersonal communication, In M. Dainton & E.D. Zelley (Eds.), *Communication theory for professional life: A practical* introduction (pp. 50-73). Thousand Oaks, CA: Sage.

Deetz, S. (1992). *Democracy in an age of corporate colonization: Developments in communication and the politics of everyday life.* Albany, State University of New York: SUNY Press.

Deetz, S., & Mumby, D. K. (1990). Power, discourse, and the workplace: Reclaiming the critical tradition. *Communication Yearbook, 13,* 18–47.

Donath, J. S. (1999). Identity and deception in the virtual community. In M. A. Smith & P. Kollock (Ed.), *Communities in cyberspace.* (pp. 29-59) New York: Routledge.

Duck, S. (1985). Social and personal relationships. In M. L. Knapp & G. R. Miller (Eds.), *Handbook of interpersonal communication* (pp. 665–686). Beverly Hills, CA: Sage.

Eisenberger, R., Armeli, S., Rexwinkel, B., Lynch, P. D., & Rhoades, L. (2001). Reciprocation of perceived organizational support. *Journal of Applied Psychology, 86,* 42-51.

Emerson, R. (1972), Exchange theory, Part I: A psychological basis for social exchange. In J. Berger, M. Zelditch, & B. Anderson (Eds.), *Sociological Theories in Progress.* (pp. 38–57). Boston: Houghton-Mifflin.

Emerson, R. M. (1981). Social exchange theory. In M. Rosenberg & R. H. Turner. (Eds.). *Social psychology: Sociological perspectives* (pp. 30–65). New York: Basic Books.

Facebook.com (2010). Statistics. Retrieved: May 29, 2010, from http://www.facebook.com/press/info.php?statistics

Fisher, W. (1978). *Perspectives on human communication.* New York: Macmillan.

Foa, U.G., & Foa, E.B. (1974) *Societal Structures of the Mind.* Springfield, IL: Charles C. Thomas.

Fredin, E. & David, P. (1998) Browsing and the hypermedia interaction cycle: a model of self-efficacy and goal dynamics. *Journalism and Mass Communication Quarterly, 75,* 35-54.

Friedkin, N.E., (1980). A test of structural features of Granovetter's strength of weak ties theory. *Social Networks,* 2 (4), 11-22.

Gergen, K. (1985). The social constructionist movement in modern psychology. *American Psychologist.* 40(3). 266-275.

Granovetter, M. (1973). The strength of weak ties. *American Journal of Sociology, 78* (6), 1360-1380.

Hall, A. D., & Fagen, R. E. (1968). Definition of system. In W. Buckley (Ed.), *Modern systems research for the behavioral scientist.* (pp. 81–92). Chicago: Aldine.

Harré, R. (1979). *Social being: A theory for a social psychology II.* Oxford: Blackwell.

Henderson, S., & Gilding, M. (2004). 'I've Never Clicked this Much with Anyone in My Life': Trust and hyperpersonal communication in online friendships. *New Media Society,* 6(4), 487-506.

Herring, S. (1999). Interactional coherence in CMC. *Journal of Computer-Mediated Communication,* 4(4), np.

Homans, G. (1958). Social behavior as exchange. *American Journal of Sociology.* 63 (May 6), 597–606.

Huberman, B. A., Romero, D. M., & Wu, F. (2008). Social networks that matter: Twitter under the microscope. Retrieved October 9, 2009 from http://www.hpl.hp.com/research/scl/papers/twitter/twitter.pdf

Java, A., Song, X., Finin, T., & Tseng, B. (2007). *Why we twitter: understanding microblogging usage and communities.* Paper presented at the 9th WebKDD and 1st SNA-KDD 2007 workshop on Web mining and social network analysis, San Jose, CA.

Joinson, A. N. (2001). Self-disclosure in computer-mediated communication: The role of self-awareness and visual anonymity. *European Journal of Social Psychology, 31*(2), 177–192.

Kalman, Y. M., Ravid, G., Raban, D. R., & Rafaeli, S. (2006). Pauses and response latencies: A chronemic analysis of asynchronous CMC. *Journal of Computer-Mediated Communication, 12*(1), article 1. Retrieved October 9, 2009 from http://jcmc.indiana.edu/vol12/issue1/kalman.html

Katz, E., Blumler, J. G., & Gurevitch, M. (1974). Utilization of mass communication by the individual. In J. G. Blumler & E. Katz (Eds.), *The uses of mass communications: Current perspectives on gratifications research* (pp. 19–32). Beverly Hills, CA: Sage.

Katz, E., Gurevitch, M., & Haas, H. (1973). On the use of the mass media for important things. *American Sociological Review, 38*, 164-181.

Kelley, H.H., & Thibault, J.W. (1978). *Interpersonal relations: A theory of interdependence.* New York: John Wiley.

Knapp, M. (1984). *Interpersonal communication and human relationships.* Boston: Allyn & Bacon.

Krishnamurthy, B., Gill, P., & Arlitt, M. (2008). *A few chirps about twitter.* Paper presented at the first workshop on Online social networks, Aachen, Germany.

Lasswell, H.D. (1927) *Propaganda technique in the World War* (1927; Reprinted with a new introduction, 1971). New York: Knopf.

Lazarsfeld, P.F., & Stanton, F. (1944). *Radio research 1942-43.* New York: Duell, Sloan & Pearce.

Lea, M., & Spears, R. (1992). Paralanguage and social perception in computer-mediated communication. *Journal of Organizational Computing, 2*, 21.

Lin, N., Ensel,W. M., & Vaughn, J. C. (1981). Social resources and strength of ties: Structural factors in occupational status attainment." *American Sociological Review. 46*, 393-405.

Littlejohn, S.W. (2002). *Theories of human communication.* (7th ed.). Belmont, CA: Wadsworth.

McQuail, D. (1983). *Mass Communication Theory* (1st ed.). London: Sage.

McQuail, D., Blumler, J. G., & Brown, J. (1972). The television audience: A revised perspective. In D. McQuail (Ed.), *Sociology of Mass Communication* (pp. 135–165). Harmondsworth, UK: Penguin.

Meyers, R. A., Seibold, D. R., & Shoham, M. D. (July, 2007). *Communicative influence in groups: A review and critique of theoretical perspectives and models.* Paper presented at the Conference of the Interdisciplinary Network for Group Research, Lansing, MI.

Miller, G.R., & Sunnafrank, M.J. (1982). All is for one but one is not for all: A conceptual perspective of interpersonal communication. In F.E.X. Dance (Ed.), *Human communication theory: Comparative essays* (pp. 220–242). New York: Harper & Row.

Miller, K. (2004). *Communication theories: Perspectives, processes, and contexts.* New York: McGraw Hill.

Miller, V. (2008). New media, networking and phatic culture. *Convergence, 14*(4), 14.

Mischaud, E. (2007). *Twitter: Expressions of the whole self–An investigation into user appropriation of a web-based communications platform.* London: London School of Economics and Political Science.

Molm, L., Schaefer, D., & Collett, J. (2007). The value of reciprocity. *Social Psychology Quarterly.* 70 (2): 199–217.

Monge, P.R. (1973). Theory construction in the study of human communication: The system paradigm. *Journal of Communication, 23*, 5–16.

Nielsen Mobile. (2008). http://blog.nielsen.com/nielsenwire/online_mobile/in-us-text-messaging-tops-mobile-phone-calling/

Palmgreen, P., & Rayburn, J.D. (1982). Gratifications sought and media exposure: An expectancy-value model. *Communication Research, 9*, 561–580.

Parks, M. R., & Floyd, K. (1996). Making friends in cyberspace. *Journal of Communication, 46*(1), 21.

Peters, O., Almekinders, J., van Buren, R., Snippers, R. & Wessels, J. (2003). *Motives for SMS Use.* In: 53rd Annual Conference of the International Communication Association, May 23-27, 2003 , San Diego, CA.

Pew Internet & American Life Project (2004). How Americans use instant messaging. Retrieved October 9, 2009, from http://www.pewinternet.org/Reports/2004/How-Americans-Use-Instant-Messaging.aspx

Pew Internet & American Life Project (2008). Mobile access to data and information. Retrieved October 9, 2009, from http://www.pewinternet.org/Press-Releases/2008/Mobile-Access-to-Data-and Information. aspx

Pew Internet & American Life Project (2009). Twitter and Status Updating, Fall 2009. Retrieved October 9, 2009, from http://www.pewinternet.org/Reports/2009/17-Twitter-and-Status-Updating-Fall-2009.aspx

Postmes, T., Spears, R., & Lea, M. (1998). Breaching or building social boundaries? SIDE-effects of computer-mediated communication. *Communication Research,* 25, 689-715.

Putnam, R. D. (2000). *Bowling alone: The collapse and revival of the American community.* New York: Simon & Schuster.

Rapoport, A. (1968). Definition of system. In W. Buckley (Ed.). *Modern systems research for the behavioral scientist* (pp. xiii–xxv). Chicago: Aldine.

Rogers, E. (1979). Network analysis of the diffusion of innovations. In P. W. Holland & S. Leinhardt (Ed.), *Perspectives on Social Network Research,* (pp. 137-64). New York: Academic Press.

Ruggiero, T. E. (2000). Uses and gratifications theory in the 21st century. *Mass Communication and Society,* 3(1), 35.

Sabourin, T.C (2006). Theories and metatheories to explain family communication: An overview. In L. Turner & R. West (Eds.), *The family communication sourcebook* (pp. 43-69). Thousand Oaks, CA: Sage.

Shi-xu. (2005). A *cultural approach to discourse.* New York: Palgrave Macmillan.

Stamp, G, H. (2004). Theories of family relationships and a family relationships theoretical model. In A. L. Vangelisti (Ed.), The handbook of family communication (pp. 1-30). Mahwah, NJ: Lawrence Erlbaum Associates.

TechCrunch (2010) from http://techcrunch.com/2010/04/21/facebook-500-million-visitors-comscore/

Thibault, J.W., & Kelley, H.H. (1959). *The social psychology of groups.* New York: Wiley.

Tidwell, L. C., & Walther, J. B. (2002). Computer-mediated communication effects on disclosure, impressions, and interpersonal evaluations. *Human Communication Research, 28*(3), 32.

Von Bertalanffy, L. (1968). *General systems theory.* New York: Braziller.

Von Bertalanffy, L. (1972). *General systems theory.* New York: Braziller.

Walther, J. B. (1992). Interpersonal effects in computer-meditated interaction; a relational perspective. *Communication Research.* 19, 52-90. Retrieved on July 20, 2010 from http://find.galegroup.com/gps/infomark.do?&contentSet=IAC-Documents&type=retrieve & tabID=T002&prodId=IPS&docId=A11871962&source=gale&srcprod=ITOF&user GroupName=ksstate_wichita&version=1.0

Walther, J. B. (1996). Computer-mediated communication: Impersonal, interpersonal, and hyperpersonal interaction. *Communication Research.* 23, 3-43.

Walther, J. B., & D'Addario, K. P. (2001). The impacts of emoticons on message interpretation in computer-mediated communication. *Social Science Computer Review, 19,* 323-345.

Walther, J. B., Slovacek, C., & Tidwell, L. C. (2001). Is a picture worth a thousand words? Photographic images in long term and short term virtual teams. *Communication Research, 28,* 105-134.

Walther, J. B., & Tidwell, L. C. (1995). Nonverbal cues in computer-mediated communication, and the effect of chronemics on relational communication. *Journal of Organizational Computing, 5,* 355-378.

Watzlawick, P., Beavin, J.H., & Jackson, D.D. (1967). *Pragmatics of human communication.* New York: W.W. Norton.

Weiner, N. (1948) *Cybernetics: Or the control and communication in the animal and the machine.* Cambridge, MA: MIT Press.

Werner, C.M., & Haggard, L. M. (1985). Temporal qualities of interpersonal relationships. In M.L. Knapp & G.R. Miller (Eds.), *Handbook of Interpersonal Communication* (pp. 59–99). Beverly Hills, CA: Sage.

Wright, K. B. (2004). On-line relational maintenance strategies and perceptions of partners within exclusively internet-based and primarily internet-based relationships. *Communication Studies, 55*(2), 15.

PART 2

Processes and Goals in Computer-Mediated Communication in Personal Relationships

CHAPTER FIVE

Communication Competence and Apprehension during CMC in Online and Face-to-Face Relationships

W. Scott Sanders

Patricia Amason

Communication using mediated channels is a common and permanent fixture in the lives of many Americans. Internet usage in the United States alone increased by 32% from an average of 53 million adult users per day to over 70 million (Pew Internet and American Life Project [PIALP] 2005). Based on the results of the PIÁLP it is estimated that 63% of adult Americans, or 128 million people, use the Internet allowing for communication among users through applications such as email, instant messaging, and social networking sites. IM was identified as the most highly selected medium for communicating among teens (Lenhart, Madden, & Hitlin, 2005) and email is the single most popular Internet activity with more time devoted to email than to any other single activity online (PIALP). More importantly, surpassing the telephone, the internet has become the most popular mediated channel for relational management (Baym, Zhang, & Lin, 2004). The PIALP study first conducted in 2000 demonstrated that Internet communication was linked to greater frequency of interaction among friends. Moreover, the results of the PIALP study the following year (2001) showed that IM often is used as a means of initiating dating relationships as well as in their dissolution (see also Lenhart, Madden, & Hitlin, 2005). Moreover, persons use computer-mediated communication (CMC) channels in the day-to-day maintenance of existing relationships (Rabby & Walther, 2003). Coming in a variety of types centered on numerous interests, social networking sites such as Facebook and MySpace are used by millions who integrate them into their daily internet use (Ellison, Steinfield, & Lampe, 2008). Recent PIALP data collected in 2008 reveals that the percentage of adult online users who have profiles on social networking sites increased from 8% in 2005 to 35% (PIALP, 2009).

CMC research focuses on how the characteristics of an online environment shape communication. Much research in this area is devoted to exploring the development of new relationships between strangers via computer networks. Relatively little research, however, examines the impact of CMC on relationships formed and primarily grounded in face-to-face (FtF) interaction. This is significant because many people communicate FtF prior to interacting online. Early models used to explain relationship development online often assume visual anonymity exists between users and that little to no information is exchanged through other media (Postmes, Spears, & Lea, 1998; Walther, 1993, 1996).

As these conditions do not apply to relationships that are initially FtF, these models may not accurately describe how people in FtF relationships use CMC as a supplement to or as a substitution for FtF interaction. Although a strong research and cultural bias indicates preferences towards relational maintenance via FtF interactions (see O'Sullivan, 2000), persons whose friendships and romantic relationships were established primarily through FtF interaction indicate they indeed also communicate on a substantial basis using CMC channels (Sanders & Amason, 2006).

Evidence presented in extant research indicates that people who interact online form deeply personal relationships (Parks & Floyd, 1996; Sanders & Amason, 2006) featuring moderate to high online self-disclosure and a great deal of breadth in the topics discussed online. Spitzberg (2006) contends that one of the major contributions of CMC within social contexts is the use of the media for relational management as it provides alternative and "empowering" (p. 635) means of initiating as well as maintaining existing relationships, in addition to serving as a potential channel for their dissolution (see also Hovick, Meyers, & Timmerman, 2003; McCown, Fischer, Page, & Homant, 2001). Additionally, CMC provides a significant outlet for social contact for persons facing problems associated with social anxiety and loneliness (Patterson & Gojdycz, 2000). CMC channels thus provide a mechanism for discourse that may provide a unique aid in the self-presentation process for persons who interact online with partners in primarily FtF relationships.

Self-Presentation

When people interact with others they do so in an attempt to accomplish interpersonal goals. These goals influence how they choose to interpret events and how they present information to others (Schlenker & Weigold, 1992). Self-presentation, or impression management, occurs when persons seek to control how others perceive them (Leary, 1995). The early research on impression management suffered from the belief that consciously regulating information about oneself to create a desired impression was inherently dishonest. Although blatant deception is undeniably an element in some self-presentations, the majority of self-presentations simply involve the selection of true information that will help actors achieve their goals. However, this tailoring or omission of information to help actors achieve their goals is a normal and constant process during social interaction and therefore does not constitute deception (Leary, 1995; Schlenker & Weigold, 1992).

People have multiple motivations for regulating information about themselves in the conduct of impression management. First, people control information in a manner that maximizes their self-esteem. When they attempt to present an ideal self to others, they are not only attempting to create a positive impression for another person but also to regulate their own emotions (Schlenker & Weigold, 1992). Second, people regulate information about themselves in an effort to elicit feedback promoting accurate self-knowledge (Trope, 1986). Finally, people desire consistency between their self-concepts and the real world (Swann, 1990). They seek confirmatory feedback from their audiences to verify their existing self-conceptions and will even present themselves in such a manner as to confirm negative self-beliefs. Thus, persons use CMC channels in their efforts to mold the perceptions their relational partners have of them and ultimately positive impression management as outcomes of the interactions.

Mediated Impression Management

O'Sullivan (2000) contends that at times people strategically select mediated channels while shaping the information they wish to convey or to maintain more control over communication outcomes (Kelly, Keaten, Larsen, & West, 2004). Mediated channels offer a greater degree of ambiguity as opposed to the relative clarity available in FtF interaction that can be used to the channel selector's advantage. O'Sullivan (2000) found that when the situation was negatively valenced more people preferred mediated channels than when the situation was positively valenced. For example, a person might choose a channel such as email or a letter to obscure unattractive or embarrassing aspects of information they are attempting to convey. On the contrary, people might prefer FtF interaction when revealing positive information because they are present to accept positive feedback and other rewards.

O'Sullivan (1996; 2000) proposes an impression management model that explains how people choose between mediated channels and FtF as a means of meeting self-presentational goals. The underlying premise behind this model is that the channel selector's goals for an interaction are to minimize costs and maximize relational rewards and benefits. When selecting channels for an interaction, users consider the characteristics of the channel, what message the channel itself presents, and whether they possess the technical skills to effectively use the channel. Moreover, the valence of the information and whether the information is salient to the channel selector or the receiver also influence channel selection. In general, information that threatens the preferred self-presentation of channel selectors or that of their partners is downplayed, while information that supports a preferred self-presentation will be emphasized. O'Sullivan (2000) found that people demonstrated the greatest preference for mediated channels when information involved the channel selector's self-presentation and was negatively valenced.

Self-Disclosure

Self-disclosure occurs when one shares private thoughts and feelings with another (Jourard, 1971). Self-disclosures can be divided into two broad categories—personal self-disclosures and relational self-disclosures. Personal self-disclosures simply refer to the revelation of facts or information about oneself. However, a relational self-disclosure functions as metacommunication because it clarifies what an individual thinks or feels about a specific interaction or the relationship between the discloser and the recipient (Baxter, 1987). Relational self-disclosure is important because it is an explicit and direct strategy for gathering information about how the discloser's partner perceives the relationship. It is likely to be employed only under certain circumstances because the explicitness of relational self-disclosure makes it the most risky of the available information-seeking strategies.

Relational self-disclosure often is considered to contribute to the maintenance of relationships (Canary & Stafford, 1994). Sanders and Amason (2006) found that persons in primarily FtF relationships who used CMC channels reported greater perceived relational closeness and higher levels of intimacy in self-disclosures than those who did not select CMC channels. Similar results are reported by Tidwell and Walther (2002) and Whittty (2002). According to the respondents in the PIALP (2000) study, they felt they had greater opportunities for frankness and open disclosure of negative or unpleasant information while interacting through email rather than FtF.

Persons' self-disclosures are oriented towards some type of interpersonal goal accomplishment (Derlega, Metts, Petronio, & Margulis, 1993) such as presenting a positive, likable self-image or attempting to define their relationship with the recipient of the self-disclosure. However, self-disclosure can fulfill many other goals that a person defines when choosing to self-disclose. For example, self-disclosure can also be used as an information-seeking strategy because it is heavily shaped by the norm of reciprocity. The level of intimacy of these responses is usually equivalent to the intimacy level of the original self-disclosure. This traditional explanation for reciprocity states that people wish to maintain equality in their relationships and reciprocate so that both parties share the rewards and risks that are involved in self-disclosure (Derlega et al., 1993). One's abilities to determine how much, when, and to who to self-disclose are tied significantly to how competently they adapt their communication strategies and execute them considering the goals they wish to accomplish in the interaction.

Regardless of their motivation, everyone engages in self-presentation, either subconsciously or purposefully, by controlling what information is self-disclosed as a means of influencing how they are perceived by others. However, because self-presentations are tailored specifically to the audience and the situation, they are influenced by the amount of privacy the actor enjoys. Self-presentation is impacted by persons' comfort in communicating across a range of situations and channels and in managing their personal relationships. Their perceived communication competence also may impact self-presentation strategies in sustaining their FtF relationship and their choice of using CMC channels.

Communication Competence

Much has been written about what constitutes competent communication. Often, the notion of communication competence is described as persons' abilities to demonstrate skills, either innate or developed, to accomplish communicative goals (Spitzberg, 1993) or "to choose among available communicative behaviors" in the attempt to accomplish "interpersonal goals during an encounter while maintaining the face and line" of "fellow interactants within the constraints of the situation" (Wiemann, 1977, p. 198). Moreover, communication competence describes "the evaluative impression of the quality of interaction" (Spitzberg & Cupach, 2002, p. 575).

Distinctive criteria are provided for determining the extent to which a communicator demonstrates competence (Spitzberg, 1993; Spitzberg & Cupach, 2002). First, competent communicators are *clear* in making their intentions known (McCroskey, 1982a; Powers & Lowry, 1984), thus the target understands the meanings intended by the speaker (see Spitzberg and Cupach, 2002, for a full discussion of this construct). Second in the list of criteria involves the degree to which communicators are *satisfied* at the end of the encounter, or what is described by Hecht (1978) as obtaining a sense of fulfillment upon sending or receiving information resulting in a positive psychological state. Additionally, one may gain a sense of satisfaction when less positive expectations resulted in more positive outcomes due to the communication strategies employed or when faced with negative circumstances the person used the best strategies they had available such as when dealing with a highly conflictual situation. The third of the criteria that must be met for competent outcomes associated with communication skills is that messages must be delivered in the most *efficient* method possible. Efficiency is described as the extent to which skills are selected among those that are the most useful in achieving desired outcomes with the least amount of time and effort, complexity, and investment of resources (Spitzberg & Cupach, 2002).

Furthermore, among the criteria for communication skill competency is that speakers also must be *effective* in accomplishing their desired goal (see, for example, Berger, 1997; O'Keefe & McCornack, 1987; Spitzberg & Cupach, 1984). Communicators must be skilled at selecting strategies that result in satisfying outcomes delivered in the most efficient manner possible. The fifth of the criteria targets the need to adapt messages to the dictates of the *situation*. Therefore communicators must use appropriate communication strategies taking into account audience and situational needs (Spitzberg & Cupach, 1984; Wiemann & Bradac, 1985). Finally, competent communicators use their skills to adapt their messages in order to achieve communication goals appropriately but also from within the person's *ethical* or moral code. "This moral code . . . envisions a world in which ideal speech situations could empower all and provide respect and voice to each person regardless of station or stereotype" (Spitzberg & Cupach, 2002, p. 583). Thus, the moral code is not focused so much on the positive outcomes of the interaction (described by Spitzberg and Cupach as the "ends" of communication), but rather the moral code is directed to "what good communication *is*" (p. 583)–the overall quality of the message.

Communication skills are used in a performative manner as they are employed as persons attempt to achieve prescribed goals. The success at which individuals engage communication skills is a key element for demonstrating communication competence. These skills vary from those related to altercentrism, composure, and coordination to expressiveness. Competencies are identified as observable from the contextual to the macro level (see Spitzberg & Cupach, 2002, for the entire taxonomy of interpersonal skills addressing types of communication competencies).

Of great importance in identifying whether competent communication occurs is the degree to which the contextual demands are considered in the selection of communicative behaviors and how well the behaviors are consistent with features of the context in question (Spitzberg & Brunner, 1991). Thus, competent communicators adapt their behaviors not only to improve their attempts to attain prescribed goals, but also to adhere to situational constraints. Selecting an appropriate and effective channel for a given situation demonstrates a degree of communication competence. This may be difficult for persons to accomplish when competence levels are low or when faced with reticence in communication situations or when highly apprehensive individuals are involved across various communication situations. The amount and quality of relational self-disclosure may be influenced by the interactants' degrees of communication apprehension such as in the case of a highly apprehensive person who may choose a CMC channel rather than to disclose information FtF that the receiver may interpret negatively. The following section discusses communication apprehension and in particular, the link between communication apprehension and selection of CMC channels in maintaining primarily FtF relationships.

Communication Apprehension

Communication apprehension, "an individual's level of fear or anxiety with either real or anticipated communication with another person or persons" (McCroskey, 1977, p. 78), is a condition affecting many. Persons with high levels of communication apprehension experience anxiety and may avoid communication when an interaction is perceived to outweigh any potential benefits (McCroskey, 1970); avoidance then often leads to lowered self-esteem. Thus, highly apprehensive persons find self-disclosure threatening and fail to engage in this strategy of relational maintenance. It may lessen the level of apprehension if a CMC channel is selected over disclosing information FtF.

Undoubtedly communication apprehension has been one of the most carefully examined individual difference variables within the field of communication. It can broadly be defined as anxiety experienced by an individual during or in anticipation of a communication event (McCroskey, 1970, 1977, 1995). Communication apprehension research has been conducted under a number of distinct but related constructs such as (un)willingness to communicate, reticence, and shyness. Burgoon (1976) defined unwillingness-to-communicate as "a chronic tendency to avoid and/or devalue oral communication and to view the communication situation as relatively unrewarding" (p. 60). Similarly, Keaten and Kelly (2000) argue that reticent individuals "avoid communication because they believe it is better to remain silent than to risk appearing foolish" (p. 168). Cheek and Buss (1981) define shyness as "one's reaction to being with strangers or casual acquaintances: tension, concern, feelings of awkwardness and discomfort, and both gaze aversion and inhibition of normally expected social behavior" (p. 330). Although these definitions differ from one another slightly on affective and behavioral components, they tap into the same underlying cognitive predisposition. As other researchers have pointed out (Daly, 2002; Kelly, 1982; Segrin & Givertz, 2003), while each of these constructs possesses important theoretical distinctions, they share much in common and ignoring their mutually informative research findings for semantic reasons serves little purpose. Therefore the term communication apprehension will be used in this review to refer to the overarching trait that underlies these constructs except where theoretical distinctions imply that findings may not easily generalize to other related constructs.

Situational and trait-based communication apprehension. Communication apprehension traditionally was viewed as a continuum with state communication apprehension on one pole and trait-based communication apprehension at the other. State communication apprehension, occurring when individuals find themselves in particular types of communication encounters, varies according to the context and the characteristics of an individual's partner (Richmond, 1978). Furthermore, it is triggered only by specific contexts. An example of state communication apprehension occurs when a person fears giving a speech but has no apprehension interacting with others in ordinary daily encounters. In context with the mediated world, a person may experience disclosing personal information to a friend FtF, but is less concerned for potential threats to "face" if the disclosure occurs online.

In contrast, the effects of trait-based communication apprehension are more far reaching. Trait-based communication apprehension is viewed as a personality trait and is likely to occur in any communication context. Trait-based communication apprehension is considered to be a learned trait that was conditioned and reinforced during childhood communication. However, rather than being linked to a few specific personality dimensions, communication apprehension has a broad relationship to an individual's entire personality. For example, McCroskey, Sorensen, and Daly (1976) found that communication apprehension was positively correlated with anxiety, dogmatism, and an external locus of control. It also is negatively correlated with emotional maturity, dominance, adventurousness, confidence, self-control, tolerance for ambiguity, and the need to achieve.

McCroskey et al. (1976) concluded that a complete knowledge of an individual's personality could to a large degree indicate variance in communication apprehension. Additionally, they believed that the combined shared traits of highly apprehensive individuals resulted in persons being "withdrawn, socially maladaptive individual[s] who have little chance for success in contemporary society" (McCroskey et al., 1976, p. 379).

Recently, an alternative approach to trait-based communication apprehension was proposed. The communibiological paradigm views the trait of communication apprehension as the manifestation of genetic characteristics that are present from birth (Beatty, McCroskey, & Heisel, 1998). A biological origin for communication apprehension is not a new concept; early research by McCroskey (1977) acknowledged the possibility of a biological root. Furthermore, it follows that socialization and learning do not play a vital role in the development of communication apprehension if these genetic characteristics are present at birth, prior to the socialization process. Beatty et al. (1998) proposed that the environment contributed no more than 20% of the influence towards the development of trait communication apprehension. According to the communibiological paradigm the cause of high communication apprehension rests in the limbic system of the brain controlling emotion. One part of the limbic system is the *behavioral inhibition system* (BIS) which responds to negative stimuli. The outcome of the activation of the BIS is anxiety. Differences in the sensitivity of the BIS account for the variance in the amount of communication apprehension individuals experience. People with high levels of communication apprehension have a lower tolerance for stimulation than those low in communication apprehension. However, even high apprehensives can be prompted to engage in communication given the appropriate motivation. The *behavioral activation system* (BAS) responds to positive, rewarding stimuli and potentially can be aroused to an extent that it will override the BIS of highly apprehensive individuals. In short, actual communication behavior is the result of the collective influences of these two systems.

Self-Disclosure and Communication Apprehension

Communication apprehension has several effects upon self-disclosure. Individuals high in communication apprehension perceive that they disclose less information overall, disclose more negative information, and are less honest when self-disclosing than individuals low in communication apprehension (McCroskey & Richmond, 1977). Furthermore, the self-disclosures of highly apprehensive individuals have less depth than those of their low apprehensive counterparts (Wheeless, Nesser, & McCroskey, 1986). McCroskey and Richmond (1977) conclude that people with high levels of communication apprehension may avoid disclosing negative information by restricting the amount and frequency of self-disclosure. However, it should be noted that those high in communication apprehension may not actually disclose more negative information but merely perceive that they do so. This may be a reflection of the lower self-esteem typical to highly apprehensive individuals. Communication apprehension is just one of many variables that can influence patterns of self-disclosure. Regardless of whether it is viewed as trait-based or as having a biological origin, it can have a crippling effect on communication in general and by extension the levels of self-disclosure.

Perceived communication competence is linked to levels of reticence and feelings of discomfort while persons communicate FtF in difficult conversations (Keaten, Kelly, Leffingwell, & McCoy, 2006; Kelly & Keaten, 2005; Roberts, Smith, & Pollack, 2000). Highly reticent individuals are more likely to use IM rather than converse FtF as it enables a greater sense of situational control and reduction of inhibition and anxiety and allows for opportunities to prepare messages in advance (Kelly, Keaten, Hazel, & Williams, 2007). It is important to study online self-disclosure in FtF relationships where CMC channels are used in relational maintenance because deviations from established models are likely to change how self-disclosure functions in these hybrid relationships. Self-disclosure may be hindered by a sense of self-consciousness, anxiety, reticence, or unwillingness to communicate or it may be a communication strategy employed to overcome these barriers. Fenigstein, Scheier, and Buss (1975) describe three dimensions of self-consciousness: private, where the self is the object or stimulus; public, and social. Self-consciousness may impact one's competence in successfully accomplishing communication goals. Social self-consciousness has similar characteristics to McCroskey's (1970, 1977, 1995) notion of communication apprehension.

Computer-Mediated Communication Apprehension as a Distinct Construct

There is some evidence suggesting that computer-mediated communication apprehension is distinct from communication apprehension and other constructs dealing with anxiety. Traditionally, communication apprehension research was concerned with oral communication and much of its research has revolved around public speaking, classroom settings, and interpersonal relationships (McCroskey, 1982b). Likewise, writing apprehension, which addresses the avoidance of composition and the fear of having one's writing evaluated, was developed primarily in the context of educational performance and testing (Daly & Miller, 1975; Daly & Shamo, 1978). Computer anxiety involves "fear of impending interaction with a computer that is disproportionate to the actual threat presented by the computer" (Howard, Murphy, & Thomas, 1986, p. 630). None of these constructs may adequately reflect the nature of computer-mediated communication. CMC is fundamentally different from the FtF communication of conventional communication apprehension research because the nonverbal cues conveying much of the interaction's information are constrained. Indeed, some researchers hold that nonverbal cues can contain as much as 60% of the interaction's information (Birdwhistell, 1955). Likewise, the off-the-cuff nature of instant messaging, SMS, and email does not reflect the carefully composed responses found in an academic context. The conversational nature of this text may preclude writing apprehension because users do not have the same concerns regarding evaluation (Scott & Timmerman, 2005). Finally, computer anxiety has been critiqued for reflecting users' unease with the technical aspects of computing such as coding and computation (Leso & Peck, 1992; Yeaman, 1992). However, it may not accurately measure the anxieties of today's communication technology users when many computer users' experiences are defined by graphical user interfaces and require little coding.

Research findings support a separate construct focusing on the anxieties produced by social interaction via technology. Scott and Rockwell (1997) found that while communication apprehension was a significant predictor for the likelihood to use oral technologies such as telephony, it did not predict many common internet activities such as email, electronic discussion groups, or computer conferencing. Furthermore, writing apprehension was not correlated with likelihood to use any new technologies. Patterson and Gojdycz (2000) found similar results when communication and writing apprehension failed to predict email, chat, and worldwide web usage but computer anxiety predicted all three. Additionally, computer anxiety was shown to be both positively related to communication apprehension and writing apprehension. In response to the difficulties CA has had in predicting technology use, Scott and Timmerman (2005) proposed a separate CMC apprehension variable that measured apprehension during online interaction. They found that CMC apprehension predicted additional variance in frequency of use for communication technologies beyond computer anxiety, communication apprehension, and writing apprehension. Furthermore, it was most strongly related to text-based communication technologies and conferencing tools rather than telephony. Other researchers show promising results tapping into this construct by measuring affect for communication technologies (see Wrench & Punyanunt-Carter, 2007). Given that conventional constructs do not seem to adequately describe the experience of using CMC and that research findings suggest that a separate construct of CMC apprehension is capable of providing greater explanatory power for new communication technology use, it is appropriate for researchers to focus further research efforts on CMC apprehension rather than communication apprehension.

Spitzberg (2006) offers a model of computer-mediated communication competence modeled after earlier views of interpersonal communication competence where competence is determined by the motivation, knowledge, and skills of the interactant in relation to the demands of the context and desired outcomes of the discourse. One important element in self-presentation is self-disclosure because it allows an individual some ability to shape and control self-relevant knowledge. Communication competence plays a significant role in how one engages in self-presentation.

Spitzberg (2006) states that "FtF and CMC interaction are more similar than they are different (p. 652) as they both involve the use of relevant interpersonal communication skills. Over 100 separate communication skills are extracted from the literature, yet they neatly fit within four categories: attentiveness, composure, coordination, and expressiveness (Spitzberg & Cupach, 2002). In an interaction, these skill sets are driven by motivations or the goals underlying the purpose of the interaction. Thus, the interaction may be influenced by positive motives such as comfort and confidence. In the case of shy, reticent, or apprehensive individuals, they may be motivated to employ CMC channels as a means of managing the interaction and their uncertainties simultaneously whereas the selection of FtF interaction is more threatening. CMC users were found to be more likely to choose these CMC channels where they anticipated positive results from the channel use. Underlying ones skills and motivations, persons must possess knowledge of the appropriate skills to employ given the nature of the goals and the demands of the situation (Spitzberg, 2006). However, the various ways in which the communication competence model may apply to CMC interactions needs support as very limited research exists to offer such support.

CMC *Apprehension and Channel Selection*

Much of the research efforts for CMC apprehension have focused on the effect of apprehension on media selection. Numerous studies report that highly apprehensive individuals experience reduced anxiety when communicating online (see Mazur, Burns, & Emmers-Sommers, 2000; Roberts et al., 2000; Stritzke, Nguyen, & Durkin, 2004). McKenna and Bargh (2000) found that highly apprehensive individuals were more likely to form online relationships than individuals low in apprehension. Likewise, research by Caplan (2005, 2007) suggests that highly apprehensive individuals may have an affinity for online communication because it affords them greater control over their self-presentation. Papacharissi and Rubin (2000) found that internet communication was a functional alternative to FtF communication for individuals who exhibited anxiety and failed to satisfy their interpersonal needs. In contrast, Birnie and Horvath (2002) concluded that the internet did not function in a compensatory manner for highly anxious individuals by helping them expand their contacts. However, they did find that shy individuals made more intimate self-disclosures through this channel than FtF.

There currently are two explanations for the relationship between CMC apprehension and channel selection. First, it is possible that mediated channels serve to create a safe environment for apprehensive individuals to interact and self-disclose by limiting the nonverbal information that is conveyed between conversational partners (High & Caplan, 2009, Strizke et al., 2004). Individuals who experience social anxiety appreciate the ability to plan ahead (Arkin & Grove, 1990) and may seek low threat contexts for self-disclosure (Leary & Kowalski, 1995). Many online communication channels, such as email and IM, are asynchronous allowing users to carefully formulate messages. Furthermore, the text-based nature of these channels may dampen both the interpersonally damaging non-verbal cues that an apprehensive individual may display as well as any disconfirming responses from their conversational partner. High and Caplan (2009) found that the more socially anxious a CMC user was the less they were perceived to be so by their partner. Additionally, partner satisfaction with the interaction increased with user apprehension. Although these effects are small, they may be indicative of highly motivated users taking advantage of the buffering effect of CMC to create a desired self-presentation. Peter and Valkenberg (2006) found that highly apprehensive adolescents appreciated the enhanced control online interactions afforded them and perceived internet communication to be more intimate and addressing a greater number of topics than FtF communication. In short, the lack of disconfirming feedback leads individuals to feel more secure and assists in the creation of positive self-presentation.

Research suggesting that computer-mediated communication buffers face-threatening interactions highlights the importance of presence research to CMC apprehension. Early social presence research, such as social presence theory (Short, Williams, & Christie, 1976) and media richness theory (Daft & Lengel, 1984), classified media by their ability to transmit social context cues and nonverbal behaviors believed to contribute to the experience of social presence. The inability to provide users with these cues was believed to impair social presence, thus, discouraging interaction and relationship development. More recent conceptualizations of social presence define it as the simulation of another intelligence when "users feel that a form, behavior, or sensory experience indicates the presence of another intelligence" (Biocca, 1997). Biocca (1997) contends that social presence is felt to the extent that users have access to the intentions and sensory impressions of others. Communication apprehension research suggests that there is a threshold of stimulation at which highly apprehensive individuals begin to feel anxiety. If channel selection is not arbitrary but is a goal-driven process, then channel selection is an effort to manage anxiety by selecting a channel that does not meet the stimulation threshold for a particular interaction. Indeed, the focus of presence research on sensory perception and evolutionary psychology (see Lee, 2004, for a review) seems to mesh quite neatly with the communibiological approach and may help refine our understanding of specific variables within channels that contribute to anxiety.

The second explanation for the influence of communication apprehension on channel preference holds that users strategically select communication channels on the basis of self-perceived competencies (Kelly and Keaten, 2007). Specifically, the asynchronous nature of many text-based channels of communication, such as email and IM, provide users greater opportunities to plan messages and to exert control over the interaction resulting in higher levels of self-perceived skill. There are a handful of studies that investigate the connection between social skill and channel selection. Keaten and Kelly (2008) found that perceptions of competence in email use were associated with the ability to prepare and control communication. Kelly et al. (2007) noted that high levels of reticence were associated with "the perception that [instant messaging] reduces both anxiety and inhibition, provides more opportunity to prepare as well as control communication, and less agreement that face-to-face communication is more meaningful and enables more emotional exchange" (p. 17). However, they also found that in scenarios requiring a high degree of social skill the reductions in anxiety produced by online communication failed to compensate for general skills deficits. Similarly, Hertel, Schroer, Batinic, and Naumann (2008) found that compared to introverts, high extraverts tend to prefer rich media capable of conveying nonverbal information. Furthermore, the relationship between personality and channel preference was mediated by social skill and was most notable in a scenario involving high social threat. Finally, Keaton and Kelly (2008) found that the individuals who reported greater competence using email than FtF communication had greater fear of negative evaluations, were more likely to report that email increased their control during an interaction, and were more likely to use email and less likely to use FtF communication during a difficult interpersonal encounter.

The connection between interpersonally challenging interactions and channel preferences implies that individuals select communication media based upon the circumstances. Indeed, O'Sullivan (2000) notes that communication channel selection is influenced not only by possessing the requisite social skills to effectively use a particular channel but also by the self-presentational concerns rooted in an interaction's goals. Keaten and Kelly (2008) proposed revising models of communication competence to include channel selection which they contend is determined primarily by stable dispositional inclinations. Specifically, communication apprehension creates a persistent preference for particular channels and communicators must consider their own preferences in conjunction with situational factors to achieve competent outcomes. Keaten and Kelly (2008) observe that channel selection may create a dilemma between effectiveness and appropriateness of communication that are considered to be benchmarks of competence. For example, while a highly apprehensive individual may feel more in control ending a romantic relationship via mediated communication, the strong social norms for FtF communication in romantic relationships mean it is probably not appropriate. A competent communicator will be able to consider the inherent tensions between these goals and make a channel selection so as to best achieve their goals in the interaction. In short, this approach holds that individuals have a default preference but that their selection of a communication channel is biased by situational factors.

Conclusion

There are two major conclusions that can be drawn from this review. First, a vast majority of research supports the contention that highly apprehensive individuals prefer interaction via mediated channels more strongly than non-apprehensive users. However, the work being conducted does not address how users choose between the ever increasing number of mediated channels available to them. Under the current paradigm of communication apprehension research, IM, email, and cell phone texting all provide similar advantages in managing self-presentation as compared to FtF communication. Despite this Leung (2007) found that the heavier users of cell phone texting were low in communication apprehension. This contrasts with the finding reported by Kelly, Keaten, and Palmer (2003) of no difference in the amount or type of CMC usage. Furthermore, given the dynamic nature of technology, researchers should focus less on the differences between FtF and specific mediated channels and more on technological variables that form the basis for this preference.

Second, there is no reason why the buffering effect of CMC and the influence of CMC on self-perceptions of communication competence need to be mutually exclusive. The safe environment created by CMC where users have the opportunities to try out strategies and refine interpersonal skills may contribute to the development of competence. Indeed, the practice and frequent use of communication skills is fundamental to being able to successfully employ them when necessary (Greene, 2003). Research over the past decade suggests that communication apprehension is an individual difference rooted in biology. Indeed, Beatty et al. (1998) state that as much as 80% of communication apprehension may be determined by genetics. Clearly if genetics plays such a strong role in the manifestation of communication apprehension cognitive and affective treatments may never fully ameliorate it (see Segrin & Givertz, 2003). Teaching users to take advantage of various communication channels may go a long way towards alleviating communication apprehension because the limited cues of text-based media means that highly apprehensive individuals may not reach the stimulation threshold at which anxiety becomes debilitating. However, communication competence research reminds us that the benefits accrued by channel selection must be balanced against the social backlash that may occur due to using an inappropriate channel. A more nuanced understanding of how dispositional preferences may be biased by situational factors would be useful in designing a course of treatment.

In conclusion, CMC dramatically alters the patterns of interaction for individuals high in communication apprehension. Specifically, text-based communication channels put highly apprehensive individuals at ease and may lead them to communicate more frequently and reveal more intimate information. However, researchers are only beginning to develop models that incorporate the strategic selection of communication channels. As this research continues it is important to clarify not only the nature of communication apprehension but also technological variables that moderate communication via CMC.

References

Arkin, R. M., & Grove T. (1990). Shyness, sociability, and patterns of everyday affiliation. *Journal of Social and Personal Relationships*, 7, 273–81.

Baxter, L. A. (1987). Self-disclosure and relationship disengagement. In V. J. Derlega & J. H. Berg (Eds.), *Self-disclosure: Theory, research, and therapy* (pp. 155–174). New York: Plenum.

Baym, N. K., Zhang, Y. B., & Lin, M. C. (2004). Social interactions across media. *New Media & Society*, 6, 299–318.

Beatty, M. J., McCroskey, J. C., & Heisel A. D. (1998). Communication apprehension as temperamental expression: A communibiological paradigm. *Communication Monographs*, *65*, 197–219.

Berger, C. R. (1997). *Planning strategic interaction: Attaining goals through communication*. Mahwah, NJ: Lawrence Earlbaum.

Biocca, F. (1997). The cyborg's dilemma: Progressive embodiment in virtual environments. *Journal of Computer-Mediated Communication*, *3*(2), 0-0.

Birdwhistell, R. L. (1955). Background to kinesics. *Etc.*, *13*, 10–18.

Birnie, S. A., & Horvath, P. (2002). Psychological predictors of Internet social communication. *Journal of Computer Mediated Communication*, 7(4), 0-0.

Burgoon, J.K., 1976. The unwillingness to communicate scale: Development and validation. *Communication Monographs*, *43*, 60–69.

Canary, D. J., & Stafford, L. (1994). Maintaining relationships through strategic and routine interactions. In D. J. Canary & L. Stafford (Eds.), *Communication and relationship maintenance* (pp. 3022). New York: Academic Press.

Caplan, S.E. (2005). A social skill account of problematic internet use. *Journal of Communication*, *55*, 721–736.

Caplan, S.E. (2007). Relations among loneliness, social anxiety, and problematic internet use. *Cyberpsychology and Behavior*, *10*, 234–241.

Cheek, J. M., & Buss, A. H. (1981). Shyness and sociability. *Journal of Personality and Social Psychology*, *41*(2), 330–339.

Daft, R. L., & Lengel, R. H. (1984). Information richness: a new approach to managerial behavior and organizational design. *Research in Organizational Behavior*, 6, 191–233.

Daly J. .A. (2002). Personality and interpersonal communication. In M.L. Knapp and J.A. Daly (Eds.), *Handbook of interpersonal communication* (3rd ed., pp. 133-180). Thousand Oaks, CA: Sage.

Daly, J. A., & Miller, M. D. (1975). The empirical development of an instrument to measure writing apprehension. *Research in the Teaching of English*, 9, 242–249

Daly, J. A., & Shamo, W. (1978). Academic decisions as a function of writing apprehension. *Research in the Teaching of English*, *12*(2), 119–126.

Derlega, V., Metts, S., Petronio, S., & Margulis, S. (1993). *Self-disclosure*. Newbury Park, CA: Sage.

Ellison, N. B., Steinfield, C., & Lampe, C. (2008). The benefits of Facebook "friends": Social capital and college students' use of online social network sites. *Journal of Computer-Mediated Communication*, *12*, 1143–1168.

Feningstein, A., Scheier, M. F., & Buss, A. H. (1975). Public and private self-consciousness: Assessment and theory. *Journal of Consulting and Clinical Psychology*, *43*, 522-527.

Greene, J. O. (2003). Models of adult communication skill acquisition: Practice and the course of performance improvement. In J.O. Greene and B. R. Burleson (Eds.), *Handbook of communication and social interaction skills* (pp. 51–91). Mahwah, NJ: Lawrence Erlbaum.

Hecht, M. L. (1978). The conceptualization and measurement of interpersonal communication satisfaction. *Human Communication Research*, *4*, 253–264.

Hertel, G., Schroer, J., Batinic, B., & Naumann, S. (2008). Do shy people prefer to send email? Personality effects on communication media preferences in threatening and nonthreatening situations. *Social Psychology, 39*, 231–243.

High, A. C., & Caplan, S. E. (2009). Social anxiety and computer-mediated communication during initial interactions: Implications for the hyperpersonal perspective. *Computers in Human Behavior, 25*, 475–482.

Hovick, S. R. A., Meyers, R. A., & Timmerman, C. E. (2003). E-mail communication in workplace romantic relationships. *Communication Studies, 54*, 468–482.

Howard,G. S., Murphy, C. M., & Thomas,G. E. (1986). *Computer anxiety considerations for design of introductory computer courses*. In Proceedings of the 1986 Annual Meeting of the Decision Sciences Institute (pp. 630–632). Atlanta, GA: Decision Sciences Institute.

Jourard, S. (1971). *The transparent self* (2nd ed.). New York: Van Nostrand Reinhold.

Keaten, J. A., & Kelly, L. (2000). Reticence: An affirmation and revision. *Communication Education, 49*, 165–177.

Keaten, J. A. & Kelly, L. (2008). "Re: We really need to talk": Affect for communication channels, competence, and fear of negative evaluation. *CommunicationQuarterly, 56*, 407–426.

Keaten, J. A., Kelly, L., Leffingwell, D., & McCoy, R. (2006). *"Re: We really need to talk": Affect for communication channels, competence, and fear of negative evaluation*. Paper presented at the annual convention of the National Communication Association, Chicago, IL.

Kelly, L. (1982). A rose by any other name is still a rose: A comparative analysis of reticence, communication apprehension, unwillingness to communicate and shyness. *Human Communication Research, 8*, 99–113.

Kelly, L., & Keaten, J. (2000). Treating communication anxiety: Implications of the communibiological paradigm. *Communication Education, 49*, 45-57.

Kelly, L., & Keaten, J. A. (2005). *Explaining the appeal of email to reticent individuals: Development of the Affect for Communication Channels Scale*. Paper presented at the annual convention of the National Communication Association, Boston, MA.

Kelly, L., & Keaten, J. A. (2007). Development of the affect for communication channels scale. *Journal of Communication, 57*, 349–365.

Kelly, L., Keaten, J. A., Hazel, M., & Williams, J. A. (2007). *Effects of reticence and affect for communication channels on usage of instant messaging and self-perceived competence*. Paper presented at the annual convention of the National Communication Association, Chicago, IL.

Kelly, L., Keaten, J. A., Larsen, J., & West, C. (2004). *The impact of reticence on use of computer-mediated communication II: A qualitative study*. Paper presented at the annual convention of the National Communication Association, Chicago, IL.

Kelly, L., Keaten, J. A., & Palmer, D. L. (2003). *The impact of reticence on use of computer-mediated communication*. Paper presented at the annual convention of the National Communication Association, Miami Beach, FL.

Leary, M. (1995). *Self-presentation: Impression management and interpersonal behavior*. Madison, WI: Brown & Benchmark.

Leary, M. R., & Kowalski, R. M. (1995). *Social anxiety*. New York: Guilford.

Lee, K. M. (2004). Why presence occurs: Evolutionary psychology, media equation, and presence. *Presence: Teleoperators and Virtual Environments, 13*, 494–505.

Lenhart, A., Madden, M., & Hitlin, P. (2005). *Teens and technology: Youth are leading the transition to a fully wired and mobile nation*. Washington, DC: Pew Internet & American Life Project.

Leso, T., & Peck, K. L. (1992). Computer anxiety and different types of computer courses. *Journal of Educational Computing Research, 8*, 469–476.

Leung, L. (2007). Unwillingness-to-communicate and college students' motives in SMS mobile messaging. *Telematics and Informatics, 24*, 115-129.

Mazur, M. A., Burns, R. J., & Emmers-Sommers, T. M. (2000). Perceptions of relational Interdependence on online relationships: The effects of communication apprehension and introversion. *Communication Research Reports, 17*, 397–406.

McCown, J. A., Fischer, D., Page, R., & Homant, M. (2001). Internet relationships: People who meet people. *CyberPsychology & Behavior, 4*, 593–596.

McCroskey, J. C. (1970). Measures of communication-bound anxiety. *Speech Monographs, 37*, 269–277.

McCroskey, J. C. (1977). Oral communication apprehension: A summary of recent theory and research. *Human Communication Research, 4*, 78–96.

McCroskey, J. C. (1978). Validity of the PRCA as an index of oral communication apprehension. *Communication Monographs, 45*, 192-203.

McCroskey, J. C. (1982a). *An introduction to rhetorical communication* (4th ed.). Englewood Cliffs, NJ: Prentice Hall.

McCroskey, J. C. (1982b). Communication competence and performance: A research and pedagogical perspective. *Communication Education, 31*, 1–8.

McCroskey, J. C. (1995). Personal report of communication apprehension (prca-24). Retrieved October 10, 2009, from http://www.jamescmccroskey.com/measures/prca24.htm

McCroskey, J. C., & Richmond, V. P. (1977). Communication apprehension as a predictor of self-disclosure. *Communication Quarterly, 25*, 40–43.

McCroskey, J. C., Sorensen, G., & Daly J. A. (1976). Personality correlates of communication apprehension: A research note . *Human Communication Research, 2*, 376–380.

McKenna, K. Y. A., & Bargh, J. A. (2000). Plan 9 from cyberspace: The implication of the Internet for personality and social psychology. *Personality and SocialPsychology Review, 4*, 57–75.

O'Keefe, B. J., & McCornack, S. A. (1987). Message design logic and message goal structure: Effects on perceptions of message quality in regulative communication situations. *Human Communication Research, 14*, 68–92.

O'Sullivan, P. B. (1996, November). *An impression management framework for the study of mediated communication.* Paper resented at the annual conference of Speech Communication Association, San Diego, CA.

O'Sullivan, P. B. (2000). What you don't know won't hurt me: Impression management functions of communication channels in relationships. *Human CommunicationResearch, 26*(3), 403–431.

Papacharissi, Z., & Rubin, A. M. (2000). Predictors of internet use. *The Journal of Broadcasting and Electronic Media, 44*, 175–196.

Parks, M. R., & Floyd, K. (1996). Making friends in cyberspace. *Journal of Communication, 46*, 80–97.

Patterson, B. R., & Gojdycz, T. K. (2000). The relationship between computer-mediated communication and communication related anxieties. *Communication Research Reports, 17*, 278–287.

Peter, J., & Valkenberg, P. M. (2006). Research note: Individual differences in perceptions of internet communication. *European Journal of Communication, 21*, 213–226.

Pew Internet & American Life Project. (2001). Online communities. Retrieved October 10, 2009, from http://www.pewinternet.org/Shared-Content/Data-Sets/2001/Online-Communities-2001- Survey-Data.aspx

Pew Internet & American Life Project. (2005). The internet: Mainstreaming of online life. Retrieved October 10, 2009, from http://www.pewinternet.org/pdfs/Internet_Status_ 2005.pdf

Pew Internet & American Life Project. (2009). Adults and social networking sites. Retrieved October 10, 2009, from http://www.pewinternet.org/topics/Social-Networking.aspx

Postmes, T., Spears, R., & Lea M. (1998). Breaching or building social boundaries? Side-effects of computer-mediated communication. *Communication Research, 25*(6), 689–715.

Powers, W. G., & Lowry, D. N. (1984). Basic communication fidelity: A fundamental approach. In R. N. Bostrom (Ed.), *Competence in communication: A multidisciplinary approach* (pp. 57–73). Beverly Hills, CA: Sage.

Rabby, M. K., & Walther, J. B. (2003). Computer-mediated communication effects on relationship formation and maintenance. In D. J. Canary & M. Dainton (Eds.), *Maintaining relationships through communication: Relational, contextual, and cultural variations* (pp. 141–162). Mahwah, NJ: Lawrence Erlbaum.

Roberts, L. D., Smith, L. M., & Pollock, C. M. (2000). 'U r a lot bolder on the net.' In W. R.Crozier (Ed.), *Shyness: Development, consolidation and change* (pp. 121–138). New York: Routledge.

Richmond, V. P. (1978). The relationship between trait and state communication apprehension and interpersonal perceptions during acquaintance stages. *Human Communication Research, 4*, 338–349.

Sanders, W. S., & Amason, P. (2006). *The breadth and depth of computer-mediated self-disclosure in pre-existing and primarily face-to-face relationships.* Paper presented at the annual convention of the National Communication Association, San Antonio, TX.

Scott, C. R., & Rockwell, S. C. (1997). The effect of communication, writing, and technology apprehension on likelihood to use new communication technologies. *Communication Education, 46*, 44–62.

Scott, C. R., & Timmerman, C. E. (2005). Relating computer, communication, and computer-mediated communication apprehensions to new communication technology use in the workplace. *Communication Research, 32*(6), 683–725.

Schlenker, B. R., & Weigold, M. F. (1992). Interpersonal processes involving impression regulation and management. *Annual Review of Psychology, 43*, 133–168.

Segrin, C., & Givertz, M. (2003). Methods of social skills training and development. In J.O. Greene and B. R. Burleson (Eds.), *Handbook of communication and social interaction skills.* Mahwah, NJ: Lawrence Erlbaum.

Short, J., Williams, E., & Christie, B. (1976). *The social psychology of telecommunications.* London: John Wiley.

Spitzberg, B. H. (1993). The dialectics of (in)competence. *Journal of Personal and Social Relationships, 10*, 137–158.

Spitzberg, B. H. (2006). Preliminary development of a model and measure of computer-mediated communication (CMC) competence. *Journal of Computer-Mediated Communication, 11*, 629–666.

Spitzberg, B. H., & Brunner, C. C. (1991). Toward a theoretical integration of context and competence inference research. *Western Journal of Speech Communication, 55*, 28–46.

Spitzberg, B. H., & Cupach, W. R. (1984). *Interpersonal communication competence.* Beverly Hills, CA: Sage.

Spitzberg, B. H., & Cupach, W. R. (2002). Interpersonal skills. In M. L. Knapp, & J. A. Daly (Eds.), *Handbook of interpersonal communication* (3rd ed., pp. 564–611). Thousand Oaks, CA: Sage.

Stritzke, W. G. K., Nguyen, A., & Durkin, K. (2004). Shyness and computer-mediated communication: A self-presentational theory perspective. *Media Psychology, 6*, 1–22.

Swann, W. B., Jr. (1990). To be adored or to be known: The interplay of self-enhancement and self-verification. In R. M. Sorentino and E. T. Higgins (Eds.), *Handbook of motivation and cognition.* (pp. 208–250). New York: Guilford.

Tidwell, L. C., & Walther, J. B. (2002). Computer-mediated communication effects on disclosure, impressions, and interpersonal evaluations: getting to know one another a bit at a time. *Human Communication Research, 28*, 317–348.

Trope, Y. (1986). Self-enhancement, self-assessment, and achievement behavior. In R M. Sorrentino & E. T. Higgins (Eds.), *Handbook of motivation and cognition* (pp. 350–378). New York: Guilford.

Walther, J. B. (1993). Impression development in computer-mediated interaction. *Western Journal of Communication, 57*, 381–398.

Walther, J. B. (1996). Computer mediated communication: Impersonal, interpersonal, and hyperpersonal interaction. *Communication Research, 23*, 3–43.

Wheeless, L. R., Nesser, K., & McCroskey J. C. (1986). The relationships of self-disclosure and disclosiveness to high and low communication apprehension. *Communication Research Reports, 3*, 129-134.

Whitty, M. T. (2002). Liar, liar! An examination of how open, supportive, and honest people are in chat rooms. *Computers in Human Behavior, 18*, 343–353.

Wiemann, J. M. (1977). Explication and test of a model of communicative competence. *Human Communication Research, 3*, 195–213.

Wiemann, J. M. & Bradac, J. J.-(1985). Review essay: The many guises of communicative competence. *Journal of Language and Social Psychology, 4*, 131-138.

Wrench, J. S., & Punyanunt-Carter, N. M. (2007). The relationship between computer mediated-communication competence, apprehension, self-efficacy, perceived confidence, and social presence. *Southern Communication Journal, 72*, 355–378.

Yeaman, A. R. J. (1992).Seven myths of computerism. *TechTrends, 37*, 22–26.

CHAPTER SIX

Relational Maintenance and CMC

Stephanie Tom Tong[1]

Joseph B. Walther[1, 2]

Is Facebook the way to maintain old friendships? Is Twitter the latest development in far-flung family members' efforts to stay in touch? Can economics theory explain why the mundane musings that make new media so very trivial are the very reason they are so successful relationally? The use of Internet technologies in the context of relationship maintenance—supporting and nurturing existing relationships over time—will be a major focus of research both with respect to comprehending the uses and functions which many social technologies provide, and also with respect to the most significant shift in how relational maintenance is conducted since the diffusion of the telephone. Since relational maintenance activities are performed more often than relationship initiation or termination (Duck, 1988), relational maintenance has become a popular research topic within the last two decades. The application of various Internet-based social technologies has led to extensive use of computer-mediated communication (CMC) in the maintenance of relationships among friends, family members, and intimates. Because CMC is relatively new, and because of the consistent emergence of novel Internet applications that become appropriated for relational maintenance, there is a relatively little literature examining CMC's role in the maintenance process. Additionally, little theoretical development has guided relational maintenance in offline contexts, making comparisons to electronic relational maintenance all the more difficult. This chapter attempts to help bridge the gap by reviewing existing research and by advancing speculative assertions about the application of CMC tools to various relational maintenance activities in the context of several relationship types, as well as to revisit and extend theoretical notions about social exchange processes in order to apply them to the affordances that computer-mediated relationship maintenance offers.

This chapter begins by briefly explicating extant perspectives on relational maintenance, and extending these into mediated contexts. One goal of this chapter is to identify current trends surrounding contemporary CMC technologies such as email, blogs, social network sites (such as Facebook), and microblogging via Twitter, and describe how these various technologies have been shown, or may be suggested, to apply to relational maintenance behaviors across a sample of relationship types (e.g., families, friendships, romantic partners). Another goal is to discuss how technology is being used to serve both proximal relationships—relationships among intimates and families who cohabitate or are in frequent physical contact, but use CMC to supplement relational maintenance—as well as long-distance relationships of various kinds. An additional focus of the chapter will be the extension of the equity theory approach to relational maintenance through an application of a transaction cost perspective on the effects of social technology in ongoing relationships, with specific suggestions about technology's support of presence, tie signs, and mundane conversation. Lastly, the chapter will conclude with some cautions and recommendations for future research in terms of studying CMC applications and their effects on relational outcomes, as well as the utility of new frameworks in understanding these applications and the relationships to which they apply.

Relational Maintenance: Definitions, Contexts, and Types

Definitions. Several perspectives on and definitions of relational maintenance exist within the interpersonal communication literature (see for review Dindia, 2003) and within its long research tradition it is often difficult to find a consistent definition of relational maintenance. Duck (1988) was the first to point out the most basic definition of relational maintenance, which is simply preserving a relationship's existence. In that definition, no assertions are made about the state of the relationship, the feelings of either party, or the amount or performance of relational maintenance behaviors. As relational maintenance scholarship expanded, others began to speculate on some of these analytic issues. Some defined relational maintenance as those behaviors that are necessary to keep a relationship stable or consistent; in other words, maintaining a certain level of intimacy or closeness (Dindia & Canary, 1993). In contrast, other perspectives suggested that relational maintenance requires not only consistency, but also a movement towards a mutually satisfying end state.

Contexts. Although scholars debate these definitional terms generally, many have focused their research on identifying which communication behaviors are used most successfully to maintain *specific types* of relationships. That is, many scholars adopt a particular theoretical perspective, but apply it to only one context (i.e., familial maintenance) in an effort to center their research efforts. While empirically convenient, it has resulted in a lopsided research agenda, with much work examining romantic *couples* (Bell, Daly, & Gonzalez, 1987; Bell, Schumm, Knot, & Ender, 1999; Dainton, 2000; Jerny-Davis et al., 2005; Rabby, 2007; Stafford & Canary, 1991; Stafford & Merolla, 2007), and less on *friendships* (Shklovski, Kraut, & Cummings, 2008; Wang & Anderson, 2007; Wright, 2004) or *families* (Harwood, 2000; Holladay & Seipke, 2007; Trice, 2002).

For the purposes of this chapter, we consider the performance of behaviors, which sustain both the existence of the relationship and satisfaction of each partner, to be important in the enactment of relational maintenance. Our definition includes all types of relationships, ranging from friends and families to romantic couples. As such, we will address the types of behaviors performed by partners in a variety of relationship contexts, the perceived equality between partners, the costs and benefits of behavioral performance, and, most importantly, the ways in which technology is changing the way in which we conceptualize and understand the field of relational maintenance.

Behavioral Typologies

Over the years, many scholars have developed typologies in an effort to categorize the range of relational maintenance behaviors, specifically among romantic couples (e.g., Ayres, 1983; Bell et al., 1987; Baxter & Dindia, 1990; Stafford & Canary, 1991), and many of these have been used to try and identify which of the behaviors is most important in predicting relational satisfaction. These typologies refer to those behaviors that are deliberate and strategic in nature, although some research has also begun to focus on those "routine" behaviors (Duck, 1988) that are done unconsciously as part of a routine set of relational maintenance activities (Dainton, 1998).

Perhaps the most commonly used typology in the current research is that developed by Stafford and Canary (1991). The developers of this typology asked romantic couples to respond to questionnaire items regarding relational maintenance strategies, and through factor analysis found five dimensions of relational maintenance behaviors:

1. *Positivity* (being cheerful)
2. *Openness* (being direct through self-disclosure and discussion of the relationship)
3. *Assurances* (stressing commitment, love, demonstrating faithfulness)
4. *Sharing tasks* (helping equally with tasks facing the couple)
5. *Networks* (time spent with common friends/acquaintances)

The application of this typology has indeed expanded into CMC arenas (Rabby, 2007; Wright, 2004); however, the findings from these offline relational maintenance studies do not necessarily replicate in online contexts. Furthermore, technological tools are changing the way these behaviors are performed, which suggests that this typology may need to be updated or reconceptualized.

Mediated Relational Maintenance: Relationships, Technological Tools, and Trends

Two continuing classifications exist in the study of relational maintenance via CMC: First is the difference between how those in *long-distance* (geographically and physically separated) and *proximate* (co-located) relationships use technology. Although scholars are beginning to examine the utility of media in both geographically dispersed and proximal relationships (Rabby, 2007), much more progress has been made regarding the way long-distance partners mitigate effects of the geotemporal divide using CMC (see for review Stafford, 2005). As many as seven million people report being in long-distance romantic relationships (LDRRs; Center for Long Distance Relationships, 2009). Since these couples cannot engage in face-to-face (FtF) communication on a daily basis, many of them rely on CMC as a means of communication (Stafford, 2005). Although the necessity of everyday FtF communication for relational satisfaction and intimacy between partners is a long-standing theoretical and paradigmatic assumption held by many relational maintenance scholars (Duck & Pittman, 1994), many scholars have come to recognize that the low cost, convenience, and asynchrony offer strong advantages that allow partners to use a variety of mediated channels to perform relational maintenance (Boneva, Kraut, & Frohlich, 2002). As technological tools have developed in form and function, and dispersed among a larger population of users, the adoption of technology for relational maintenance purposes in many different relationship settings is not surprising. But what are the effects on these relationships of a substitution of CMC for FtF interaction?

Stephen (1986) was among the first to recognize that because geographic separation requires partners to communicate via some technological medium, their communication is affected by the loss of nonverbal cues present in FtF communication, placing greater importance on verbal communication. It was long thought that the reduced cues of CMC would be a poor substitute for FtF communication and would lead to task-related, impersonal communications (Culnan & Markus, 1986). Paradoxically, Stephen (1986) found that geographically separated couples experienced positive levels of affection despite their mediated communication. An early study by Gunn and Gunn (2000) compared the effects of CMC versus other forms of communication in existing LDRRs and found that those who used CMC to stay in touch with their partners reported greater levels of love and intimacy than those who did not: "people who were online preferred their long-distance relationships to their local (unmediated) relationships, whereas people who were not online preferred their local relationships to their long-distance (letter- or telephone-based) relationships" (p. 2). In addition, Dainton and Aylor (2002) found a positive correlation between CMC use and trust within long-distance romantic relationships. The results of such studies challenge the necessity of FtF communication, and question how CMC may lead to greater satisfaction among geographically separated romantic partners.

Research across a variety of communication contexts has indicated that CMC has the potential to improve relational communication, relative to FtF interaction, and the lessons of those efforts may generalize to relational maintenance as well. The hyperpersonal model of online communication (Walther, 1996), in particular, offers a conceptual basis with which to understand how online communication may be more desirable than parallel FtF effects. The very lack of nonverbal cues in CMC, and the greater control that language offers in message construction (especially when it is asynchronous and editable), affords communicators several means that inflate relational interactions: selective self-presentation, partner idealization, editing and attention advantages, and mutually enhancing reciprocal feedback are parallel processes that improve relational communication in CMC (Walther, 1996). These factors have been investigated in a variety of settings, from online groups and business to online dating (see for review Walther & Ramirez, 2009). For instance, Walther (1997) studied college students in groups, members of which were geographically separated between England and the United States, who used CMC to communicate. Those who worked with partners over a longer period of time rated their CMC-only team members as more physically and socially attractive (despite never having seen them), and also reported greater intimacy and affection with them, than with those with whom they communicated FtF. The more editing that individuals put into their asynchronous electronic messages, the more affection and immediacy such messages reflect (Walther, 2007).

Little research has examined hyperpersonal dynamics within relationships that have their genesis offline but become geographically dispersed, with a few exceptions. Even before email was the written medium of choice, Stafford and Reske (1990) discovered that the frequency with which couples sent letters to one another was associated with higher ratings of love and marital adjustment, compared to the use of telephones and FtF meetings. These authors suggested that written communication leads partners to idealize one another; they forget the difficult aspects of FtF conversation and cohabitation when they use relatively restricted media to maintain their relationships at a distance. So too with email: As Rabby and Walther (2002, p. 154) suggested, "When one communicates largely through email he or she loses the sense of that partner's bad manners, slow speech, frequent cursing, and other questionable habits." Out of sight may be out of mind, and what remains in sight are crafted, edited messages. Recent work by Human and Lane (2008), examined platonic friends whose relationships migrated from offline to online status. They found that CMC preserves and enhances these partners' "lingering" and "fictive relational memories" of one another, at a distance. That is, virtual contact without visual cues reminds a partner of how the other looks and acts, subject to biased inaccuracies in these reconstructions. The degree and conditions under which CMC enhances rather than diminishes relational maintenance, and in what forms, is the subject of current research. In the following section, we review some existing findings about the use of various CMC technologies for relational maintenance.

Email

As Internet access became more widespread in the 1990s, it became clear that email would be one of the first forms of CMC to be used for relational maintenance. Stafford, Kline, and Dimmick (1999) reported on the use of email for relational maintenance when they found that 61% of randomly sampled interview subjects reported using their home computers to "keep in touch with friends, family, or relatives" (p. 663) who were across the country or globe. As technological adoption and access has increased, this trend has escalated. In 2002, Horrigan and

Rainie found that 72% of U.S. college students reported using the Internet to communicate with friends both on and off campus.

Friends. Other research has tested the ability of individuals to maintain psychological closeness using mediated channels with friends after moving away (Shklovski et al., 2008). Although after the residential move friends experienced a decrease in overall amount or frequency of communication, results showed that decreases in FtF communication did not diminish psychological closeness, whereas changes in frequency of phone calls and emails had a profound impact. Results showed that increases and decreases in phone calls were correlated to similar increases and decreases in closeness. However, effects were only found for *decreases* in emails. Specifically when email frequency decreased, closeness did as well, but no effects were found for email increases. The authors therefore concluded: "While phone calls seemed to contribute to relational growth, email communication may only serve to maintain it. Email may keep a relationship alive, while phone calls may help it to grow deeper" (p. 812). Thus Shklovski et al. (2008) characterized email as providing a "hygienic" function rather than constituting a proactive or strategic relational maintenance behavior, in that it requires a threshold level of activity, below which relational maintenance declines.

Families. Research has also focused on the ways in which individuals use email to maintain relationships with family members who have moved away from home to pursue school or work opportunities. Horrigan and Rainie (2002) interviewed a sample 1,500 Internet users and found that 84% use email for maintaining communication with family members, 70% emailing family members for advice, and 63% emailing to express worries. Such trends suggest that CMC is a medium in which individuals feel comfortable communicating serious issues. More specifically, a consistent line of research has examined the ways in which college students use mediated channels to maintain communication with geographically separated grandparents (Harwood, 2000; Holladay & Seipke, 2007) and parents (Trice, 2002). These findings are reviewed as follows.

The intrinsic advantages that mediated channels provide for grandparent-grandchild communication were evident in research conducted by Holladay and Seipke (2007). Their findings showed that grandparents who were geographically farther than two-hour-driving distance away from their grandchildren relied on telephone and email more heavily than face-to-face visits to stay in touch with their grandchildren. Overall, 58% of grandparents reported regular use of email with their grandchildren. Although some of these respondents indicated that they initiated the email correspondence more frequently (33.3% of sample), the majority (50.6%) indicated that the initiation and contact frequency was mutually determined by both parties, while others indicated that their grandchildren initiated contact more frequently (6.2%). This suggests that the medium is not only being used for relational maintenance by tech-savvy teenagers, but also their octogenarian counterparts.

Other research has tried to uncover the use of email between college students and their parents. Examining a sample of 578 emails that college freshmen sent to their parents, Trice (2002) found that email was a frequent form of relational contact. Students sent an average of six emails to their parents during a five-day week. It also seems that as periods of stress during the college semester increased, so did the emails ($r = .42$, males; $r = .54$, females). Furthermore, his analysis found that in some emails (22%) were students disclosing or requesting advice in what were deemed "frequently studied content areas in adaptation to college"—financial, social, or academic areas. However, the majority of these emails (78%) contained other relational maintenance messages. These results imply that between family members, both serious (e.g., open disclosures, requests for advice, social support) as well as mundane exchanges (e.g., news from home, reports of events at school) are communicated through email.

Couples. Among romantic couples, one reason that CMC helps facilitate relational maintenance and increased satisfaction may be its ubiquity. A line of current research has focused on couples in which at least one spouse is a deployed member of the U.S. military. When email became available as early as 1990, it became (and still remains) the most frequently used form of communication among deployed soldiers. Many soldiers and their spouses like email because it "provides speed, relative privacy, decentralization, and personal communication" (Booth et al., 2007). Because it is a much cheaper financial alternative than the telephone, email is often used by military families, but as a supplement, rather than a replacement channel for contact (Ender & Segal, 1996).

Web 2.0: Interpersonal Relational Maintenance in Public Spaces

Whereas email has become a widely used medium for relational maintenance due to its early development and widespread adoption, recent years have seen the explosion of "Web 2.0" applications that have increased in popularity and usage. These applications are making significant inroads in relationship maintenance activities. Applications such as blogs and social network sites like Facebook[3] offer unique combinations of communication elements. As a group, these new forms of participative web technologies not only allow individuals to display messages for, or broadcast to, a large but potentially selective number of people, they also allow for relational partners to add their own relational maintenance responses. Social network sites provide users their own pre-formatted web pages on which the owner of that profile describes himself or herself, and lists his or her proclivities, and a microblog—providing relatively short updates—about what that person is doing or feeling. The sites also feature participatory spaces on the profile where an individual's friends—anyone with whom a user has indicated a connection with—can post messages to the profile owner. In most cases, all these messages, by both the profile owners and their friends, are publicly available for other friends to see. Blogs are sometimes called "online diaries," but unlike traditional diaries, they are open to other people to read. Blogs consist of self-authored content that is posted in a (semi)public web space. Additionally, many blogs allow readers to post comments or replies to entries, creating an interactive space different than traditional offline diaries.

These and other new technologies bring together four important attributes that enhance their utility for relational maintenance and other social functions. First, they are asynchronous. Like email, messages are transmitted or published only after they are finished being constructed. This gives their users time to craft and edit their messages to achieve a desired effect. As we mentioned before, research in other arenas shows that time spent editing enhances relational qualities of electronic messages. Second, messages posted on these systems can be widely disseminated, but are also limited in their reach. Like traditional websites, the default setting on social network sites and blogs is for open access: Anyone in the world can see what an individual has posted online. Unlike traditional websites, both technologies feature various privacy settings, which allow users to restrict access to their postings to specific groups, networks, or individuals. These restrictions can include (or exclude) various family members, or strangers, or other parameters. Blogs in particular are quite frequently made accessible to small groups of family and friends. Within these networks of personal connections, however, relational maintenance and other messages can be shared so that everyone who has access to one's site can see what the author wishes them to see. Third, these systems foster participation, feedback, and interactivity when relationship partners initiate messages that are posted on an individual's site, or reply to an author's posting. Finally, these systems can embed multimedia sources from other websites. Individuals can share photos of themselves or other things, video clips that they find amusing, or interactive programs—such as quizzes—along with their own results to which partners can compare. The sharing of these elements suggests the passing of virtual tokens among relationship partners, and may function as an (asynchronous) shared activity.

Using these technologies, interpersonal relational maintenance often happens in a publicly viewable space (e.g., Facebook walls, blog entries, Twitter updates). It is this blend of public and private communication with which Web 2.0 technologies transform the way we communicate, enact, and interpret relational maintenance behaviors. The exact nature of how these new technologies are changing our actions (or perhaps how old, learned behaviors are being retrofitted into new technologies) has not yet been the focus of much formal research, but recent efforts are uncovering some surprising ways in which people use these media for maintenance purposes.

Facebook. Since its development in 2004, Facebook has become a worldwide social networking phenomenon. With over 500 million active members at the time of this writing (Facebook Press Room, 2010), it is one of the most heavily trafficked websites on the Internet. Allowing such things as profile creation, photo and video sharing, message postings, and status updates, Facebook's interactive nature made it a popular interface to seen and be seen, leading many scholars to speculate early on about the possibility of connecting users to novel social networks and new contacts. But more recent research concludes that users are more interested in maintaining ties with *existing* offline contacts rather than forging new ones. In their survey of college Facebook users, Lampe, Ellison, and Steinfield (2006) found that the most frequently reported use of Facebook was to "keep in touch with an old friend or someone I knew in high school" (p. 169). Similarly, Joinson's (2008) survey found that a frequently reported use among Facebook users was "maintaining relationships with people you don't get to see very often" (p. 1030). Relational maintenance communication often takes place on the profile-owner's "status updates," where they can inform their audience about activities, accomplishments, or musings. Relational maintenance also occurs on a profile-owner's "Wall"—a publicly viewable space on which one's friends can write messages. Relational maintenance on Facebook includes everything[4] from the mundane ("Hi, how are you?") to encouraging ("Hang in there! You can do it!"), and affectionate ("Hey girl, hey. Miss you around here"). Other messages facilitate other offline actions, such as coordinating a face-to-face meeting ("I'm coming this weekend. Are you available Saturday night?"). Such examples clearly reflect that Facebook is being used for relational maintenance activities.

Facebook has expanded its user base beyond its original focus—members of academic institutions—so that anyone who can access the Internet can now have a Facebook account. Opening up the membership to include "nonacademic" individuals has allowed parents and grandparents to connect to their children. In fact, self-help and how-to guides have been launched by those who realize Facebook's potential as a parenting and relational maintenance tool (e.g., www.facebookforparents.org). As Facebook continues to grow, its application as a relational maintenance tool will be further solidified among its users.

Blogs. As of 2004, a relatively small proportion of Internet users maintained a blog (8 million), but as many as 32 million people reported reading blogs on a regular basis, and 14.4 million reported posting content to blogs (Rainie, 2005). Many blog authors seem aware of the public Internet environment, and tend to blend private thoughts and mass communications together into a single communicative entity (see Nardi, Schiano, & Gumbrecht, 2004).

Research on blog users has confirmed that users do employ them to maintain relationships (Nardi et al., 2004). Although blogging has received a great deal of attention due to its capacities to foster "citizen journalism" and to provide an alternative to traditional mass media as a source of news and information, "A national phone survey of bloggers finds that most are focused on describing their personal experiences to a relatively small audience of readers and that only a small proportion focus their coverage on politics, media, government, or technology," according to Lenhart and Fox (2006); 59% of bloggers revealed that staying in touch with friends and family is a major or minor reason that they blog. And among bloggers, Stefanone and Jang (2007) found that those who were more extraverted were prone to use them for relational maintenance purposes. They also found that, like email and networking sites, blogs were more likely to be used by those who had larger social networks consisting of "strong ties," suggesting that such personal information is intended for close friends and family to maintain online contact with those who they do not get to see often offline.

Traditional topics in blogs consist of personal updates on day-to-day activities, or personal thoughts (Herring, 2004), and the "one-to-many" broadcast quality of blogs makes it a useful tool for relational maintenance. From a convenience perspective, one blog post could save a lot of time, allowing individuals to connect with their entire social network about personal matters much more frequently. Functioning similarly to the hygienic "annual holiday Christmas card" that many families send, blogs allow a more regular form of relational maintenance communication, at a reduced price, both economically and temporally. Moreover, the interactivity of blogs gives readers the opportunity to post their own responses, or perhaps begin a multimember interpersonal dialogue, providing a greater advantage for relational maintenance—something not possible in other types of one-to-many communications, such as cards or letters.

Twitter & Microblogging

Can relational maintenance be communicated in 140 typographical characters or less? That is precisely what 1.2 million users who use Twitter (and other applications like it) do every day. Twitter's interface limits messages to 140 characters, and therefore the reports, reflections, and narratives that users can post are necessarily brief; this characteristic is what earns Twitter its characterization as a "microblogging" system. Accessible via any web browser, or via wireless mobile devices, Twitter allows users to post and read updates from other subscribers about their day-to-day activities in the form of text messages known as "Tweets." Individuals can "follow" another's Twitter posts by specifying the people from whom the system will reveal the messages those individuals write, whenever they write them; or to transmit those individual's tweets to one's mobile phone as a text message. The persons being "followed" are able to reject the requests of anyone wishing to receive their tweet, allowing any Twitter user to define his or her own audience. According to Lenhart's (2009) survey of Twitter users, most users tend to be young adults, but it seems that microblogging is becoming more mainstream with the average user's age on the rise.

Although many users employ Twitter to share news about events they see or hear, and observations about things they witness, there is growing recognition of its capacity for interpersonal purposes. Twitter seems almost tailor-made for contacting relational partners, as seen in the system's very definition on Twitter.com: "Twitter is a service for friends, family, and co-workers to communicate and stay connected through the exchange of quick, frequent answers to one simple question: What are you doing?" (Twitter, 2009). The *openness and disclosures* that satisfy this question are not limited to one's activities, but include users' discussion of their evaluations and feelings. The convenience of microblogging and the rapidity with which communication can take place make it a tool that many have adopted for interpersonal purposes. As such, relational maintenance applications have been informally described among coworkers (Zhao & Rosson, 2009) and family members (Slatalla, 2008), although no empirical studies to date have documented such behaviors.

Theoretical Perspectives: Offline and Online

Although descriptive trends and usage patterns of CMC for relational maintenance have been documented in recent research, major theoretical perspectives reflected in the current offline relational maintenance literature have only recently been examined within an online arena (Dainton & Aylor, 2000; Gunn & Gunn, 2000; Rabby, 2007; Ruppel, 2009; Wright, 2004). We review these theoretical perspectives and the implications and shortcomings of their translation into online environments.

Examining Traditional Relational Maintenance in Online Channels

Several studies have examined how the various dimensions of relational maintenance behavior—positivity, openness, assurances, sharing tasks, and networks—are affected by CMC.

Openness and Disclosure. Several studies have taken the Stafford and Canary typology and applied it to CMC maintenance with interesting outcomes. In traditional offline research, trends have shown *openness* to be negatively associated with relational characteristics such as satisfaction, and commitment (Dainton, 2000; Stafford, 2003; Stafford, Dainton, & Haas, 2000). However, Wright's (2004) examination of those in primarily Internet-based (PIB) and exclusively Internet-based (EIB) relationships found that openness was the most frequently performed behavior for both types of relationships, followed by positivity. Additionally, the use of these two strategies was positively associated with the quality of their communication.

Similar results were obtained in Rabby's (2007) study which examined the relational maintenance behaviors of different types of CMC-using couples. *Real worlders* were those couples who had initially met offline, and continued to do so; *Cyber emigrants* are couples who had met offline, but continued their relationship online after geographical separation; *Pinocchios* consisted of couples who met initially online, but now interact offline; and *Virtuals* met online and continued to rely on CMC as their primary form of communication. Rabby (2007) found that among virtuals, *openness* was the most frequently performed strategy: "Without the capacity to engage in activities and to share in social events that most couples do, these relationships exist entirely in the communication that they engage in with each other. Their relationships are maintained through self-disclosure and sharing information with their partners" (pp. 331–332). In contrast to offline research, openness played an important role in sustaining relational commitment among online partners.

This discrepancy raises questions about the role of self-disclosure and openness in online and offline relationships. Some speculate that culturally held belief that self-disclosure and openness are necessary for relational success reflects a social-desirability bias, in which individuals overestimate their use of openness (Stafford, 2003; Weigel & Ballard-Reisch, 1999). It may also be that the content of self-disclosures in online and offline relationships is different. Because disclosure is so important in online relationships, content may reflect both positive and negative aspects of the relationship. By contrast, those that are co-located and communicate face-to-face may not feel the need to disclose information on a daily basis, instead using this form of communication only for problem-solving, negative disclosure, or troublesome issues. Although disclosures would seem to be less likely to take place online among co-located couples, some research has shown that the communication of sensitive or hurtful information and bad news can be done more easily through online channels (McGlone & Batchelor, 2003; Tong & Walther, in press). Regardless, however, openness and disclosures function much more broadly for online individuals who must use these to convey positive (affection, intimacy) and negative (disapproval, criticism) affect alike. Thus the blanket claim that openness is negatively associated with relational commitment and success deserves re-examination by CMC researchers.

New Activities: Sharing Tasks and Social Networks. In a recent diary study (Jerney-Davis et al., 2005) military spouses were asked to record their communication activities with their deployed spouse. These communications were then content-analyzed and coded for the above strategies. Interestingly, the authors found that the categories in the original Stafford and Canary typology needed to be expanded. Because the original definitions of *sharing tasks* and *networks* necessitated physical presence, it seemed nearly impossible for separated spouses to be able to perform these behaviors. To remedy this, the authors extended the category to any "discussion of social networking," "interactions with persons outside the relationship," or "ordinary tasks that needed to be completed" (p. 13).

Furthermore, with tools such as online shopping, social networking sites, and photo-sharing sites on the rise, it is interesting to speculate on the utility of these websites for relational maintenance and the ways that the definitions within behavioral typologies would have to evolve. For example, imagine that a husband and wife want to send a birthday present to a cousin. The husband is away on business in Japan and the wife is at home in the United States. They communicate via email (because the time difference makes phone calls too difficult to coordinate), they shop for a birthday present online, and have it shipped to their cousin all without ever being physically co-located. Should such behaviors be considered "sharing tasks"? Using social networking sites, couples can check in with mutual acquaintances, each "spending time" with important friends and family within their mutual social network, albeit independently. Individuals can provide "status" feeds to update others with what they are doing (e.g., "Arrived back to the dorm safe and sound. I love you"). Although not physically co-located, intimate, open communicative maintenance is taking place. Furthermore, these messages are no longer considered a privately conveyed interpersonal communication between two individuals, but rather a blend of interpersonal-mass-media messages broadcast to many, but intended for few (see Walther et al., 2010). It is yet unknown if these dynamics change the way relational maintenance behaviors are performed.

New Theoretical Directions: Costs and Benefits Transformed

One of the most useful approaches to relational maintenance has been the application of equity theory, in explaining an individual's motivations to perform relational maintenance activities and the benefits derived from doing so. In terms of relational maintenance, equity refers to the fact that both relational partners desire equitable ratios of behavioral input, or costs, relative to positive relational outcomes, or benefits. When the cost-to-reward ratio reaches a balanced state for both partners, equity theory suggests they will both be satisfied (Hatfield, Traupmann, Sprecher, Utne, & Hay, 1985). If one partner perceives that his or her relational maintenance costs exceed the outcomes he or she receives, that individual is said to be *under-benefited.* Individuals who believe their rewards exceed their relational maintenance costs, however, are *over-benefited.* When this balance is threatened, equity theory predicts that each will attempt to restore the equity of rewards and costs. Although this has been a useful framework for several relational maintenance researchers (Canary & Stafford, 1994; Dainton & Lendzinski, 2008; Dainton, Zelley, & Langan, 2003), it is worth noting how technology is changing the amount of effort, or cost, of performing strategic and nonstrategic relational maintenance behaviors. New communication technologies are significantly reducing the cost to transmit and receive relational maintenance messages, for both geographically separated partners as well as those who live together but are separated during portions of the day. Information technology researchers have identified the advantages of various "lightweight tools" for communication, which reduce the resource costs required to maintain communicative connections with others. The aspect that makes them "lightweight" refers to the relatively low effort required to look up, initiate a connection to, and create messages for, another person, as well as how many operations a program or device requires in order to use it. Primarily text-based and involving no programming or code, "they are also cognitively lightweight for users," according to Churchill and Bly (1999, p. 2): "once an account has been set up, a user can begin interacting with others who are present in the (system) with only a small amount of instruction." These characteristics were originally applied to instant messaging (Nardi, Whittaker, & Bradner, 2000) and other real-time chat systems, but the characteristics enumerated above certainly describe a growing number of web-based systems such as the ones we have discussed in this chapter.

The cost reduction comes as lightweight Internet-based systems are present in the same machine that an individual uses for work or education, when it incurs no additional financial resources to contact relational partners (as is the case when computers or other devices and their Internet connections are already in place), and since some technologies allow individuals to contact whole hosts of relational partners with a single message (as do blogs, Facebook, or Twitter). It is far simpler to transmit relational maintenance messages under these conditions than it is to do so with numerous and sequential dyadic partners, using phones or FtF encounters. The reduction of costs implies that the scarcity of communicative opportunities is reduced, and more frequent relational maintenance communication should be expected when technology is available among relational partners. If relational maintenance provides pleasant distractions from work or study, even more the case that its frequency should be greater in the presence of technological facilitation. As Nardi et al. (2000) documented among office workers with Instant Messenger (IM) systems on their computers,

> A frequent use of IM was to keep in touch with friends and family while at work. These interactions were often very brief, like, "Hi, Hon!" Such interactions seemed to provide a moment of respite in a busy day, a sort of "pat on the shoulder"...IM injected playfulness and intimacy...allowing (workers) to connect to loved ones in quick but meaningful ways. (p. 82)

The field of economics (the home of equity theory's origins) offers a good deal of research on the transaction costs perspective (Williamson, 1975), which can also be applied to the impact of lightweight tools on the rewards and costs of relational maintenance. Transaction costs were originally the costs accompanying the conduct of an economic exchange, such as a broker's fee that accompanies buying stock, but the concept has been extended to include the time, effort, and other resources that are required to scan the environment for alternatives, assess alternatives, procure something, pay for it, transport it, etc. In transaction cost analyses of organization communication, information technology provides numerous factors that reduce the cost of coordinating activities and exchanges (Malone, Yates, & Benjamin, 1987). From this perspective we may say that the net reward value of communicating relational maintenance with others increases as the transaction costs for exchanging relational maintenance messages decrease. The effect of new communication technologies on relational transaction costs is fairly clear: technology reduces the initial expense of communicating maintenance because it is present, lightweight, and asynchronous. Furthermore, social technologies such as blogs or microblogs that allow an individual to broadcast relational maintenance to many partners further reduce transaction costs on a per partner basis. Whether receivers interpret broadcasted messages as less relationally genuine than dyadic messages is a question for future research. Research on "cheap talk" in the economics literature—where messages require few resources to exchange—has explored the value of such modes in coordinating independent actors (see for review Wildman, 2008). Further research on equity theory and relational maintenance more generally, from a relational transaction costs perspective, may help expand our understanding of both relational maintenance and technology's role in its conduct.

Moreover, as a result of these persistently present lightweight tools, new relational maintenance functions and types are emerging. We suggest three types, two of which have been suggested in the communication technology literature but which have not been tied to relational maintenance before—presence and tie signs—and a communication function discussed in recent relational dynamics research but previously unconnected to social technology—mundane conversation.

Presence. Communication technologies allow relational partners to maintain a sense of presence with one another even when they are physically separated. Although presence has been defined in a variety of ways (see Lee, 2006), we use the term here to mean that partners are at least mildly cognizant of one another and feel as though they are in present or potential interpersonal contact. Presence implies a sense of emotional connection and psychological propinquity, which research has shown can be achieved even through the leanest of communicative systems, especially when other more complex communication alternatives are inopportune (Walther & Bazarova, 2008). Seeing an email message arrive in one's inbox, or watching occasional status updates on Facebook, or an interesting Tweet may all revive a sense of connection to the partner who sent it.

Tie signs. Second, several of these systems indicate for others to see that partners share a relationship. Facebook displays who one's apparent friends are, in a way that other friends can see. Twitter shows who one follows and who is following someone. Although these are not particularly dynamic displays, and require no effort to maintain once they are established, their persistent display on the systems' interfaces reminds individuals and partners, as well as observers, who knows whom. It should be noted that the meaning of being a friend within Facebook is quite weaker than in the traditional sense of a friend, and the number of friends one has in Facebook bears a curvilinear relationship to perceived social desirability by others (Tong, Van Der Heide, Langwell, & Walther, 2008). Even this disparaging finding, however, reinforces the notion that these publicly displayed sociometric linkages have meaning.

These visible clues about sociometric and relational links, and the messages people leave for one another in a semi-publicly visible manner online, may function as "tie signs." Better known in the study of nonverbal communication, with theoretical roots in sociology and anthropology (Goffman, 1971; Morris, 1971; see for review Afifi & Johnson, 2005), the concept of tie signs refers to displays of mutual belonging through physical behaviors or adornments. Wedding rings, wearing someone else's clothing (like a varsity jacket), or a tattoo with someone's name provide offline examples of tie signs. These kinds of symbolic links translate readily into contemporary media. Facebook wall posts or Twitter Tweets, for example, convey not only the content of the message; they also signify that the sender and receiver are connected, which has led Donath and boyd (2004) to describe social network sites as creating "public displays of connection." Such a link notifies observers about who has a relationship, as well as reinforces that relationship for its members, since tie signs "typically serve as relational indicators to the dyad using the behaviors and to their audience" (Afifi & Johnson, 2005, p. 190).

Mundane sharing of activities and experiences. One of the chief bewilderments, if not complaints, about Twitter's Tweets and Facebook's status updates is the incredibly mundane quality of many, if not most, such postings. And yet, relational communication research has started to recognize that mundane conversations have a lot to do with the sustenance of relationships, their character, or even their demise (Duck, 2005). Therefore somewhat ironically, we argue that it may be the tendency to use new social technologies for the mundane expression of activities and the sharing of mundane personal observations that make them the most useful in the process of relationship maintenance. To post messages on social technology systems, to responsive or imagined relational partners, may be to engage in "self-validation, social comparison with other people, and emotional integration (i.e., ratification of one's own emotional responses to events)" (Duck, 2005, p. 211)—the functions which mundane conversations appear to provide in offline interaction—and implicitly to invite participation in, concurrence about, or confirmation of an individual's forays into these important psychological processes. Beyond the shared physical activities, it is the mundane events and interpretations of day-to-day life that geographically separated relational partners do not ordinarily share, and the expense of traditional long-distance communication make it too costly to re-enact them during brief encounters. Among relational partners who reside together but are separated during part of the day—couples or families, for instance—they can probably ill afford to interrupt one another's professional activities to describe trivial musings ad hoc, but the asynchronous capabilities of email, Facebook, and Twitter mean that mundane updates will wait until the receiver seeks diversion or connection. Communication technology mitigates cost and normalizes discussing these otherwise unremarkable events (as the system asks, "what are you doing right now?") as well as inviting reflection and responses. Moreover, in offline interaction, discussion of mundane events is the commonplace not only of intimate relationships but of non-intimate relations as well (Duck, Rutt, Hurst, & Strejc, 1991). Therefore social technologies that allow individuals to broadcast mundane narratives and reflections to both highly intimate and less intimate partners provide more relational maintenance "bang" for the message-sending "buck."

The sharing of mundane discussions has not frequently been a focus of relational maintenance behavior typologies. New technology may make this activity more salient. On that basis analysts may be inclined to suggest that the communicational sharing of mundane observations is a new form of relational maintenance which technology has wrought. Such a conclusion would be an error of causal inference. Despite its basis in traditional relational communication, however, considering discussion of the mundane as a relational maintenance activity that technology makes more prevalent may be a useful approach to understanding technology's role in the relational maintenance process.

Conclusion

With the increase in Internet use across a variety of relational contexts, it is important for relational maintenance researchers to continue considering the intersection of technology and interpersonal communication. The careful attention that was paid to the initial examination of offline relational maintenance behaviors over the past two decades must be expanded if we are to understand whether and how new technologies are changing and extending the performance of relational maintenance. Future research should broaden its scope, discovering how media are used in a variety of ongoing relationship types, in both geographically separated and geographically collocated settings. Along with this expansion, we argue that a rigorous research agenda should be established and pursued by researchers. Simply applying old relational maintenance typologies to new mediated environments will not advance explanatory or predictive power. New factor analytic studies conducted to see what emergent dimensions of relational maintenance may surface as a result of additional channels will be of little utility unless such work is preceded with thoughtful conceptual notions about what is different, and why, with specific respect to computer-mediated relational maintenance. Other research that simply charts the usage trends of media for interpersonal functions, although interesting, only provides descriptive information about communicative behaviors, and these kinds of studies, too, beg for more thorough examination and explanation.

Research should offer careful attention to a number of tensions that mediated relational maintenance presents. The importance of both routine and strategic communication online, the novel ways in which media can foster both the remarkable and the mundane, and the nature of private relational maintenance in public spaces, for instance, are all dualities that merit consideration. Do family members find Tweets to be valuable relational maintenance communications or mundane expressive patter? Does expressing a problem in one's Facebook status update beget provisions of social support through Facebook wall posts by one's friends? Does the joint conduct of mutual relationship tasks coordinated and conducted across continents and time zones occur with more or less conflict than when they are done FtF? Are some relational maintenance functions especially well suited to mediated channels, and more effective or satisfying this way than by FtF means? And what are the implications for relationship success, stability, or termination?

Notes

1 Department of Communication, Michigan State University
2 Department of Telecommunication, Information Studies & Media, Michigan State University
3 Although MySpace is often mentioned as having a vast user base, recent research by Parks (2010) has found that most MySpace visitors cease using the system after a relatively short period of time, and that many users who return make "friends" with musical groups and other organizational entities, suggesting that use of MySpace for relational maintenance is relatively unusual.
4 All wall post examples were taken from real Facebook profiles.

References

Afifi, W. A., & Johnson, M. L. (2005). The nature and function of tie-signs. In V. Manusov (Ed.), *The sourcebook of nonverbal measures: Going beyond words* (pp. 189–198). Mahwah, NJ: Lawrence Erlbaum.

Ayers. J. (1983). Strategies to maintain relationships: Their identification and perceived usage. *Communication Quarterly, 31*, 62-67.

Baxter, L. A., & Dindia, K. (1990). Marital partners perceptions of marital maintenance strategies. *Journal of Social and Personal Relationships, 7*, 187-208.

Bell, D. B., Schumm, W. R., Knott, B., & Ender, M. G. (1999). The desert fax: A research note on calling home from Somalia. *Armed Forces & Society, 25*, 509-521.

Bell, R. A., Daly, J. A., & Gonzalez, M. C. (1987). Affinity-maintenance in marriage and its relationship to women's marital satisfaction. *Journal of Marriage and the Family, 49*, 445-454.

Boneva, B., Kraut, R., & Frohlich, D. (2001). Using e-mail for personal relationships: The difference gender makes. *American Behavioral Scientist, 45*, 530-549.

Booth, B., Segal, M. W., Bell, D. B., Martin, J. A., Ender, D. G., & Rohall, D. E. (2007). What we know about army families: 2007 update. Retrieved June 10, 2009, from http://www.army.mil/fmwrc/documents/research/whatweknow2007.pdf

Canary, D. J., & Stafford, L. (1994). Strategic and routine interaction. In D. J. Canary & L. Stafford (Eds.), *Communication and relational maintenance* (pp. 3–22). San Diego, CA: All Academic.

Center for Long Distance Relationships. (2009). Retrieved July 26, 2009, from http://www.longdistancerelationships.net/index.html

Churchill, E. F., & Bly, S. (1999, November). *It's all in the words: Supporting work activities with lightweight tools.* Paper presented at the conference on Groups 99, Phoenix, AZ.

Culnan, M. J., & Markus, M. L. (1987). Information technologies. In F. M. Jablin, L. L. Putnam, K. H. Roberts, & L. W. Porter (Eds.), *Handbook of organizational communication: An interdisciplinary perspective* (pp. 420-443). Newbury Park, CA: Sage Publications.

Dainton, M. (1998). Everyday interaction in marital relationships: Variations in relative importance and event duration. *Communication Reports, 11*, 101-110.

Dainton, M., & Aylor, B. (2000). Patterns of communication channel use in the maintenance of long-distance relationships. *Communication Research Reports, 19*, 118–129.

Dainton, M., & Lendzinski, J. (2008, November). *The frequency, valence, and importance of relationship maintenance behaviors.* Paper presented at the annual conference of the National Communication Association, Chicago.

Dainton, M., Zelley, E., & Langan, E. (2003). Maintaining friendships throughout the lifespan. In D. J. Canary & M. Dainton (Eds.), *Maintaining relationships through communication* (pp. 79–102). Mahwah, NJ: Lawrence Erlbaum.

Dindia, K. (2003). Definitions and perspectives on relational maintenance communication. In D. J. Canary & M. Dainton (Eds.), *Maintaining relationships through communication: Relational, contextual, and cultural variations* (pp. 51–77). Mahwah, NJ: Lawrence Erlbaum.

Dindia, K., & Canary, D. J. (1993). Definitions and theoretical perspectives on relational maintenance. *Journal of Social and Personal Relationships, 10*, 163-173.

Donath, J., & boyd, d. (2004). Public displays of connection. *BT Technology Journal, 22*(4), 71–82.

Duck, S. (1988). *Relating to others*. Milton Keynes, UK: Open University Press.

Duck, S. (2005). How do you tell someone you're letting go? *The Psychologist, 18*(4), 210–213.

Duck, S., & Pittman, G. (1994). Social and personal relationships. In M. L. Knapp & G. R. Miller (Eds.), *Handbook of interpersonal communication* (2nd ed., pp. 676–695). Thousand Oaks, CA: Sage.

Duck, S., Rutt, D. J., Hurst, M. H., & Strejc, H. (1991). Some evident truths about conversations in everyday relationships: All communications are not created equal. *Human Communication Research, 18*, 228–267.

Ender, M. G., & Segal, D. R. (1996). V(E)-mail to the foxhole: Isolation, (tele)communication, and forward deployed soldiers. *Journal of Political and Military Sociology, 24*, 83–104.

Facebook Press Room: Statistics. (2009). Retrieved July 21, 2010, from http://www.facebook.com/press.php

Goffman, E. (1971). *Relations in public*. New York: Basic Books.

Gunn, D. O., & Gunn, C. W. (2000, September). *The quality of electronically maintained relationships*. Paper presented at the annual conference of the Association of Internet Researchers, Lawrence, KS.

Harwood, J. (2000). Communication media use in the grandparent-grandchild relationship. *Journal of Communication, 50*, 56-78.

Hatfield, E., Traupmann, J., Sprecher, S., Utne, M., & Hay, J. (1985). Equity and intimate relations: Recent research. In W. Ickes (Ed.), *Compatible and incompatible relationships* (pp. 91-117). New York: Springer.

Holladay, S. J., & Seipke, H. L. (2007). Communication between grandparents and grandchildren in geographically separated relationships. *Communication Studies, 58*, 281-297.

Horrigan, J. B., & Rainie, L. (2002). Getting serious online. Pew Internet & American Life Project. Retrieved July 29, 2009, from http://www.online-publishers.org/media/202_W_PIP_Getting_Serious_Online3ng.pdf

Human, R., & Lane, D. (2008, November). *Virtually friends in cyberspace: Explaining the migration from FtF to CMC relationships with electronic functional propinquity theory*. Paper presented at the annual meeting of the National Communication Association, San Diego, CA.

Jerney-Davis, M., Kim, R., Kim, I., Raphael, D., Kawamura, A., & Lau, J. (2005, May). *Relational maintenance during deployment: Communication between spouses*. Paper presented at the annual meeting of the International Communication Association, New York.

Joinson, A. N. (2008, April). "Looking at," "Looking up," or "Keeping up with" people? Motives and uses of Facebook. *Proceedings of CHI 2008*, Florence, Italy, 1027–1036.

Lampe, C., Ellison, N. B., & Steinfield, C. (2006). A face(book) in the crowd: Social searching vs. social browsing. *Proceedings of CSCW 2006*, Banff, Alberta, Canada.

Lee, K. M. (2006). Presence, explicated. *Communication Theory, 14*, 27–50.

Lenhart, A. (2009). *Twitter and status updating.* Pew Internet & American Life Project. Retrieved July 18, 2009, from http://www.pewinternet.org/Experts/~/link.aspx?_id=6C747837133C4A54A4D0351E2683478B&_z=z

Lenhart, A., & Fox, S. (2006, July 19). *Bloggers.* Pew Internet & American Life Project. Retrieved July 18, 2009, from http://www.pewinternet.org/Reports/2006/Bloggers.aspx

Malone, T. W., Yates, J., & Benjamin, R. I. (1987). Electronic markets and electronic hierarchies. *Communications of the ACM, 30,* 484–497.

McGlone, M. S., & Batchelor, J. A. (2003). Looking out for number one: Euphemism and face. *Journal of Communication, 53,* 251-264.

Morris, D. (1971). *Intimate behavior.* New York: Random House.

Nardi, B. A., Schiano, D. J., & Gumbrecht, M. (2004). Blogging as social activity, or, would you let 900 million people read your diary? *Proceedings of CSCW 2004,* Chicago, IL.

Nardi, B. A., Whittaker, S., & Bradner, E. (2000). Interaction and outeraction: Instant messaging in action. *Proceedings of the ACM Conference on Computer-Supported Cooperative Work* (pp. 79–88). New York: ACM.

Parks, M. R. (2010). Social network sites as virtual communities. In Z. Papacharissi (Ed.), *A networked self: Identity, community and culture on social network sites* (pp. 105-123). New York: Routledge.

Rabby, M. K. (2007). Relational maintenance and the influence of commitment in online and offline relationships. *Communication Studies, 58,* 315–337.

Rabby, M., & Walther, J. B. (2002). Computer-mediated communication impacts on relationship formation and maintenance. In D. Canary & M. Dainton (Eds.), *Maintaining relationships through communication: Relational, contextual, and cultural variations* (pp. 141–162). Mahwah, NJ: Lawrence Erlbaum.

Rainie, L. (2005). *The state of blogging.* Pew Internet & American Life Project. Retrieved July 18, 2009, from http://www.pewinternet.org/Reports/2005/The-State-of-Blogging.aspx

Shklovski, I., Kraut, R., & Cummings, J. (2008, April). *Keeping in touch by technology: Maintaining friendships after a residential move.* Paper presented at the ACM conference on Computers and Human Interaction, Florence, Italy.

Slatalla, M. (2008). Cyberfamilias: If you can't let go, Twitter. *The New York Times.* Retrieved July 29, 2009, from http://www.nytimes.com/2008/02/14/fashion/14Cyber.html?pagewanted=1&_r=1

Stafford, L. (2003). Maintaining romantic relationships: A summary and analysis of one research program. In D. J. Canary & M. Dainton (Eds.), *Maintaining relationships through communication: Relational, contextual, and cultural variations* (pp. 51–77). Mahwah, NJ: Lawrence Erlbaum.

Stafford, L. (2005). *Maintaining long-distance and cross-residential relationships.* Mahwah, NJ: Lawrence Erlbaum.

Stafford, L., & Canary, D. J. (1991). Maintenance strategies and romantic relationship type, gender, and relational characteristics. *Journal of Social and Personal Relationships, 8,* 217-242.

Stafford, L., Dainton, M., & Haas, S. (2000). Measuring routine and strategic relational maintenance: Scale revision, sex versus gender roles, and the prediction of relational characteristics. *Communication Monographs, 67,* 306-323.

Stafford, L., Kline, S. L., & Dimmick, J. (1999). Home e-mail: Relational maintenance and gratification opportunities. *Journal of Broadcasting and Electronic Media, 43,* 659-669.

Stafford, L., & Merolla, A. J. (2007). Idealization, reunions, and stability in long-distance dating relationships. *Journal of Social and Personal Relationships, 24,* 37–54.

Stafford, L., & Reske, J. R. (1990, May). Idealization and communication in long-distance premarital relationships. *Family Relations, 39,* 274–279.

Stephen, T. (1986). Communication and interdependence in geographically separated relationships. *Human Communication Research, 13,* 191–210.

Tong, S. T., Van Der Heide, B., Langwell, L., & Walther, J. B. (2008). Too much of a good thing? The relationship between number of friends and interpersonal impressions on Facebook. *Journal of Computer-Mediated Communication, 13,* 531–549.

Tong, S. T., & Walther, J. B. (in press). Just say "No thanks": Romantic rejection in computer-mediated communication. *Journal of Social and Personal Relationships.*

Trice, A. D. (2002). First semester college students' email to parents: Frequency and content related to parenting style. *College Student Journal, 36,* 328–334.

Twitter (2009). *Twitter.* Retrieved July 29, 2009 from http://twitter.com

Walther, J. B. (1996). Computer-mediated communication: Impersonal, interpersonal, and hyperpersonal interaction. *Communication Research, 23,* 3–43.

Walther, J. B. (1997). Group and interpersonal effects in international computer-mediated collaboration. *Human Communication Research, 23,* 342–369.

Walther, J. B. (2007). Selective self-presentation in computer-mediated communication: Hyperpersonal dimensions of technology, language, and cognition. *Computers in Human Behavior, 23,* 2538–2557.

Walther, J. B., & Bazarova, N. (2008). Validation and application of electronic propinquity theory to computer-mediated communication in groups. *Communication Research, 35,* 622–645.

Walther, J. B., Carr, C., Choi, S., DeAndrea, D., Kim, J., Tong, S. T., & Van Der Heide, B. (2010). Interaction of interpersonal, peer, and media influence sources online: A research agenda for technology convergence. In Z. Papacharissi (Ed.), *A networked self: Identity, community and culture on social network sites* (pp. 17-38). New York: Routledge.

Walther, J. B., & Ramirez, A., Jr. (2009). New technologies and new directions in online relating. In S.W. Smith & S. R. Wilson (Eds.), *New directions in interpersonal communication research* (pp. 264-284). Thousand Oaks, CA: Sage.

Wang, H., & Andersen, P. A. (2007, May). *Computer-mediated communication in relationship maintenance: An examination of self-disclosure in long-distance friendships.* Paper presented at the annual meeting of the International Communication Association, San Francisco.

Weigel, D. J., & Ballard-Reisch, D. S. (1999). All marriages are not maintained equally: Marital type, marital quality, and the use of maintenance behaviors. *Personal Relationships, 6,* 291–303.

Wildman, S. S. (2008). Communication and economics: Two imperial disciplines and too little collaboration. *Journal of Communication, 58,* 693–706.

Williamson, O. E. (1975). *Markets and hierarchies: Analysis and antitrust implications, a study of the economics of internal organization.* New York: Free Press.

Wright, K. B. (2004). Online maintenance strategies and perceptions of partners within exclusively Internet-based and primarily Internet-based relationships. *Communication Studies, 55,* 239–253.

Zhao, D., & Rosson, M. B. (2009). How and why people Twitter: The role that microblogging plays in informal communication at work. *Proceedings of the ACM conference on Supporting Group Work,* Sanibel Island, FL.

CHAPTER SEVEN

Locating Computer-Mediated Social Support within Online Communication Environments

Andrew C. High

Denise H. Solomon

Social support encompasses the comfort, assistance, and reassurance that people experience as a function of social relationships. Social support enables people to cope with a multitude of personal, physical, social, or mental stressors and experience important physical and psychological benefits (i.e., Burleson & MacGeorge, 2002; Cohen & Wills, 1985; Cunningham & Barbee, 2000). To date, the communication of social support has been predominantly studied as an activity that unfolds in face-to-face (FtF) interaction. Although FtF communication is an important source of social support, this focus neglects the fact that social interaction unfolds in a variety of communication modalities and mediated discourse may even be preferred or required in some circumstances (Walther & Parks, 2002). In this chapter, we consider computer-mediated communication (CMC) as a means of seeking and receiving social support.

Computer-mediated communication refers to interaction between two or more people that is enabled by the use of computer technology. Whereas FtF communication requires temporal or spatial proximity, CMC enables people to exchange messages asynchronously and across great physical distances using the Internet and mediating technology. For these reasons, CMC can be a powerful tool for the communication of social support. Research on mediated social support has examined online support groups (e.g., Davison, Pennebaker, & Dickinson, 2000; Hildingh, Fridlund, & Segesten, 1995; Winzelberg, 1997; Wright, 2000, 2002; Wright & Bell, 2003) and has compared CMC to FtF communication experiences (e.g., Adams, Roch, & Ayman, 2005; Burgoon, Bonito, Ramirez, Dunbar, Kam, & Fischer, 2002; O'Sullivan, 2000; Tidwell & Walther, 2002; Walther, Slovacek, & Tidwell, 2001). Our goal is to build upon this work by considering how features of CMC environments, in general, shape the communication of social support within a variety of specific CMC contexts.

In this chapter, we propose to examine how different modes of CMC provide fundamentally different contexts for social support. Both social support (Burleson, 1994; Burleson & MacGeorge, 2002; Xu & Burleson, 2001) and CMC are multifaceted phenomena. We advance a conceptual framework that appreciates the nuances in both social support and CMC and illustrates how they map onto each other. To this end, we begin by describing types of social support. Next, we discuss several dimensions that shape mediated interactions. Then, we consider some issues that emerge when social support is situated within six distinct CMC environments.

Explicating Social Support

Burleson, Albrecht, Goldsmith, and Sarason (1994, p. xviii) claimed that "social support should be studied as communication because it is ultimately conveyed through messages directed by one individual to another in the context of a relationship that is created and sustained through interaction." With the emergence of a communication perspective on social support, researchers began to appreciate the centrality of messages exchanged between people, the dynamics of interaction, and the relational consequences of support episodes (Burleson & MacGeorge, 2002). The types of social support recognized in previous research range from sharing thoughts (Hildingh et al., 1995, p. 225) to promoting healthy habits (Callaghan & Morrissey, 1993); not surprisingly, messages with different content often lead to divergent support experiences (Burleson & MacGeorge, 2002; Burleson & Samter, 1990; Cutrona & Russell, 1990; Hale, Tighe, & Mongeau, 1997). In the paragraphs that follow, we describe a variety of ways in which people communicate social support.

Emotional support encompasses messages that address a target's emotional state. Kohn (1996) described this type of support as any effort at ventilating, managing, or suppressing an emotional reaction to an incident. Other research has conceptualized emotional support as openly disclosed, genuine feelings of caring (Burleson & MacGeorge, 2002). Burleson and Goldsmith (1998) explicitly recognized the centrality of care, concern, and acceptance in their conceptualization of emotional support. Likewise, Albrecht and Adelman (1987) suggested that emotional support should convey understanding for what a person is feeling. In general, then, emotional support involves promoting a positive affective experience for a distressed individual.

Whereas emotional support addresses the sentiments of distressed individuals, *informational support* focuses on advising distressed people (Burleson & MacGeorge, 2002). Cobb (1976, p. 300) conceptualized this construct as information leading a person to believe that he or she is "a member of a network of mutual obligation." Informational support is often operationalized as attempts to provide people with practical information that will help remedy their problems. Although informational support might not directly provide a solution, its content should enable a distressed individual to become a self-sufficient problem solver.

Although emotional and informational support have received the most research attention (Burleson & MacGeorge, 2002), other types of support merit mention. For example, Xu and Burleson's (2001) typology of social support also includes esteem, tangible, and network support. Providers of *esteem support* reaffirm people's identities and remind the targets of support that they are valuable and worthwhile.

Tangible support involves giving practical, material aid, which allows distressed people to concentrate on more troubling aspects of their lives. *Network support* expands a distressed individual's supportive options by either initiating new social contacts or providing novel support resources (Xu & Burleson, 2001).

Person-centered messages have also been examined in the context of social support (Burleson, 1982). Person-centeredness is the extent to which a message "reflects an awareness of and adaptation to the affective, subjective, and relational aspects of communication contexts" (Burleson, 1987, p. 305). Highly person-centered messages recognize people's feelings, often by employing evaluatively neutral messages to help them articulate, elaborate, and understand their emotions or the situation (Burleson & MacGeorge, 2002). These messages are sophisticated utterances because they acknowledge several different factors, including the distressed person, the social situation, the process of communication, and people's emotional and cognitive states. Prior research suggests that highly person-centered messages result in more effective social support than messages low in person-centeredness (e.g., Burleson, 2008, 2009; Holmstrom, Burleson, & Jones, 2005; Jones, 2004, 2005).

Social support is also conveyed by messages that help people change their appraisals of stressful situations. *Cognitive reappraisal* involves facilitating an individual's expression, elaboration, and clarification of distress-relevant thoughts and feelings. Effective cognitive reappraisal is demanding; it requires that participants engage in a detailed conversation about a stressful situation, focus on distressing thoughts and feelings, and create a personal narrative by assembling, clarifying, and processing thoughts and feelings associated with an event (Burleson & Goldsmith, 1998). When people work through a difficult experience in this manner, they are able to reframe traumatic events and achieve physical and psychological relief (Pennebaker, 1992, 1997).

This brief review highlights the variety of interpersonal messages that convey social support. Some forms of support are largely instrumental, such as providing people with tangible aid or connecting them with someone else who can provide help. Other forms of support—bolstering a person's identity, attending to emotions, and offering person-centered messages—are more personal and potentially face-threatening to the support recipient. Person-centered messages and cognitive reappraisal are types of support that may be especially taxing, given the attention and time they require to implement. This conceptualization of support messages as varied in form and function provides the foundation for our thinking about computer-mediated social support.

Explicating Computer-Mediated Communication

At its inception, scholars treated CMC as a singular modality that was only applicable for impersonal, task-oriented communication (Hiltz, Johnson, & Turoff, 1986; Parks & Floyd, 1996). In reality, CMC unfolds in a variety of contexts, such as online support groups, public discussion boards, chat rooms, mediated social networks, instant messaging, and virtual worlds. Moreover, CMC venues can be characterized in terms of a diverse set of underlying features (Sundar, 2008). These features determine not only the structure of a CMC venue, but also the normative communication practices that occur therein. The following paragraphs describe eight dimensions on which CMC venues can vary: synchronicity, anonymity, customization, processual interactivity, degree of social presence, number of users, homophily of users, and source perceptions. Although not exhaustive, these dimensions highlight characteristics of CMC environments that may be especially relevant to understanding people's experience of computer-mediated social support.

Synchronicity refers to the degree to which the exchange of messages is immediate rather than delayed. CMC environments are synchronous when communication occurs in real time with immediate feedback. For example, chat rooms, instant messaging programs, and virtual communities all enable people to synchronously communicate with others. In these contexts, the pacing of a conversation mirrors the response time of an FtF interaction. On the other hand, public discussion boards, and social networking sites are all environments in which time elapses between the posting of a message and its reply. The level of synchronicity not only helps define mediated venues, but it also influences the communication that occurs therein.

CMC environments also differ in the level of *anonymity* they offer. Anonymity represents the level of personal, individuating information transmitted by a given channel. Public discussion boards represent an especially anonymous mode of online communication in which users interact with relatively unknown others. On the other hand, instant messenger conversations, wherein users often disclose detailed information about themselves with people they know, are a CMC context with low anonymity. Certain CMC venues allow users to input personal content; however, other venues do not possess such capabilities. Thus, mediated locales fall along a continuum of anonymity as determined by the norms and features of a specific environment.

Customization is a third dimension on which mediated locales differ. In the domain of CMC, customization is the degree to which a mediated environment "modifies itself according to specific user input or user navigation and then provides information that is tailored to the user as a unique individual" (Kalyanaraman & Sundar, 2006, p. 112). In other words, users of customizable sites have the ability to manipulate the sites to match their individual needs and preferences. Some researchers have posited that customization promotes involvement with information, thereby encouraging in-depth processing of message content (cf. Petty & Cacioppo, 1986). Consistent with this reasoning, Kalyanaraman and Sundar (2006) observed that high levels of customization on political websites resulted in more positive impressions of featured candidates. To the extent that these findings generalize across contexts, customization should influence communicative exchanges in different CMC contexts.

CMC venues also vary in the degree to which they allow *processual interactivity*. Stromer-Galley (2004, p. 392) defined this type of interactivity as "interaction that occurs between two or more people communicating with each other, in which subsequent messages consist of responses to prior messages in a contingent fashion" (see also Bucy, 2004). Processual interactivity can have beneficial social effects, such as increased gregariousness and civic participation (Bucy, 2004; Shah, Cho, Eveland, & Kwak, 2005). On the other hand, extreme interactivity can be detrimental, resulting in fragmentation, individualization, a lack of shared experiences, and selfishness (Bucy, 2004). Given the consequences it has for CMC, processual interactivity is a meaningful index of differences between CMC environments.

A channel's *degree of social presence* is another factor with the potential to influence communication outcomes (Short, Williams, & Christie, 1976). Presence occurs when a mediated interaction does not seem mediated (Lombard & Ditton, 1997); channels high in social presence provide a salient impression of other interactants, whereas channels low in social presence transmit only superficial impressions of other communicators. Mediated contexts that enable people to communicate in real time, in shared spaces, through a variety of media, typically yield high levels of social presence, whereas those that limit fluid interaction between communicators yield correspondingly limited social presence. In turn, the degree of social presence promoted by a CMC infrastructure affects the quality of mediated communicative exchanges.

Different CMC venues also attract and allow different numbers of users. In turn, the number of users interacting within a given mediated environment can influence the communication experience. The vast number of interpersonal contacts available online has been touted as an important advantage of CMC, relative to FtF communication (Turner, Grube, & Meyers, 2001; Wright, 2000). In particular, a greater number of contributors increases the amount of resources and opinions transmitted in a given venue. In addition, having a larger pool of potential interactants increases the prevalence of weak tie networks, which involve frequent communication coupled with nonintimate interpersonal bonds (Granovetter, 1973). Because CMC venues vary in the number of users who participate in exchanges, the *number of users* in a CMC venue is an important feature that distinguishes online communication environments.

CMC venues also vary in the extent to which perceptions of *homophily*, or feelings of similarity, develop among users (Wright, 2000, 2002). Users of CMC sometimes come to see fellow participants as similar to themselves, based on their shared interests and time spent conversing via CMC. In turn, these perceptions can promote satisfaction with online communication and heighten perceptions of emotional support (Cline, 1999; Wright, 2000). After sufficient interaction and disclosure, some people even believe members of online groups are more similar to them than their offline acquaintances (Wright, 2000). In this manner, perceptions of homophily are an important element of mediated venues of interpersonal communication.

The ease with which people can create meaningful contributions to mediated discourse generates different *source perceptions* in online venues. Sundar and Nass (2000) distinguished conceptions of source as it pertains to the mediated world by enumerating several different types of CMC source. For example, a message's creator reflects traditional notions of a source. In addition, CMC channels represent technological sources. Receivers of mediated content can even perceive themselves to be sources in light of their ability to browse and select content for consumption (see also Reeves & Nass, 1996). Thus, the perceived source of information obtained through CMC constitutes a final dimension that captures variation among environments for online interaction.

In this section, we elaborated upon underlying dimensions that distinguish various CMC contexts. Specifically, we identified synchronicity, anonymity, customization, interactivity, social presence, number of users, homophily of users, and source perceptions as characteristics that can influence communication experiences with particular online venues. Although this list does not encompass the full range of dimensions that characterize technology-enabled communication experiences (see Sundar, 2008), these dimensions capture variation in CMC venues that can influence the provision of mediated social support. With this foundation in place, the following section examines the qualities of five CMC environments and discusses how those features shape the type of social support communication likely to unfold within each context.

Social Support within Five CMC Contexts

Thus far in this chapter, we have described different forms of supportive communication and considered ways in which CMC environments vary. In this section, we draw upon these features to describe five different CMC contexts that provide venues for computer-mediated social support. Specifically, we discuss CMC support groups, public discussion boards, chat rooms, instant messaging, and virtual worlds to demonstrate how the unique qualities of particular contexts give rise to particular types of supportive communication.

Online Support Groups

Mediated support groups, which represent the most widely studied venue of CMC social support, are electronic venues in which people post topical threads regarding the issues associated with different stressors. CMC support groups are novel contexts because users frequently disclose their most private feelings and thoughts to a group of relative strangers. For example, Davison et al. (2000) documented the existence of online support groups helping people deal with ailments like multiple sclerosis, diabetes, chronic fatigue syndrome, and depression. Once posted, topical threads exist on the discussion board to be viewed or replied to by any member of the support group. In fact, one study showed that 36% of caregivers found advice or support from other people in online support groups (Madden & Fox, 2006). Thus, online support groups represent an important venue of social support with tangible benefits for many people.

Online support groups, in general, share several features that shape the communication occurring therein. For example, online support groups typically involve asynchronous communication, but processual interactivity develops overtime through connected posts and responses. Although people often share detailed personal information, CMC support group contributors are typically anonymous. And, based on the shared experiences that bring people to the online support group, members develop perceptions of homophily; in fact, people sometimes perceive fellow members of CMC support groups to be more similar to them than their offline counterparts (Wright, 2000). These features distinguish the discourse of online support groups from what occurs in other CMC venues.

The relative anonymity of online support groups, compared to FtF contexts, is a key force shaping supportive communication in this venue. As Galagher, Sproull, and Kiesler (1998, p. 497) maintained:

> Confidentiality regarding the face-to-face group's proceedings may be expected, but one's physical presence and the possibility of encountering others in one's community create a risk of unwanted public exposure. Furthermore, these groups often exert social pressure on members to participate actively and to disclose their thoughts and feelings. Small size, local geography, and social pressure make these groups less private, less anonymous, and more conformist than are electronic social support groups.

Shielded by the anonymity that characterizes online support groups, participants can seek and provide forms of social support that require emotional openness and private disclosure. Accordingly, online social groups should be especially conducive to emotional support, esteem support, and person-centered messages.

The perceived similarity among contributors to online support groups should also affect people's support experience. Attitude and behavioral homophily are positively correlated with satisfaction and perceived emotional support in both CMC and FtF situations (Cline, 1999; Wright, 2000). Moreover, Wright (2000, p. 47) argued that communication partners in online support groups gain credibility from the "perception that the provider of support has been through similar circumstances, has had similar problems and engaged in similar behaviors, and has similar attitudes and beliefs about the conditions he or she is facing." The identification of common situations, emotions, or stressors provides members of online support groups with a sense of similarity that can lead to important communicative and supportive benefits. As in the case of anonymity, feelings of homophily facilitate the expression of sensitive forms of support, including emotional and esteem support and person-centered messages. In this manner, feelings of homophily could contribute to the sensitive, supportive interactions documented in online support contexts (i.e., Braithwaite, Waldron, & Finn, 1999).

Public Discussion Boards

Whereas CMC support groups are essentially private venues, public discussion boards are, by definition, open environments available to any Internet user interested in commenting on a specific issue. Public discussion boards are dedicated to everything from sports teams and televisions shows to types of cuisine and hobbies. Members voluntarily post correspondence organized into threads. Although the members of these groups are predominantly strangers, a common topical interest unites the members and provides a focal point for conversation.

Discussion boards often exist as asynchronous public and impersonal gathering places, but they can develop into familiar, even intimate, communities through continued posting and interaction. These venues contain high levels of processual interactivity because users are responsive to each other's comments. In addition, the members of public discussion boards consist of anonymous users who interact based on a common topical interest. The number of individuals who post on a specific topic can vary widely, but discussion boards are typically defined by a large and unrestricted number of users. And, although the topical focus of a discussion board represents the venue's de facto purpose, participants can post information about a broad variety of topics; in this sense, individual participants are typically the perceived source of information shared within public discussion boards.

The anonymity of online discussion boards can facilitate supportiveness through the development of weak tie networks. Unlike committed relationships, Adelman, Parks, and Albrecht (1987) asserted that weak tie networks are effective at providing access to diverse information and encouraging disclosure of sensitive information because people are not concerned that the information will reach closer relationship partners. In fact, weak tie associations can promote more positive support interactions than family ties because expressions of support are not based on obligations or expectations (Nussbaum, 1994). The benefits of weak ties are magnified in discussion boards when people reveal more personal information than they normally would in parallel FtF settings (Wright, 2000). Because of their ability to access information and promote interpersonal comfort, weak tie networks encourage informational support, as well as sensitive forms of support, such as emotional support and person-centered messages.

The number of members who participate in public discussion boards also affects the quality of support people derive. A large discussion board membership increases the likelihood that an individual will find someone who can provide useful information or direct them to relevant resources (Turner et al., 2001; Wright, 2000). In fact, Walther and Boyd (2001) highlighted continuous access to large stores of information as a major benefit of turning to CMC for social support. Accordingly, informational and network support are heightened in public discussion boards, especially those with large, active memberships.

With the advent of discussion boards, on which anyone can post a message, users possess an increased opportunity to be an initiator of wide-spread electronic discourse. Besides themselves, other users, professional writers, and even robotic computer systems can all be sources of electronic content. Interestingly, Sundar and Nass (2000) reported that users liked content more, perceived it was higher quality, and believed it was more representative when they thought other users produced the content, rather than when they were told it was developed by editors or a computer. Accordingly, people could be attracted to the support offered in public discussion boards, regardless of its precise form, simply because it was produced by fellow users. These findings suggest that the actual users and their contributions to a public discussion board could be important features that contribute to the perceived quality of social support.

Chat Rooms

Like public discussion boards, chat rooms are virtual public spaces dedicated to providing an open forum for topical or social discussion. Chat rooms are dedicated to a large variety of topics ranging from places for singles to meet, to venues for fans of specific musical genres, to settings for the enactment of virtual book clubs. Once inside a chat room, people type comments into a textbox and send them to the room's inhabitants. Every contributor's comments are visible to everyone else in a common chat window that refreshes as new comments are generated. Users are free to come and go as they please, venturing into several chat rooms until they encounter one in which they are comfortable and desire to interact.

Chat rooms maintain rather high levels of processual interactivity, which stems from the fact that they enable synchronous interaction. In fact, chat room discussions resemble the rate of message exchange experienced in FtF discussions. Like the other venues discussed thus far in this chapter, chat room participants typically maintain their anonymity. These main features of chat rooms—anonymity, processual interactivity, and synchronicity—shape the social support that occurs within them.

As previously discussed, the anonymity that characterizes chat rooms allows people to feel comfortable disclosing stressful thoughts and feelings (see also McKenna, Green, & Gleason, 2002). In general, CMC reduces self-presentational anxiety, social risk, and face threat (Caplan, 2003), and the Internet's anonymity enables people to reach high levels of disclosure without anxiety or fear (Anolli, Villani, & Riva, 2005). In this manner, the anonymity of chat rooms enables forms of social support that could be face-threatening in FtF situations. In particular, emotional support, esteem support, and person-centered messages are likely to occur in chat rooms.

The high levels of processual interactivity and synchronicity that define chat rooms should also benefit the social support process by drawing people into mediated conversations. The constant disclosure, feedback, and openness that constitute a chatroom creates a welcoming environment for conversation that facilitates the expression of a person's inner feelings (Bargh, McKenna, & Fitzsimons, 2002; Caplan & Turner, 2007). In fact, Tidwell and Walther (2002) found that CMC interactants produced a higher proportion of self-disclosures than did strangers interacting FtF. The ability to obtain immediate responses may also facilitate informational and network support, because people can request immediate help and provide feedback that helps support providers tailor their advice to the user's specific needs. In combination, then, processual interactivity and synchronous communication allow involving conversational forms of support, such as cognitive reappraisal or person-centered messages, as well as instrumental assistance in the form of informational and network support.

Instant Messaging

Instant messaging (IM) is strictly text-based communication that allows users to send and receive short messages via specialized chat programs. Besides conversing, IMers are able to create short profiles and informative "away messages," as well as browse other people's profiles. People from younger age groups are even more likely to use instant messaging programs than email (Shiu & Lenhart, 2004). Thus, IM is a widely used CMC context, which is likely to become more popular in the future.

Unlike the other forms of CMC discussed in this chapter, IM exchanges are dyadic, private conversations that commonly occur between people with a prior relationship. In fact, researchers have determined that people's IM networks are relatively modest, with 66% of people regularly IMing only between one and five people (Shiu & Lenhart, 2004). IM interactions are typically characterized by a lack of anonymity, synchronous information exchange, and high processual interactivity. The rapid message exchange and prior knowledge of IM interactants also yields high levels of social presence. Processual interactivity and social presence are the features of this CMC venue that are especially relevant to the communication of social support.

IMing is an ideal example of processual interactivity; however, there is still some debate as to whether this form of interactivity benefits or curtails relationship development. Some scholars have touted processual interactivity as enhancing sociability (Bucy, 2004; Shah et al., 2005), whereas others contend that interactivity is socially debilitating because it promotes fragmentation and individualization (Bucy, 2004). Because of the debate surrounding interactivity, it is not clear whether it is a benefit or detriment to the social support process. Certainly, individuation and selfishness are incompatible with notions of emotional and esteem social support; however, civic participation seems to promote network support. Interactivity could also be intrinsically supportive, because just maintaining a conversation with a distressed person could generate esteem or network support and confirm that the distressed person is worthy of another's attention.

Although the degree of social presence in IM conversations is high relative to other CMC environments, IM provides a less immediate form of interaction relative to FtF discussions. O'Sullivan (2000) argued that CMC exerts a "buffer effect," such that it reduces the face threats that are present within difficult FtF interactions. Consistent with this reasoning, O'Sullivan (2000) found that people prefer mediated channels when they are concerned with their self-presentation and when they believe that ambiguity would favor their comments. Thus, the engagement facilitated by IM, coupled with this buffering quality, may provide unique support benefits to individuals uncomfortable with FtF communication about stressful experiences.

In particular, men could prefer the reduced social presence and perceived security of IM when seeking and expressing support. Although men believe they are more skilled supporters, women are more comfortable and more effective in support situations (Hale et al., 1997; Kunkel & Burleson, 1999; Sarason, Sarason, Hacker, & Basham, 1985). For men, then, IM might provide a place where they can synchronously express feelings that they cannot articulate FtF. Shiu and Lenhart (2004) found that males (29%) are more likely than females (19%) to IM someone who is in the same location, despite the fact that men and women are otherwise equivalent users of IM (Shiu & Lenhart, 2004). More generally, we expect that IM conversations allow people to provide esteem or emotional support, as well as sensitive person-centered messages, in a manner that capitalizes on both the heightened social presence relative to other CMC venues, and the decreased social presence relative to FtF encounters.

Virtual Worlds

Virtual worlds, such as Secondlife or Habbo Hotel, represent a unique segment of the online landscape. In these environments, people manipulate avatars, traverse virtual terrain, and synchronously interact with other users. The goal of virtual worlds is to provide an alternate form of reality that accurately mirrors FtF life in many respects. Besides social pursuits, many virtual worlds contain a commercial component whereby users are able to purchase and sell virtual commodities to other users through the exchange of actual currency. Secondlife, perhaps the largest and most publicized of the virtual worlds, boasts an international membership of several million individuals. As the site's homepage declares, it is "a vast digital continent, teeming with people, entertainment, experiences and opportunity."

Because of their multimodal capabilities, interactants can create a variety of social interactions in virtual worlds. In these contexts, people can interact with others through synchronous chat, video, audio, or other asynchronous means. Furthermore, people possess the capability to manipulate their avatar's appearance, their behavior, and their virtual surroundings to suit their needs, personalities, or goals. Accordingly, virtual worlds have very high levels of customization relative to other mediated contexts. Moreover, the potential for rapid conversational exchange, processual interactivity, and the ability to control avatars enable people to create a strong sense of social presence. The interpersonal, social, and psychological benefits inherent in virtual worlds make them a useful resource for many people.

The high levels of customization in virtual worlds can facilitate selective self-presentation. The ability to strategically edit physical features is especially pronounced in virtual worlds, where people interact using a customized avatar. Some scholars have proclaimed that without visual information, message senders can selectively present themselves by masking or editing undesirable and uncontrollable cues while simultaneously magnifying preferred cues (Walther, 1996, 1997). As Walther (1996, p. 20) asserted, "such social evaluations as one is able to garner are not impeded by messy hair, lack of makeup, or normal imperfections, much less more pronounced physical distracters or disabilities." For example, enhanced selective self-presentation should facilitate the production of some types of support, perhaps particularly those types of support that are difficult to provide FtF, such as emotional support, person-centered messages, or cognitive reappraisal. People can customize both their virtual appearance and support statements to correspond with support goals. Thus, virtual world's customization embodies the benefits of selective presentation.

Virtual world's levels of social presence, which are greater than those of most CMC venues yet smaller than FtF interaction, can afford people great flexibility in editing supportive statements and managing cognitive resources. Because CMC requires people to type their responses before sending them, communicators are able to distance themselves from their thoughts to revise or abandon unfavorable messages. "The channel itself facilitates goal-enhancing messages by allowing sources far greater control over message construction than is available in FtF settings" (Walther & Parks, 2002, p. 541). Communicators can also exploit virtual worlds' social presence to redirect cognitive resources to where they can be applied most effectively (Burgoon & Walther, 1990; Walther, 1996, 1997). Some theorists have even argued that CMC users can reallocate unused cognitive resources towards message construction (Walther, 2006). If people can direct more cognitive resources to verbal message production, they should be able to thoroughly combine the contextual, emotional, and personal aspects of these messages to create person-centered messages. As Walther (1996, p.33) imparted, "CMC affords opportunities, however, to communicate as desired; an impulse that seems to be inherently human yet may be more easily enacted via technology."

In this section of the chapter, we discussed the types of social support enabled by the features of five different venues for CMC: CMC support groups, public discussion boards, chat rooms, IM, and virtual worlds. Rather than yielding identical communication outcomes, each of these venues possesses distinct features that shape the communication experiences and social support messages that transpire. In some cases, the anonymity of participants in a CMC environment was linked to the more personal communication required by esteem support, emotional support, and person-centered messages. In other venues, synchrony and processual interactivity make possible forms of support, like cognitive reappraisal, that rely on an ongoing exchange of messages between communicators. Likewise, the number of users, homophily of users, and source perceptions were argued to affect the prevalence of different forms of support. This discussion reveals how the array of CMC environments vary in ways that have consequences for computer-mediated social support.

Closing Thoughts

McLuhan (1964) claimed that "the medium is the message" in that the meanings attached to any symbols are driven by qualities of the channel through which messages are communicated. Relatedly, Meyrowitz (1985) asserted that different media have the ability to drastically shape people's social relations. Whereas these two scholars imply that the media, rather than the content they carry, should be the focus of study, we highlight how qualities of both communication channels and support messages affect experiences of computer-mediated social support. Like others before us, we assume that a communication channel strongly influences the form of the messages it conveys (e.g., Pingree, Wiemann, & Hawkins, 1988). In particular, we have argued that variations in the features that characterize CMC venues influence the social support processes that unfold within them. In addition, our analysis illustrated how characteristics of support messages themselves determine whether they are likely to occur within specific online environments. In this way, we propose that it is the interplay between features of support messages and characteristics of CMC venues that determines the experience of computer mediated social support.

Our approach in this chapter also emphasized differences between CMC contexts, rather than the issues that surround CMC, in general. Walther (1992, p. 82) cautioned against generalizing findings across different online channels, and he advocated a focus on "the theoretical underpinnings regarding communication functions in any context" against which "differences due to channel attributes will become more precise, interesting, and may possibly be employed with greater discretion and utility." To this end, we identified dimensions that capture variation among CMC environments, and we used these underlying features to clarify the kinds of social support likely to occur in different contexts. Thus, we recognize differences between specific CMC venues, while aligning those differences in terms of underlying and theoretically important dimensions.

Just as CMC needs to be examined in ways that attend to underlying similarities and nuanced differences, so must social support be appreciated as a multifaceted phenomenon that takes many forms. In this chapter, we emphasized different types of support that are conveyed through communication; some of these are defined by the type of help provided (e.g., information vs. attention to emotion) and others (e.g., cognitive reappraisal) are better understood as a communication process. Social support can also be conveyed implicitly by a person's membership in a social network or via a sense of community with similar others. As we look forward to future research on computer-mediated social support, we anticipate that the most theoretically and practically important insight will come from research that considers the variety of ways in which social relationships can be supportive.

Davison et al. (2000, p. 210) asserted that "the social connections enabled by the advent of the Internet constitute a new forum of social support that has unknown, and largely unstudied potential." Computer-mediated communication coexists with face-to-face interaction as the prominent modes for giving and receiving social support. Given its prevalence and impact, research is needed to shed light on the experience of computer-mediated social support. The approach in this chapter provides a framework for undertaking that research in ways that attend to the complexities of both computer-mediated communication and social support.

References

Adams, S. J., Roch, S. G., & Ayman, R. (2005). Communication medium and member familiarity: The effects on decision time, accuracy, and satisfaction. *Small Group Research, 36*, 321–353.

Adelman, M. B., Parks, M. R., & Albrecth, T. L. (1987). Beyond close relationships: Support in weak ties. In T. L. Albrecht & M. B. Adelman (Eds.), *Communicating social support* (pp. 126–147). Newbury Park, CA: Sage.

Albrecht, T. L., & Adelman, M. B. (1987). Communicating social support: A theoretical perspective. In T. L. Albrecht & M. B. Adelman (Eds.), *Communicating social support* (pp. 18–39). Newbury Park, CA: Sage.

Anolli, L., Villani, D., & Riva, G. (2005). Personality of people using chat: An on-line research. *CyberPsychology and Behavior, 8*, 89–95.

Bargh, J. A., McKenna, K. Y. A., & Fitzsimons, G. M. (2002). Can you see the real me? Activation and expression of the "true self" on the Internet. *Journal of Social Issues, 58*, 33–48.

Braithwaite, D. O., Waldron, V. R., & Finn, J. (1999). Communication of social support in computer-mediated groups for people with disabilities. *Health Communication, 11*, 123–151.

Bucy, E. P. (2004). Interactivity in society: Locating an elusive concept. *The Information Society, 20*, 373–383.

Burgoon, J. K., Bonito, J. A., Ramirez, A., Jr., Dunbar, N. E., Kam, K., & Fischer, J. (2002). Testing the interactivity principle: Effects of mediation, propinquity, and verbal and nonverbal modalities in interpersonal interaction. *Journal of Communication, 52*, 657–677.

Burgoon, J. K., & Walther, J. B. (1990). Nonverbal expectancies and the evaluative consequences of violations. *Human Communication Research, 17*, 232–265.

Burleson, B. R. (1982). The development of comforting communication skills in childhood and adolescence. *Child Development, 53*, 1578–1588.

Burleson, B. R. (1987). Cognitive complexity. In J. C. McCroskey & J. A. Daly (Eds.), *Personality and interpersonal communication* (pp. 305–349). Newbury Park, CA: Sage.

Burleson, B. R. (1994). Comforting messages: Features, functions, and outcomes. In J. A. Daly & J. M. Wiemann (Eds.), *Strategic interpersonal communication* (pp.135–161). Hillsdale, NJ : Lawrence Erlbaum.

Burleson, B. R. (2008). What counts as effective emotional support?: Explorations of situational and individual differences. In M. T. Motley (Ed.), *Studies in applied interpersonal communication* (pp. 207–227). Newbury Park, CA: Sage.

Burleson, B. R. (2009). *The relationship between perceived and actual effectiveness of supportive messages: A dual-process framework.* Paper presented at the National Communication Association conference, Chicago, IL.

Burleson, B. R., Albrecht, T. L., Goldsmith, D. J., & Sarason, I. G. (1994). The communication of social support. In B. R. Burleson, T. L. Albrecht, & I. G. Sarason (Eds.), *Communication of social support: Messages, interactions, relationships, and community* (pp. xi–xxx). Thousand Oaks, CA: Sage.

Burleson, B. R., & Goldsmith, D. J. (1998). How the comforting process works: Alleviating emotional distress through conversationally induced reappraisals. In P. A. Andersen & L. K. Guerrero (Eds.), *Handbook of communication and emotion: Research, theory, applications, and contexts.* (pp. 245–280). San Diego, CA: Academic.

Burleson, B. R., & MacGeorge, E. L. (2002). Supportive Communication. In M. L. Knapp & J. A. Daly (Eds.), *Handbook of interpersonal communication* (3rd ed.) (pp. 374–424). Thousand Oaks, CA: Sage.

Burleson, B. R., & Samter, W. (1990).Effects of cognitive complexity on perceived importance of communication skills in friends. *Communication Research, 17,* 165–182.

Callaghan, P., & Morrissey, J. (1993). Social support and health: A review. *Journal of Advanced Nursing, 18,* 203–210.

Caplan, S. E. (2003). Preference for online social interaction: A theory of problematic Internet use and psychosocial well-being. *Communication Research, 30,* 625–648.

Caplan, S. E., & Turner, J. S. (2007). Bringing theory to research on computer-mediated comforting communication. *Computers in Human Behavior, 23,* 985–998.

Cline, R. J. (1999). Communication within social support groups. In L. R. Frey, D. S. Gouran, & M. S. Poole (Eds.), *Handbook of group communication theory and research* (pp. 516–538). Thousand Oaks, CA: Sage.

Cobb, S. (1976). Social support as a moderator of life stress. *Psychosomatic Medicine, 38,* 300–314.

Cohen, S., & Wills, T. A. (1985). Stress, social support, and the buffering hypothesis. *Psychological Bulletin, 98,* 310–357.

Cunningham, M. R., & Barbee, A. P. (2000). Social support. In C. Hendrick & S. S. Hendrick (Eds.), *Close relationships: A sourcebook* (pp. 272–285). Thousand Oaks, CA: Sage.

Cutrona, C. E., & Russell, D. W. (1990). Types of social support and specific stress: Toward a theory of optimal matching. In B. R. Burleson, I. G. Sarason, & G. R. Pierce (Eds.), *Social support: An interactional view* (pp. 319–366). New York: John Wiley.

Davison, K. P., Pennebaker, J. W., & Dickerson S. S. (2000). Who talks? The social psychology of illness support groups. *American Psychologist, 55,* 205–217.

Galagher, J., Sproull, L., & Kiesler, S. (1998). Legitimacy, authority, and community in electronic support groups. *Written Communication, 15,* 493–530.

Granovetter, M. S. (1973). The strength of weak ties. *American Journal of Sociology, 78,* 1360–1380.

Hale, J. L., Tighe, M. R., & Mongeau, P. A. (1997). Effects of event type and sex on comforting messages. *Communication Research Reports, 14,* 214–220.

Hildingh, C., Fridlund, B., & Segesten, K. (1995). Social support in self-help groups, as experienced by persons having coronary heart disease and their next of kin. *International Journal of Nursing Studies, 32*, 224–232.

Hiltz, R. S., Johnson, K., & Turoff, M. (1986). Experiments in group decision-making: Communication process and outcome in face-to-face versus computerized conferences. *Human Communication Research, 13*, 225–252.

Holmstrom, A. J., Burleson, B. R., & Jones, S. M. (2005). Some consequences for helpers who deliver "cold comfort: Why it's worse for women than men to be inept when providing emotional support. *Sex Roles, 53*, 153–172.

Jones, S. M. (2004). Putting the person into person-centered and immediate emotional support: Emotional change and perceived helper competence as outcomes of comforting in helping situations. *Communication Research, 31*, 338–360.

Jones, S. M. (2005). Attachment style differences and similarities in evaluations of affective communication skills and person-centered comforting messages. *Western Journal of Communication, 69*, 233–249.

Kalyanaraman, S., & Sundar, S. S. (2006). The psychological appeal of personalized online content in Web portals: Does customization affect attitudes and behavior? *Journal of Communication, 56*, 110–132.

Kohn, P. M. (1996). On coping adaptively with daily hassles. In M. Zeidner & N. S. Endler (Eds.), *Handbook of coping* (pp. 181–201). New York: John Wiley.

Kunkel, A. W., & Burleson, B. R. (1999). Assessing explanations for sex differences in emotional support: A test of different cultures and skills specialization accounts. *Human Communication Research, 25*, 307–340.

Lombard, M., & Ditton, T. B. (1997). At the heart of it all: The concept of presence. *Journal of Computer-Mediated Communication, 3*. Retrieved July 18, 2009 from http://jcmc.indiana.edu/vol3/issue2/lombard.html

Madden, M., & Fox, S. (2006). Finding answers online in sickness and in health. *Pew Internet and American Life Project.* Retrieved from http://www.pewinternet.org/pdfs/PIP_Health_Decisions_2006.pdf

McKenna, K. Y. A., Green, A. S., & Gleason, M. E. J. (2002). Relationship formation on the Internet: What's the big attraction. *Journal of Social Issues, 58*, 9–31.

McLuhan, M. (1964). *Understanding media: The extensions of man.* London: Routledge.

Meyrowitz, J, (1985). *No sense of place: The impact of electronic media on social behavior.* New York, Oxford University Press.

Nussbaum, J. F. (1994). Friendship in older adulthood. In M. L. Hummert, J. M. Wiemann, & J. F. Nussbaum (Eds.), *Interpersonal communication in older adulthood* (pp. 209–225). Thousand Oaks, CA: Sage.

O'Sullivan, P. B. (2000). What you don't know won't hurt me: Impression management functions of communication channels in relationships. *Human Communication Research, 26*, 403–431.

Parks, M. R., & Floyd, K. (1996). Making friends in cyberspace. *Journal of Communication, 46*, 80–96.

Pennebaker, J. W. (1992). Putting stress into words: Health, linguistic, and therapeutic implications. *Behavioral Research Therapy, 31*, 539–548.

Pennebaker, J. W. (1997). Writing about emotional experience as a therapeutic process. *Psychological Science, 8*, 162–166.

Petty, R. E., & Cacioppo, J. T. (1986). *Communication and persuasion: Central and peripheral routes to attitude change.* New York: Springer-Verlag.

Pingree, S., Wiemann, J. M., & Hawkins, R. P. (1988). Editor's introduction: Toward a conceptual synthesis. In R. P. Hawkins, J. M. Wiemann, & S. Pingree (Eds.), *Advancing communication science: Merging mass and interpersonal processes* (pp. 7–17). Newbury Park, CA: Sage.

Reeves, B., & Nass, C. (1996). *The media equation: How people treat computers, televisions, and new media like real people and places.* Stanford, CA: CSLI Publications and Cambridge University Press.

Sarason, B. R., Sarason, I. G., Hacker, T. A., & Basham, R. B. (1985). Concomitants of social support: Social skills, physical attractiveness, and gender. *Journal of Personality and Social Psychology, 49*, 469–480.

Shah, D. V., Cho, J., Eveland, W. P., & Kwak, N. (2005). Information and expression in a digital age: Modeling Internet effects on civic participation. *Communication Research, 32*, 531–565.

Shiu, E., & Lenhart, A. (2004). How Americans use instant messaging. *Pew Internet and American Life Project.* Retrieved July 18, 2009 from http://www.pewinternet.org/pdfs/PIP_ Instantmessage_ Report.pdf

Short, J., Williams, E., & Christie, B. (1976). *The social psychology of telecommunications.* London: John Wiley.

Stromer-Galley, J.(2004). Interactivity-as-product and interactivity-as-process. *The Information Society, 20*, 391–394.

Sundar, S. S. (2008). The MAIN model: A heuristic approach to understanding technology effects on credibility. In M. J. Metzger & A. J. Flanigan (Eds.), *Digital media, youth, and credibility* (pp. 73–100). Cambridge, MA: MIT Press.

Sundar, S. S., & Nass, C. (2000). Source orientation in human-computer interaction. *Communication Research, 27*, 683–703.

Tidwell, L. C., & Walther, J. B. (2002). Computer-mediated communication effects on disclosure, impressions, and interpersonal evaluations: Getting to know one another a bit at a time. *Human Communication Research, 28*, 317–348.

Turner, J. W., Grube, J. A., & Meyers, J. (2001). Developing an optimal match within online communities: An exploration of CMC support communities and traditional support. *Journal of Communication, 51*, 231–251.

Walther, J. B. (1992). Interpersonal effects in computer-mediated interaction: A relational perspective. *Communication Research, 19*, 52–90.

Walther, J. B. (1996). Computer-mediated communication: Impersonal, interpersonal, and hyperpersonal interaction. *Communication Research, 23*, 3–43.

Walther, J. B. (1997). Group and interpersonal effects in international computer-mediated collaboration. *Human Communication Research, 23*, 342–369.

Walther, J. B. (2006). Nonverbal dynamics in computer-mediated communication, or :(and the Net :('s with you, :) and you :) alone. In V. Manusov & M. L. Patterson (Eds.), *Handbook of nonverbal communication* (pp 461–480). Thousand Oaks, CA: Sage.

Walther, J. B., & Boyd, S. (2001). Attraction to computer-mediated social support. In C. A. Lin & D. Atkins (Eds.), *Communication technology and society: Audience adoption and uses of the new media* (pp. 133–167). New York: Hampton.

Walther, J. B., & Parks, M. R. (2002). Cues filtered out, cues filtered in: Computer-mediated communication relationships. In M. L. Knapp, J. A. Daly, & G. R. Miller (Eds.), *The handbook of Interpersonal Communication* (3rd. ed.) (pp. 529-563). Thousand Oaks, CA: Sage.

Walther, J. B., Slovacek, C. L., & Tidwell, L. C. (2001). Is a picture worth a thousand words? Photographic images in long-term and short-term computer-mediated communication. *Communication Research, 28*, 105–134.

Winzelberg, A. (1997). The analysis of an electronic support group for individuals with eating disorders. *Computers in Human Behavior, 13*, 393–407.

Wright, K. B. (2000). Perceptions of on-line support providers: An examination of perceived homophily, source credibility, communication, and social support within on-line support groups. *Communication Quarterly, 48*, 44–59.

Wright, K. B. (2002). Social support within an on-line cancer community: An assessment of emotional support, perceptions of advantages and disadvantages, and motives for using the community from a communication perspective. *Journal of Applied Communication Research, 30*, 195–209.

Wright, K. B., & Bell, S. B. (2003). Health-related support groups on the Internet: Linking empirical findings to social support and computer-mediated communication theory. *Journal of Health Psychology, 8*, 39–54.

Xu, Y., & Burleson, B. R. (2001). Effects of sex, culture, and support type on perceptions of spousal social support: An assessment of the "support gap" hypothesis in early marriage. *Human Communication Research, 24*, 535–566.

CHAPTER EIGHT

Personal Relationships and Computer-Mediated Support Groups

Kevin B. Wright

Ahlam Muhtaseb

The Internet has become a widely used resource for obtaining social support within interpersonal relationships (Walther & Boyd, 2002), particularly in the context of health concerns (Neuhauser & Kreps, 2003; Wright & Bell, 2003). One popular way in which the Internet facilitates social support is through access to computer-mediated support groups: individuals interacting in groups using the Internet and the World Wide Web to exchange social support. Websites such as Yahoo! Groups, WebMD, and the American Cancer Society, for example, offer numerous asynchronous and real-time discussion forums where individuals concerned with a specific issue share information and offer emotional assistance. An estimated 90 million Americans have participated in some type of computer-mediated support group and that 1 in 4 people seeking information about disease join such groups (Horrigan & Rainie, 2002; Levy & Strombeck, 2002).

Although the Internet provides many options for individuals seeking to supplement or replace traditional (face-to-face) sources of social support (see High & Solomon, Chapter 7, in this volume), including the maintenance of supportive relationships among people who know one another from face-to-face contexts (such as family members and friends), computer-mediated support groups provide a unique portal for the development of new supportive relationships online (e.g., individuals who "meet" in online support groups) which may augment or replace traditional sources of social support. This phenomenon has attracted the attention of social scientists and medical researchers who are interested in the benefits of computer-mediated social support groups for people with health concerns and other stressful life experiences, including important outcomes such as reduced stress and increased coping skills (Johnson, Wright, Craig, Gilchrist, Lane, & Haigh, 2008; King & Moreggi, 1998; Preece & Ghozati, 2001; Rains & Young, 2009; Wright, 2000; Wright, Rains, & Banas, 2010).

This chapter explores interpersonal issues related to the giving and receiving of social support within computer-mediated support groups. Towards that end, the chapter explores the link between social support and health outcomes, relational dilemmas surrounding the provision of social support, and advantages and disadvantages of online support groups/communities. In addition, it focuses on several theoretical frameworks that have been useful in past research in terms of understanding the nature of computer-mediated social support, and suggestions for future researchers and practitioners who are interested in understanding computer-mediated social support group relationships.

Review of Literature

While the Internet provides a surprisingly vast array of potential sources of social support for people seeking informational support, emotional support, and even tangible support online, the majority of computer-mediated support studies have focused on support within online support groups and communities. As the body of computer-mediated support group literature has developed over the past 15 years, a number of theoretical concerns and interesting findings concerning personal relationships and online social support within these groups have emerged. The following sections detail several theoretical issues related to personal relationships within computer-mediated support groups.

Defining Computer-Mediated Social Support

Computer-mediated social support (CMSS) is social support exchanged through various types of computer-mediated communication (CMC), including older forms of asynchronous and synchronous CMC, such as newsgroups, listserves, bulletin boards, instant messaging, and email, and newer forms of CMC, or "multimodal CMC" (Herring, 2002), such as combined text, voice, and video interactions. Although there are many definitions of computer-mediated social support across different studies, most researchers have acknowledged that CMSS includes informational, emotional, appraisal, and instrumental support (Barrera, 1986; Lieberman & Goldstein, 2005; Pull, 2006; Winefield, 2006).

Social Capital and Computer-Mediated Networks

An important concept related to social support is social capital. Social capital can be defined as "resources embedded in a social structure that are accessed and/or mobilized in purposive actions" (Lin, 2001, p. 29). According to Putnam, social capital refers to networks among individuals and trust, norms, and values arising from social networks (Putnam, 1995a, 1995b, 2000). According to Putnam (2000), social capital is facilitated by strong interpersonal ties, reciprocity norms, interpersonal trust, and shared values. Social capital theorists posit that there are abilities, resources, and values embedded in social networks and relationships that can potentially create emotional, informational, and instrumental benefits depending upon the types of individuals a person interacts with on a regular basis and how well he or she can capitalize on these resources.

A number of researchers have argued that the Internet has the potential to foster increased social capital, particularly among people who use it for interpersonal and community-building purposes (Drentea & Moren-Cross, 2005; Wellman, 1997; Quan-Haase, Wellman, Witte, & Hampton, 2002; Szreter, 2000). In addition, these studies have found that social capital within online support groups/communities is linked to emotional and psychological benefits to participants.

CMC *Channel Characteristics and Interpersonal Perceptions/Relationships*

Early interpersonal communication studies of CMC largely focused on the limitations of characteristics of computer-mediated communication vis-à-vis face-to-face communication. For example, media richness theory (Daft & Lengel, 1984, 1986; Daft, Lengel, & Trevino, 1987) posits that text-based CMC is "lean" media that is better suited for simple, straight forward types of communication whereas "rich" media (such as teleconferencing or face-to-face communication) is a better choice for intensive interactions, more complex informational and emotional content, and more complicated exchanges of messages between people. Other early CMC researchers suggested that computer-mediated communication reduces or filters out physical and contextual cues that make it difficult to form meaningful or intimate relationships (Kiesler & Sproull, 1992; Kiesler, Siegel, & McGuire, 1984).

Latter CMC research supported the idea that satisfying interpersonal relationships can develop through computer-mediated interactions (Parks & Floyd, 1996; Walther, 1995, 1996; Walther & Boyd, 2002). Today, evidence from a variety of research programs suggests that computer-mediated support groups are an important resource for supplementing supportive relationships, especially in cases where people are geographically separated from these traditional sources of social support.

Interpersonal Concerns Regarding Online Close Relationships

A number of researchers have attempted to explain the attraction to both face-to-face support groups and computer-mediated support groups by examining interpersonal issues that often arise in close, face-to-face supportive relationships (Walther & Boyd, 2002; Wright, 2000, 2002; Wright & Bell, 2003). The process of seeking support within relationships can be a difficult process for many individuals facing illness, who must coordinate meeting their needs while simultaneously attempting to manage relational concerns (Albrecht & Goldsmith, 2003). According to these authors, partners must not only cope with a stressor but also cope with the relational strains created by the stressor and the difficulties inherent in *coordinating* their individual coping attempts. The practice of seeking support can involve a complicated process of managing difficult individual coping needs while simultaneously attempting to manage delicate relational concerns. Findings from a variety of research programs (see Albrecht, Burleson, & Goldsmith, 1994; Barbee, Derlega, Sherburne, & Grimshaw, 1998; Brashers, Neidig, & Goldsmith, 2004; Pakenham, 1998) suggest that many individuals find it difficult to obtain appropriate support from friends and family since they may feel these potential sources of support lack experience or have limited information about certain problems. Furthermore, many people may feel uncomfortable discussing their problems with members of their close, face-to-face support network for a variety of other reasons, such as a desire to avoid feeling stigmatized, patronized, or being judged when discussing sensitive topics.

Other complicating relational concerns in social support situations may include reluctance towards receiving inappropriate support, or not wanting to appear vulnerable or incapable of handling one's own problems. In addition there are complications associated with role obligations and reciprocity issues in many relationships (Albrecht & Goldsmith, 2003; Chesler & Barbarin, 1984; Cline, 1999; LaGaipa, 1990). Role obligations refer to the idea that we sometimes feel obligated to support our loved ones even during times when we may not necessarily want to help them due to our own concerns. Role obligations in supportive encounters have been found to lead to resentment in some cases (Rook, 1995). Reciprocity issues in supportive interactions include problems that occur when one relational partner is underbenefitted (i.e., gives more support than he or she receives), and when partners are overbenefitted (i.e., receives more support than he or she can give in return). People with health problems often find themselves in a position where they are overbenefitted as people in their social network attempt to support them during their time of need. However, the inability to help others in such situations has been found to lead to feelings of inadequacy, helplessness, and demoralization in some cases.

Interpersonal Benefits and Limitations of Computer-Mediated Support Groups

Computer-mediated support groups allow people facing a wide variety of stressful situations and concerns a means to overcome a number of barriers that often present in the face-to-face world as well as a number of limitations due to channel constraints and other characteristics of the medium. This section briefly discusses some of the major interpersonal benefits and limitations of computer-mediated support groups that have been identified in the literature.

Access to multiple perspectives. One advantage of computer-mediated support groups appears to be greater access to multiple perspectives of others on stressful situations, such as facing depression or cancer, through access to an extended support network capable of providing informational, emotional, and (in some cases) instrumental support. The medium facilitates communication among a concentrated number of individuals sharing similar specific concerns. Finn and Lavitt (1994) was one of the first studies to identify interpersonal advantages to meeting online through their study of a computer-mediated sexual abuse survivors group. In particular, they found that participants were able to communicate with a wider variety of people than in the face-to-face world, and people were able to diffuse dependency needs that would normally be fulfilled through face-to-face interaction. Moreover, members were able to talk to many people and receive information from an assortment of people instead of relying on a small circle of friends or family. Later studies, such as Davison, Pennebaker, and Dickerson (2000), Pull (2006), Sharf (1997), Walther and Boyd (2002); and Wright (2002), found that members of computer-mediated support groups mentioned both the diverse experiences of group members and access to a larger network of people with more information about problems than would be possible to form in the face-to-face world as major advantages to online support groups over traditional sources of social support.

In addition to the sheer number of people that Internet support groups/communities can provide access to, these groups/communities also introduce individuals to a more diverse social network than they are typically able to access in the face-to-face world (Wright & Bell, 2003). In most face-to-face networks, individuals tend to seek support from family members and close friends. These individuals tend to be somewhat homogenous in terms of demographic, attitude, and background similarity. By contrast, online support groups/communities tend to help people transcend these similarities and introduce them to a more heterogeneous network of individuals (despite the fact that they share a common health concern) (Wellman, 1997). Many potential supportive relationships in the face-to-face world are thwarted due to perception that others are too dissimilar.

For example, in the face-to-face world, individuals tend to rely heavily on in-group/out-group differences (Giles, Mulac, Bradac, & Johnson, 1987) when comparing themselves to people who appear to be members of a different social group (e.g., based on sex, race, age, background, etc.). In the computer-mediated environment, many of these social cues are unavailable due to the reduced nonverbal information in most contexts (e.g., virtual groups/communities). People may be more likely to judge individuals on the quality of their verbal messages (i.e., postings, etc.) rather than making snap judgments based on visible social cues. As a result, participants can often receive more unique and novel viewpoints about the health issues they are facing compared to what they are able to obtain in traditional face-to-face support networks. This provides individuals with more opportunities for social comparisons with other individuals facing similar health concerns. This may help shatter perceptions of uniqueness when it comes to coping with health issues (e.g., why me?), and allow individuals to examine their own problems vis-à-vis the issues other group/community members are facing.

Lack of judgment. Another advantage for members of an online support group is the ability to discuss topics that would be difficult in the face-to-face environment, which is often a problem for people who have been traumatized by some type of disorder or illness (Cline, 1999; Finn & Lavitt, 1994; Wright & Bell, 2003). However, through sharing information, stories, feelings and emotions, members are able to develop a social network, and discuss sensitive topics without the embarrassment of revealing personal information to strangers in face-to-face encounters (Wright & Bell, 2003). Computer-mediated support groups offer individuals increased anonymity compared to traditional face-to-face sources of social support (including face-to-face support groups). This can lead to a reduction in feeling stigmatized due to having a visible health condition (as well as conditions that are not readily apparent), reduced communication apprehension in terms of initiating communication with others, and the ability to self-disclose sensitive information in a less risky environment.

Similarity. Another advantage of online support groups that has been identified in previous literature is the larger sense of similarity people feel with other members of these groups (Walther & Boyd, 2002; Wright, 2000; Wright & Bell, 2003). In many cases, people who participate in online support groups have mentioned that feeling part of a larger community of individuals who are dealing with similar concerns is an important source of validation and comfort (Braithwaite, Waldron, & Finn, 1999; Lieberman & Goldstein, 2005; Sharf, 1997). Social support researchers refer to this sense of community as social network support, and they argue that it may be an important source for companionship, validation, and a reminder that support is available when needed (Wright & Bell, 2003). Social network support may be particularly valued in terms of providing direct benefits to individuals, such as diversion, elevated mood, and reduced stress, even during times when a person is not facing an acute crisis situation (Cohen & Wills, 1985).

Convenience. Computer-mediated support groups appear to be important portals for individuals who are interested in forming supportive interpersonal relationships with people who may be difficult to conveniently locate in the face-to-face world (Eichhorn, 2008; Lieberman & Goldstein, 2005; Walstrom, 2000; Zelley, 2001). Many online support groups/communities feature both asynchronous communication (e.g., bulletin boards, email) and synchronous communication (e.g., chat rooms or chat applications) capabilities so that participants can obtain support from others in real time or post messages to the group. This provides people with access to support when they are facing immediate concerns (although the number of people using synchronous applications tends to be relatively small), or they can make comments or pose questions to the larger group/community through posting comments via bulletin board or mass email.

Other advantages. Researchers have also found that the act of expressing one's thoughts in written form (which is typically the means by which people communicate in online support groups/communities as they post messages, chat, or send emails to each other) has therapeutic value (Diamond, 2000; Weinberg, Schmale, Uken, & Wessel, 1995). Expressing thoughts in emails, bulletin boards, and chat applications appears to allow psychological distance between a person and his or her thoughts. This provides opportunities for individuals to reflect on their thoughts, re-examine them, and re-articulate them prior to sending messages to the group. Moreover, recent research suggests that act of sending affectionate messages in supportive online exchanges is related to reduced total cholesterol levels and cortisol levels (Floyd, Mikkelson, Hesse, & Pauley, 2007). Both cholesterol and cortisol are physiological products of stress, and both have been linked to heart disease and strokes among individuals facing long-term stressful situations.

This research provides an important empirical link between computer-mediated supportive communication and physical health outcomes.

Disadvantages of computer-mediated support groups. Computer-mediated support groups/communities may also have a number of disadvantages for participants. Wright (2000, 2002), in a survey of many types of health-related online support groups, identified a number of disadvantages. One problem is the relatively short-term membership that is seen in health-related groups. Participants often join online groups/communities when they are initially worried about a health problem or when they have been recently diagnosed with an illness. However, members appear to stop using these groups/communities after a few weeks. It seems that once some people feel that their initial concerns about a health issue have been addressed by the group, they decide to stop affiliating with the group and (presumably) seek support elsewhere. Such short-term membership may lead to several problems, including difficulty locating specific members and fewer "old-timers," or individuals who have been using the group/community to deal with an illness/health concern for a long period of time.

In addition, despite greater access to individuals who share similar health concerns in online support groups/communities, participants often find the lack of immediacy (associated with reduced social presence) when communicating with others frustrating. Wright (2002) also found that online support group members missed the ability to engage in haptic communication (i.e., hugs, and other expressions of supportive touch) with fellow participants.

Other disadvantages of computer-mediated support groups include off-topic remarks from participants, spam messages, privateers (people who try to use the group/community for their own selfish purposes), and flaming (i.e., antisocial behavior). These behaviors tend to increase negative perceptions of the group/community, and they may curtail membership if they occur frequently. Early computer-mediated communication researchers posited that the reduced social cues associated with the medium may encourage antisocial behaviors due to the lack of physical presence of other participants (Walther, 1996; Walther & Burgoon, 1992). In other words, it is much easier to be disruptive online since a person is in little danger of physical retaliation from other members.

Moreover, the medium may facilitate deceptive practices (see Dunbar & Jensen, Chapter 17, in this volume). The anonymity associated with online support groups/communities makes it difficult to assess who one is really communicating with. In some cases, individuals may misrepresent themselves, pretend to have an illness in order to receive attention from others, or they may be using the group/community for a variety of other reasons that are unrelated to the purpose of the group/community.

Social Support, Stress, and Well-Being

A large volume of research has provided evidence that social support tends to reduce stress, enhance people's psychological well-being, and may lead to positive health outcomes. For example, several decades of research have provided empirical support for the negative relationship between social support and perceived stress, in general (see Cobb, 1976; Cohen & Wills, 1985; Cutrona & Suhr, 1992; Franks, Cronan, & Oliver, 2004), and social support as an important mediator of stress (Chappell & Novak, 1992; Ellis & Miller, 1994; Kalliath & Morris, 2002; Tyler & Cushway, 1995). In terms of supportive communication, Shaw, Hawkins, McTavish, Pingree, and Gustafson (2006) found that insightful self-disclosure from a cancer support group had an impact on emotional well-being, although it did not affect self-reported physical health outcomes.

Studies from a variety of disciplines have consistently linked traditional sources of social support to morbidity and mortality rates (Berkman & Syme, 1979; Bruunk, 1990; Cohen, 1988; House, Landis, & Umberson, 1988; Uchino, Cacioppo, & Kiecolt-Glaser, 1996). Prolonged exposure to stress has been found to impair immune system response, increase risk for hypertension, heart disease, and stroke, as well as lead to increases in depression, tension, and nervousness (Ballieux & Heijen, 1989; Clow, 2001; Kohn, 1996).

Researchers have identified two distinct ways in which social support appears to influence health outcomes. The buffering model of social support (Cohen & Wills, 1985; Dean & Lin, 1977) posits that an individual's social network helps to shield or reduce the amount of stress he or she may experience when encountering both major crises and everyday sources of stress. The social capital and other resources associated with one's support network can help buffer or offset stressors, including those related to health concerns. The main effects model of social support asserts that there is a direct relationship between social support and health outcomes (Aneshensel & Stone, 1982). Positive interactions with one's support network in everyday relationships appear to elevate a person's mood, reduce stress, and make people more resilient to stressful situations (Berkman & Syme, 1979; Cohen, 1988).

While only a relatively small number of studies have measured psychological and physical health outcomes related to participation in computer-mediated support groups, there is some empirical evidence that these groups may provide interpersonal and health benefits for users. Early studies found that verbal and written expression about emotionally traumatic events has been linked to positive physical and mental health outcomes (see Pennebaker, 1997). Wright (2000) found that participation in a computer-mediated support community for older adults predicted reduced stress and increased coping skills. Owen, Klapow, Roth, Shuster, and Bellis (2005) found that participation within an online support group for cancer patients predicted higher quality of life and reduced stress. More recently, Wright, Rains, and Banas (2010) found that computer-mediated support group users who had more positive perceptions of the benefits of weak tie support from their group had reduced levels of stress. In addition, in a recent meta-analysis of 28 computer-mediated support group studies, Rains and Young (2009) found that across studies, participation in computer-mediated social support groups led to increased social support, decreased depression, increased quality of life, and increased self-efficacy in terms of managing health conditions.

Of course there are many mediating variables that influence the relationship between social support and health outcomes, including an individual's coping abilities (Kohn, 1996), the communication competence of both support providers and receivers (Query & James, 1989; Query & Kreps, 1996; Query & Wright, 2003), and perceptions of the support providers and type of support offered (Albrecht & Goldsmith, 2003; Burleson, 1994).

Theoretical Approaches to Computer-Mediated Social Support Groups

This section presents several theoretical frameworks that have been used to understand computer-mediated support groups/communities in prior research. These perspectives provide important insight into the interpersonal processes associated with computer-mediated social support. Additionally, these theories help provide explanations as to why several features of computer-mediated support groups may or may not appeal to participants.

Social Support and Weak Tie Network Preference

One interpersonal communication theory that has been applied to the study of computer-mediated is Granovetter's (1973) theory of weak ties (Walther & Boyd, 2002; Wright & Bell, 2003; Wright & Query, 2004). Weak tie relationships typically occur between individuals who communicate on a relatively frequent basis, but who do not consider each other to be members of their close personal network (e.g., close friends, family). While one may not think of weak ties as personal relationships, within the context of computer-mediated support groups, weak ties (as well as more developed relationships) are often important sources of social support. When seeking social support in certain stressful situations, an alternative to one's strong ties network of family and friends is the use of a weak tie support network, which was traditionally composed of one's neighbors, acquaintances, or individuals one might consult with in specific contexts, such as clergy, counselors, or members of face-to-face support groups. Today, the Internet appears to play an important role in terms of providing greater access to weak tie social support, including computer-mediated support groups (Walther & Boyd, 2002; Wright & Bell, 2003). However, studies have not investigated how weak ties that are initially encountered in computer-mediated contexts may develop into stronger ties.

Researchers have found that weak tie network members can afford a number of advantages over stronger ties when it comes to providing support in sensitive and problematic circumstances (Adelman, Parks, & Albrecht, 1987; Helgeson & Gottlieb, 2000; La Gaipa, 1990; Granovetter, 1973). According to Adelman et al.

(1987), "weak ties may provide a vital lifeline to those who lack the requisite skills or cognitive skills for intimate relationships." (p. 128).

One reason why individuals may opt for a weak tie over a strong tie support network is that weak ties often provide access to diverse points of view and information that tend not to be available within more intimate relationships (Adelman et al., 1987). Weak tie networks often provide access to diverse points of view and information that may not be available within more intimate relationships (Adelman, et al., 1987). As we have seen, many individuals form close relationships with others who are similar to them in terms of demographics, attitudes, and backgrounds (Botwin, Buss, & Shackelford, 1997). This homogeneous preference can limit the diversity of information and viewpoints obtained about topics, including health concerns. According to Adelman et al. (1987), "dependence on highly insulating social networks can prevent individuals from reaching out to needed physical and mental health professionals" (p. 129).

Access to more diverse viewpoints about health problems can provide individuals with more varied informational support about health issues, and interacting with varied types of people increases the number of social comparisons a person can make about his or her stressful situation vis-à-vis others (Adelman et al., 1987). The opportunity for more social comparisons has been found to be an integral component of face-to-face support groups (Helgeson & Gottlieb, 2000), and they often help individuals to manage uncertainty about their situation. For example, individuals facing difficult health concerns may often obtain more useful information by moving beyond their traditional strong tie support network. Using a weak tie network, such as a computer-mediated support group whose members may also have the disease or health condition, will often offer perspectives from others who are more likely to share similar feelings about their condition, even if they are dissimilar in terms of demographics, attitudes, and/or background. Furthermore, by interacting with a wider network of individuals experiencing similar problems, assessments can be made about how one is coping with a problem compared to others, further helping to reduce uncertainty and anxiety.

For a variety of reasons, strong tie support networks can be perceived as inadequate or incapable of providing satisfactory support, and a range of factors, both practical and psychological, have been shown to influence an individual's decision to pursue weak tie support networks as an alternative. For example, researchers have found that family members and friends often minimize the concerns of significant others who are seeking support for difficult health problems. In many cases, it is not uncommon for close ties to steer conversational topics away from emotional talk about problems, refrain from in-depth discussion of such topics, or avoid consequent interaction all together (Brashers et al., 2004; Dakof & Taylor, 1990; Dunkel-Schetter & Wortman, 1982; Helgeson, Cohen, Shultz, & Yasko, 2000).

In addition, studies have found that role obligations and related reciprocity issues in close ties can lead to problems with the provision of social support. Support for a loved one who is ill can lead to increased conflict, resentment, and negative feelings for both parties involved due to reluctance to form new complicated role obligations on the one hand, and feelings of guilt and shame stemming from the perceived inability to reciprocate on the other (Albrecht & Goldsmith, 2003; Chesler & Barbarin, 1984; LaGaipa, 1990; Pitula & Daugherty, 1995). According to Adelman et al. (1987), "those suffering from chronic illnesses, for example, sometimes react to support attempts by close friends and family with discomfort and anxiety because they do not believe that they will be able to reciprocate" (p. 129).

LaGaipa (1990) contends that these "social obligations may override the positive effect of companionship and social support. Such constraints may have a negative effect on a person's mental well-being that may not make up for the beneficial aspects of personal relationships" (p. 126). For example, although one may care deeply for those who are close, he or she may easily feel overburdened if a loved one becomes ill and needs a great amount of support, and the stress experienced can lead to conflict (Chesler & Barbarin, 1984). In contrast, weak tie network members, since they tend to be less emotionally attached, may be more willing to talk about difficult and/or unpleasant health concerns. According to Adelman et al. (1987), "support from weaker links will not create such intense discomfort. The expectations of weaker links are generally less extensive and more easily reciprocated" (p. 129).

As we have seen, many problems that computer-mediated support group members face often carry a social stigma (Brashers et al., 2004; Mathieson, Logan-Smith, Phillips, MacPhee, & Attia, 1996; Sullivan & Reardon, 1985), and this dehumanizing process can negatively affect the provision of social support (Bloom & Spiegel, 1984). Because members of weak tie networks do not typically share an intimate relational history, they have been found to be less likely to judge one another, and will frequently encourage one another to share concerns and feelings about living with various stigmatized health problems.

Because other computer-mediated support group members may also be contending with similar concerns, the similarity between members in terms of health concerns increases empathy and understanding of the situation, and fosters opportunities for other types of emotional support such as affirmation and validation. In addition, because of their reduced emotional attachment, weak tie network members may be more adept at providing objective feedback about health problems, and are generally more willing to discuss risky topics compared to one's strong tie network of family and friends (Adelman et al., 1987).

Social Information Processing Theory

Social information processing theory (Walther, 1996) argues that in computer-mediated communication, message senders portray themselves in a socially favorable manner to draw the attention of message receivers and foster anticipation of future interaction. Message receivers, in turn, tend to idealize the image of the sender due to overvaluing minimal, text-based cues. In addition, the asynchronous format of most computer-mediated interaction (and to some extent insynchronous formats, such as chat rooms) gives the sender and the receiver more time to edit their communication, making computer-mediated interactions more controllable and less stressful compared to the immediate feedback loop inherent in face-to-face interactions. Idealized perceptions and optimal self-presentation in the computer-mediated communication process tend to intensify in the feedback loop, and this can lead to what Walther (1996) labeled as "hyperpersonal interaction," or a more intimate and socially desirable exchange than face-to-face interactions.

Hyperpersonal interaction is enhanced when no face-to-face relationship exists, so that users construct impressions and present themselves "without the interference of environmental reality" (Walther, 1996, p. 33). Hyperpersonal interaction has been found to skew perceptions of relational partners in positive ways, and in some cases, computer-mediated relationships may exceed face-to-face interactions in terms of intensity, including within online support groups (King & Moreggi, 1998; Walther, 1996; Wright & Bell, 2003).

Optimal Matching Theory

The optimal matching model posits that an optimal match between the needs of support seekers and the resources/abilities of support providers is important in terms of coping with the many relational challenges associated with communicating social support (Goldsmith, 2004). For example, if an individual is seeking emotional support and validation for an eating disorder and he or she perceives that members of his or her support network have competently listened, expressed empathy, and acknowledged the severity of the issue, then this would be considered an example of an optimal match between the support seeker and support providers. Conversely, if an individual desires emotional support, and members of his or her support network provide unwanted advice (a negative form of informational support), then this would be considered a bad (or less than optimal) match.

Goldsmith (2004) contends that optimal matches in supportive episodes may lead to more positive perceptions of relational partners and the type(s) of support that is being offered, and this, in turn, may ultimately influence positive health outcomes. However, research drawing from this perspective has also found evidence that optimal support network patterns are dynamic and may change depending upon the severity of the crisis people are facing (Carstensen & Fredrickson, 1998; Lockenhoff & Carstensen, 2004).

While this model has been applied to a variety of face-to-face supportive contexts (see Goldsmith, 2004), relatively few researchers have used this framework to investigate computer-mediated support (Eichhorn, 2008; Turner, Grube, & Meyers, 2001; Wright & Muhtaseb, 2005). Yet, this perspective may help to provide important insights into the supportive needs of individuals who seek computer-mediated support. For example, drawing upon an optimal matching theory framework, Eichhorn (2008) found that informational support through shared experiences was the most common type of support for members of an online support group for eating disorders. Moreover, Sullivan (2003) found that men were more likely to seek informational support within online support groups whereas women were more likely to seek emotional support and validation.

Theoretical Development, Limitations of Current Research, and Future Directions

Based on the existing literature, there is clearly a need for a more comprehensive theoretical framework or frameworks that help shed light on the nature of interpersonal relationships, communication, and health benefits associated with participation in computer-mediated support groups. Such efforts are important in terms of moving beyond descriptive studies and developing a framework that can be useful in terms of understanding the processes of relational development, the communication of social support, and the relationship between these interactions and potential benefits for participants.

For example, Wright and Bell (2003) mentioned that the majority of computer-mediated support group studies have been largely descriptive in nature and have not sufficiently linked findings to a broader theoretical framework. Similarly, Eysenbach, Powell, Englesakis, Rizo, and Stern (2004) contend "the lack of measurable evidence from controlled studies is in sharp contrast to the increasing body of anecdotal and descriptive information on the self-helping processes in virtual communities" (p. 1169).

Of course, there have been a number of theories that have been applied to the study of computer-mediated support groups in a less comprehensive way. As we have seen, theories from several different academic disciplines and research areas, such as social information processing theory, weak tie network theory, social comparison theory, and the optimal matching model, have been used to explain certain processes or features of these groups. Future research would benefit from integrating overlapping features from these theories as well as integrating or combining them with theories regarding health outcomes (such as the buffering model or the direct effects model of social support).

For example, weak tie network theory appears to explain some of the motives for why people affiliate with computer-mediated support groups, particularly the need for more objective information, experience, and similarity that are more easily accessed in a heterogeneous network. However, it would be interesting to assess how perceptions of online partners (social information processing theory) affect perceptions of weak tie support benefits. Of course, it is also important to assess how weak ties may develop into stronger ties within these groups and whether or not this leads to changes in perceptions of the utility of such ties for social support. Similarly, it would be interesting to assess whether preference for weak or strong ties within these groups leads to optimal matching in terms of the type of support sought and the type of support received by participants. Assessing participant preferences/motives for using these groups may be useful in terms of identifying an optimal match between individual preferences/motives and the type(s) of support that are available within a given support group or community. Such efforts could aid in the development of support interventions where participants could receive the type of support that is most useful to them.

Most studies of computer-mediated support groups that have examined health outcomes have focused on researcher-created groups or groups led by healthcare professionals. However, according to Eysenbach et al. (2004), "Given the abundance of unmoderated peer to peer groups on the Internet, researchers must focus their efforts not only on professionally led systems, but shift their attention to consumer led, self help venues" (p. 1170). Given the sheer number of participant-led support groups on the Internet, researchers need to assess whether or not these types of groups have similar benefits as professionally led groups.

Finally, more research is needed to understand how characteristics of computer-mediated communication affect relationships between members of online support groups. As we have seen, these groups offer a number of advantages and disadvantages to participants. However, most studies have been cross-sectional, and relatively little is known about the development and maintenance of long-term supportive relationships within these groups, how contacts within the groups may become part of one's face-to-face support network, relational dilemmas in giving and receiving computer-mediated social support, and a host of other interpersonal communication concerns within this environment.

Conclusion

Computer-mediated social support groups are an important context to study supportive personal interpersonal relationships. Most of the previous research suggests that online support has potential health benefits for those individuals who use computer-mediated support groups/communities or who engage in other types of online support. However, despite some of the advantages of using these resources, there are also problems associated with them. The use of the Internet as a vehicle for obtaining social support from traditional sources (i.e., family and friends) remains an important area for future research. In addition, future research should continue to focus on the relationship between computer-mediated support and health outcomes, and researchers should attempt to disseminate findings to potential users and healthcare professionals in an effort to increase education about these sources of social support.

References

Adelman, M. B., Parks, M. R., & Albrecht, T. L. (1987). Beyond close relationships: Support in weak ties. In T. L. Albrecht & M. B. Adelman (Eds.), *Communicating social support* (pp. 126–147). Newbury Park, CA: Sage.

Albrecht, T. L., Burleson, B. R., & Goldsmith, D. J. (1994). Supportive communication. In M. Knapp & G. R. Miller (Eds.), *Handbook of interpersonal communication* (pp. 419–449). Thousand Oaks, CA: Sage.

Albrecht, T. L., & Goldsmith, D. J. (2003). Social support, social networks, and health. In T. L. Thompson, A. M. Dorsey, K. I. Miller, & R. Parrott (Eds.), *Handbook of health communication* (pp. 263–284). Mahwah, NJ: Lawrence Erlbaum.

Aneshensel, C. S., & Stone, J. D. (1982). Stress and depression: A test of the buffering model of social support. *Archives of General Psychiatry, 39,* 1392–1396.

Ballieux, R. E., & Heijen, C. J. (1989). Stress and the immune response. In H. Weiner, I. Floring, R. Murison, & D. Hellhammer (Eds.), *Frontiers of stress research* (pp. 51–55). Toronto, Canada: Huber.

Barbee, A. P., Derlega, V. J., Sherburne, S. P., & Grimshaw, A. (1998). Helpful and unhelpful forms of social support for HIV-positive individuals. In V. J. Derlega & A. P. Barbee (Eds.), *HIV & social interaction* (pp. 83–105). Thousand Oaks, CA: Sage.

Barrera, M. (1986). Distinctions between social support concepts, measures, and models. *American Journal of Community Psychology, 14,* 413–445.

Berkman, L. F., & Syme, L. S. (1979). Social networks, host resistance, and mortality: A nine-year follow-up study of Alameda County residents. *Journal of Epidemiology, 109,* 186–204.

Bloom, J. R., & Spiegel, D. (1984). The relationship of two dimensions of social support to the psychological well-being and social functioning of women with advanced breast cancer. *Social Science Medicine, 19,* 831–837.

Botwin, M.D., Buss, D.M., & Shackelford, T.K. (1997). Personality and mate preferences: Five factors in mate selection and marital satisfaction. *Journal of Personality, 65,* 107–136.

Braithwaite, D. O., Waldron, V. R., & Finn, J. (1999). Communication of social support in computer-mediated groups for people with disabilities. *Health Communication, 11,* 123–151.

Brashers, D. E., Neidig, J. L., & Goldsmith, D. J. (2004). Social support and the management of people living with HIV or AIDS. *Health Communication, 16,* 305–331.

Bruunk, B. (1990). Affiliation and helping interactions within organizations: A critical analysis of the role of social support with regard to occupational stress. In W. Stroebe & M. Hewstone (Eds.), *European review of social psychology* (Vol. 1, pp. 293–322). Chichester, UK: John Wiley.

Burleson, B. R. (1994). Comforting messages: Significance, approaches, and effects. In B. R. Burleson, T. L. Albrecht, & I. G. Sarason (Eds.), *Communication of social support: Messages, interactions, relationships and community*. Newbury Park, CA: Sage.

Carstensen, L. L., & Fredrickson, B. L. (1998). Influence of HIV status and age on cognitive representations of others. *Health Psychology, 17,* 494–503.

Chappell, N. L., & Novak, M. (1992). The role of support in alleviating stress among nursing assistants. *Gerontologist, 32,* 251–359.

Chesler, M. A., & Barbarin, O. A. (1984). Difficulties of providing help in a crisis: Relationships between parents of children with cancer and their friends. *Journal of Social Issues, 40,* 113–134.

Cline, R. J. (1999). Communication within social support groups. In L. R. Frey (Ed.), D. S. Gouran, & M. S. Poole (Assoc. Eds.), *The handbook of group communication theory and research* (pp. 516–538). Thousand Oaks, CA: Sage.

Clow, C. (2001). The physiology of stress. In F. Jones & J. Bright (Eds.), *Stress: Myth, theory, and research* (pp. 47–61). Harlow, UK: Prentice Hall.

Cobb, S. (1976). Social support as a moderator of life stress. *Psychosomatic Medicine, 38*, 300–314.

Cohen, S. (1988). Psychosocial models of the role of support in the etiology of physical disease. *Health Psychology, 7*, 269–297.

Cohen, S., & Wills, T. A. (1985). Stress, social support, and the buffering hypothesis. *Psychological Bulletin, 98*, 310-357.

Cutrona, C. E., & Suhr, J. A. (1992). Controllability of stressful events and satisfaction with spouse supportive behaviors. *Communication Research, 19*, 154–174.

Daft, R. L., & Lengel, R. H. (1984). Information richness: A new approach to managerial behavior and organizational design. *Research in Organizational Behavior, 6*, 191–233.

Daft, R. L., & Lengel, R. H. (1986). Organizational informational requirements, media richness, and structural design. *Management Science, 32*, 554–571.

Daft, R. L., Lengel, R. H., & Trevino, L. K. (1987). Message equivocality, media selection, and manager performance: Implications for information systems. *MIS Quarterly, 11*, 355–368.

Dakof, G. A., & Taylor, S. E. (1990). Victims' perceptions of social support: What is helpful from whom? *Journal of Personality and Social Psychology, 58*, 80–89.

Davison, K. P., Pennebaker, J. W., & Dickerson, S. S. (2000). Who talks? The social psychology of illness support groups. *American Psychologist, 55*, 205–217.

Dean, A., & Lin, N. (1977). The stress buffering role of social support: Problems and prospects for systematic investigation. *Journal of Health and Social Behavior*, 32, 321–341.

Diamond, J. (2000). *Narrative means to sober ends: Treating addiction and its aftermath*. New York: Guilford.

Drentea, P., & Moren-Cross, J. L. (2005). Social capital and social support on the web: The case of an Internet mother site. *Sociology of Health & Illness, 27*, 920–943.

Dunkel-Schetter, C., & Wortman, C. B. (1982). The interpersonal dynamics of cancer: Problems in social relationships and their impact on patients. In H. S. Friedman & M. R. DiMatteo (Eds.), *Interpersonal issues in health care* (pp. 69–100). New York: Academic.

Eichhorn, K. C. (2008). Soliciting and providing support over the Internet: An investigation of online eating disorder groups. *Journal of Computer-Mediated Communication, 14*, 67–78.

Ellis, B. H., & Miller, K. I. (1994). Supportive communication among nurses: Effects on commitment, burnout, and retention. *Health Communication, 6*, 77–96.

Eysenbach, G., Powell, J., Englesakis, M., Rizo, C., & Stern, A. (2004). Health related virtual communities and electronic support groups: Systematic review of the effects of online peer to peer interactions. *BMJ, 328*, 1166–1172.

Finn, J., & Lavitt, M. (1994). Computer-based self-help groups for sexual abuse survivors. *Social Work with Groups, 17*, 21–47.

Floyd, K., Mikkelson, A. C., Hesse, C., & Pauley, P. M. (2007). Affectionate writing reduces total cholesterol: Two randomized, controlled trials. *Human Communication Research, 33*, 119–142.

Franks, H. M., Cronan, T. A., & Oliver, K. (2004). Social support in women with fibromyalgia: Is quality more important than quantity? *Journal of Community Psychology, 4*, 425–438.

Giles, H., Mulac, A., Bradac, J. J., & Johnson, P. (1987). Speech accommodation theory: The next decade and beyond. *CommunicationYearbook, 10*, 13–48.

Goldsmith, D. J. (2004). *Communicating social support*. New York: Cambridge University Press.

Granovetter, M. (1973). The strength of weak ties. *American Journal of Sociology, 78*, 1360–1380.

Haythornthwaite, C., & Wellman, B. (2002). The Internet in everyday life: An introduction. In B. Wellman & C. Haythornthwaite (Eds.), *The internet in everyday life* (pp. 3-42). Oxford, UK: Blackwell.

Helgeson, V. S., Cohen, S., Schulz, R., & Yasko, J. (2000). Group support interventions for women with breast cancer: Who benefits from what? *Health Psychology, 19*, 107-114.

Helgeson, V. S., & Gottlieb, B. H. (2000). Support groups. In S. Cohen, L. G. Underwood, & B. H. Gottlieb (Eds.), *Social support measurement and intervention* (pp. 221–245). New York: Oxford University Press.

Herring, S. C. (2002). Computer-mediated communication on the Internet. In B. Cornin (Ed.), *Annual review of information science and technology* (Vol. 36, pp. 109–168). Medford, NJ: Information Today.

Horrigan, J. B., & Rainie L. (2002). *Getting serious online.* Washington, DC: Pew Internet & American Life Project.

House, J., Landis, K. R., & Umberson, D. (1988). Social relationships and health. *Science, 241,* 540–545.

Johnson, A. J., Wright, K. B., Craig, E. A., Gilchrist, E. S., Lane, L. T., & Haigh, M. M. (2008). A model for predicting stress levels and marital satisfaction for stepmothers utilizing a stress and coping approach. *Journal of Social and Personal Relationships, 25),* 119–142.

Kalliath, T., & Morris, R. (2002). Job satisfaction among nurses: A predictor of burnout levels. *Journal of Nursing Administration, 32,* 648–654.

Kiesler, S., Siegel, J., & McGuire, T. (1984). Social psychological aspects of computer-mediated communication. *American Psychologist, 39,* 1123–1134.

Kiesler, S., & Sproull, L. (1992). Group decision making and communication technology. *Organizational Behavior and Human Decision Processes, 52,* 96–123.

King, S. A., & Moreggi, D. (1998). Internet therapy and self-help groups: The pros and cons. In J. Gakenbach (Ed.), *Psychology and the Internet: Intrapersonal, interpersonal, and transpersonal implications* (pp. 77–109). San Diego, CA: Academic.

Kohn, P. M. (1996). On coping adaptively with daily hassles. In M. Zeidner & N. S. Endler (Eds.), *Handbook of coping* (pp. 181–201). New York: John Wiley.

La Gaipa, J. J. (1990). The negative effects of informal support systems. In S. Duck & R. C. Silver (Eds.), *Personal relationships and social support* (pp. 122–139). Newbury Park, CA: Sage.

Levy, J.A., & Strombeck, R. (2002). Health benefits and risks of the Internet. *Journal of Medical Systems,* 26, 495–510.

Lieberman, M. A., & Goldstein, B. A. (2005). Self-help online: An outcome evaluation of breast cancer bulletin boards. *Journal of Health Psychology, 10,* 855–862.

Lin, N. (2001). *Social capital: A theory of social structure and action.* Cambridge, UK: Cambridge University Press.

Lockenhoff, C. E., & Carstensen, L. L. (2004). Socioemotional selectivity theory, aging, and health: The increasingly delicate balance between regulating emotions and making tough choices. *Journal of Personality, 72,* 1395–1423.

Mathieson, C. M., Logan-Smith, L. L., Phillips, J., MacPhee, M., & Attia, M. L. (1996). Caring for head and neck oncology patients: Does social support lead to better quality of life? *Canadian Family Physician, 42,* 1712–1720.

Neuhauser, L., & Kreps, G. L. (2003). Rethinking communication in the e-health era. *Journal of Health Psychology, 8,* 7–23.

Owen, J. E., Klapow, J. C., Roth, D. L., Shuster, J. L., & Bellis, J. (2005). Randomized pilot of a self-guided Internet coping group for women with early-stage breast cancer. *Annals of Behavioral Medicine, 30,* 54–64.

Pakenham, K. I. (1998). Specification of social support behaviors and network dimensions along the HIV continuum for gay men. *Patient Education and Counseling 34,* 147–157.

Parks, M. R., & Floyd, K. (1996). Making friends in cyberspace. *Journal of Communication, 46,* 80–97.

Pennebaker, J. W. (1997). Writing about emotional experiences as a therapeutic process. *Psychological Science, 8,* 162–166.

Pitula, C. R., & Daugherty, S. R. (1995). Sources of social support and conflict in hospitalized depressed women. *Nursing and Health, 18,* 325–332.

Preece, J. J., & Ghozati, K. (2001). Experiencing empathy online. In R. E. Rice & J. E. Katz (Eds.), *The Internet and health communication: Experiences and expectations* (pp. 237-260). Thousand Oaks, CA: Sage.

Pull, C. (2006). Self-help Internet interventions for mental disorders. *Current Opinion in Psychiatry, 19,* 50–53.

Putnam, R. D. (1995a). Bowling alone. America's declining social capital. *Journal of Democracy, 6,* 65–78.

Putnam, R. D. (1995b). Tuning in, tuning out the strange disappearance of social capital in America. *PS: Political Science & Politics, 284,* 664–683.

Putnam, R. D. (2000). *Bowling alone: The collapse and revival of American community*. NewYork: Simon & Schuster.

Quan-Haase, A., Wellman, B., Witte, J. C., & Hampton, K. N. (2002). Capitalizing on the net: Social contact, civic engagement, and sense of community. In B. Wellman & C. Haythornthwaite (Eds.), *The internet in everyday life* (pp. 291-325). Oxford, UK: Blackwell.

Query, J. L., Jr., & James, A. C. (1989). The relationship between interpersonal communication competence and social support among elderly support groups in retirement communities. *Health Communication, 1*(3), 165–184.

Query, J. L., Jr., & Kreps, G. L. (1996). Testing the relational model for health communication competence among caregivers for individuals with Alzheimer's disease. *Journal of Health Psychology, 1 (3),* 335–351.

Query, J. L., Jr., & Wright, K. B. (2003). Assessing communication competence in an online study: Toward informing subsequent interventions among older adults with cancer, their lay caregivers, and peers. *Health Communication 15*(2), 203–218.

Rains, S. A., & Young, V. (2009). A meta-analysis of research on formal computer-mediated support groups: Examining group characteristics and health outcomes. *Human Communication Research, 35,* 309-336.

Rook, K. S. (1995). Support, companionship, and control in older adults' social networks: Implications for well-being. In J. F. Nussbaum & J. Coupland (Eds.), *Handbook of communication and aging research* (pp. 437–463). Mahwah, NJ: Lawrence Erlbaum.

Sharf, B. F. (1997). Communicating breast cancer online: Support and empowerment on the Internet. *Women & Health, 26,* 65–84.

Shaw, B. R., Hawkins, R., McTavish, F., Pingree, S., & Gustafson, D. H. (2006). Effects of insightful disclosure within computer mediated support groups on women with breast cancer. *Health Communication, 19,* 133–142.

Sullivan, C. F. (2003). Gendered cybersupport: A thematic analysis of two online cancer support groups. *Journal of Health Psychology, 8,* 83–103.

Sullivan, C. F., & Reardon, K. K. (1985). Social support satisfaction and health locus of control: Discriminators of breast cancer patients' style of coping. In M. L. McLaughlin (Ed.), *Communication yearbook* (Vol. 9, pp. 707–722). Beverly Hills, CA: Sage.

Szreter, S. (2000). Social capital, the economy, and education in historical perspective. In S. Baron, J. Field, & T. Schuller (Eds.), *Social capital: Critical perspectives* (pp. 56–77). New York: Oxford University Press.

Turner, J. W., Grube, J. A., & Meyers, J. (2001). Developing an optimal match within online communities: An exploration of CMC support communities and traditional support. *Journal of Communication,* 231–251.

Tyler, P. A., & Cushway, D. (1995). Stress in nurses: The effects of coping and social support. *Stress Medicine, 11,* 243–251.

Uchino, B. N., Cacioppo, J. T., & Kiecolt-Glaser, J. K. (1996). The relationship between social support and physiological processes: A review with emphasis on underlying mechanisms and implications for health. *Psychological Bulletin, 119,* 488–531.

Walstrom, M. K. (2000). You know who's the thinnest?: Combating and surveillance and creating safety in coping with eating disorders online. *Cyberpsychology and Behavior, 3,* 761–783.

Walther, J. B. (1995). Relational aspects of computer-mediated communication: Experimental observations over time. *Organizational Science, 6,* 186–203.

Walther, J. B. (1996). Computer-mediated communication: Impersonal, interpersonal, and hyperpersonal interaction. *Communication Research, 23,* 3–43.

Walther, J. B., & Boyd, S. (2002). Attraction to computer-mediated social support. In C. A. Lin & D. Atkin (Eds.), *Communication technology and society: Audience adoption and uses* (pp. 153–188). Cresskill, NJ: Hampton.

Walther, J. B., & Burgoon, J. (1992). Relational communication in computer-mediated interaction. *Human Communication Research, 19,* 50–88.

Weinberg, N., Schmale, J. D., Uken, J., & Wessel, K. (1995). Computer-mediated support groups. *Social Work with Groups, 17,* 43–55.

Wellman, B. (1997). An electronic group is virtually a social network. In S. Kiesler (Ed.), *Culture of the Internet* (pp. 179–205). Mahwah, NJ: Lawrence Erlbaum.

Winefield, H. R. (2006). Support provision and emotional work in an Internet support group for cancer patients. *Patient Education and Counseling, 62,* 193–197.

Wright, K. B. (2000). Perceptions of online support providers: An examination of perceived homophily, source credibility, communication and social support within online support groups. *Communication Quarterly, 48,* 44–59.

Wright, K. B. (2002). Social support within an online cancer community: An assessment of emotional support, perceptions of advantages and disadvantages, and motives for using the community. *Journal of Applied Communication Research, 3,* 195–209.

Wright, K. B., & Bell, S. B. (2003). Health-related support groups on the Internet: Linking empirical findings to social support and computer-mediated communication theory. *Journal of Health Psychology, 8,* 37–52.

Wright, K. B., & Muhtaseb, A. (May, 2005). *Perceptions of online support in health-related computer-mediated support groups.* Paper presented at the annual International Communication Association Convention, New York.

Wright, K. B., & Query, J. L., Jr. (2004). Online support and older adults: A theoretical examination of benefits and limitations of computer-mediated support networks for older adults and possible health outcomes. In J. F. Nussbaum & J. Coupland (Eds.), *Handbook of communication and aging research* (2nd ed., pp. 499–519). Mahwah, NJ: Lawrence Erlbaum.

Wright, K. B., Rains, S., & Banas, J. (2010) Weak tie support network preference and perceived life stress among participants in health-related, computer-mediated support groups. *Journal of Computer-Mediated Communication, 15,* 606-624.

Zelly, E. D. (2001, November). *Desperately seeking social support: What women with eating disorders want regarding social support from close friends.* Paper presented at the annual meeting of the National Communication Association, Atlanta, GA.

CHAPTER NINE

Online Self-Disclosure: A Review of Research

Jinsuk Kim

Kathryn Dindia

Computer-mediated environments, such as social networking sites and online dating sites, provide us with a variety of opportunities to initiate, develop, and maintain interpersonal relationships. Self-disclosure is a key factor in developing relationships in online environments as it is in face-to-face (FtF) contexts (Dindia, 2000). Researchers have been examining how individuals present themselves in online environments and the factors that influence the way individuals disclose personal information in online settings. There has been an accumulation of research on self-disclosure with respect to FtF contexts (e.g., Collins & Miller, 1994; Derlega, Metts, Petronio, & Margulis, 1993; Dindia, 2002; Dindia & Allen, 1992; Jourard, 1971); however, there has been no comprehensive review of research on how people disclose personal information in various online settings and how this disclosure affects interpersonal relationships.

The traditional definition of self-disclosure refers to only intentional, "verbal" expressions of the self, and does not include nonverbal cues, such as how people dress, as disclosure. However, this definition of self-disclosure may not be adequate for online communication. Online self-disclosure may include nonverbal communication, including pictures posted of self, which may be a conscious mechanism used to disclose self. This is not the case in FtF settings where there is less opportunity to change appearances. How we present ourselves physically, both online and offline, can be manipulated (through dress, make up, etc.); however, in online settings, depending on an individual's intention, a person's physical appearance may or may not be shared. Also, when people include their favorite links on their websites, it reveals information about themselves. This is another device that can be used, with or without intention, to reveal information about self that is not available offline. We feel that it is worthwhile to explore these possible aspects of self-disclosure on the web in addition to verbal expressions of self. Our definition of online self-disclosure extends the traditional definition of self-disclosure (verbally revealing self) to include pictures of self and favorite links posted on the web.

Considering the growth of online social interactions and their impact on people's personal and social lives, it is worthwhile to examine self-disclosure in computer-mediated communication (CMC). How do various aspects of CMC affect self-disclosure? What do we know about online self-disclosure in comparison to FtF self-disclosure? In this review, we assess the state of the literature on online self-disclosure and suggest avenues for future research, including theoretical and methodological considerations that are important to advancing our understanding of online self-disclosure. We begin with an examination of the qualities of online self-disclosure in various modes of CMC. Next, we review empirical findings on self-disclosure in CMC versus FtF settings and some of the influences on online self-disclosure such as gender, culture, age, and motivations for using CMC media.

Online Self-Disclosure

Self-Disclosure versus Self-Presentation

CMC researchers have used the term *self-presentation* and *self-disclosure* interchangeably to refer to expressions of self in online settings. While there are distinctions, conceptually and operationally, between self-disclosure and self-presentation (Schlenker, 1986), those distinctions may not be as fundamental as one supposes. Self-presentation refers to selectively presenting aspects of oneself to control how one is perceived by others and is concerned with impression management (Goffman, 1959). Self-disclosure refers to revealing personal or private information about self that is generally unknown and not available from other sources (Derlega et al., 1993; Jourard & Jaffee, 1970; Pearce & Sharp, 1973; Worthy, Gary, & Kahn, 1969). Self-disclosure always involves an element of self-presentation.

Some researchers have elaborated dimensions of self-disclosure that border on self-presentation. Cozby (1973) considered depth or intimacy as a dimension of self-disclosure. Jourard (1971) discussed honesty of self-disclosure. Pearce and Sharp (1973) implied that conscious, deliberate intent to disclose, or willingness to disclose, as well as honesty or authenticity are basic dimensions of self-disclosure.

Valence is another dimension of self-disclosure that directly bears relevance on the issue of self-disclosure versus self-presentation. Gilbert and Horenstein (1975) distinguished between intimacy (high or low) and valence of self-disclosure (positive or negative) based on the writing of Blau (1964). According to Blau, when engaging in self-disclosure, a person attempts to present qualities that make himself/herself an attractive person, especially in early stages of relationship development. Blau further explains that in initial stages of relationship development, disclosing aspects of oneself that are negative does not lead to attraction.

Gilbert and Horenstein (1975) studied the relationship of intimacy and valence of self-disclosure to interpersonal attraction. In an earlier study, Gilbert (1972) found a negative relationship between intimacy of self-disclosure and attraction, such that subjects were more attracted to a confederate in a low versus a high intimacy condition. Gilbert reasoned that it might be because of the negative information in the high-disclosure condition (a girl's negative feelings about her mother and herself). Gilbert and Horenstein (1975) hypothesized that valence of self-disclosure, not intimacy, would predict interpersonal attraction. To test this, they manipulated both intimacy and valence of self-disclosure. Results of their study indicated a main effect of valence of self-disclosure. Participants were more attracted to a confederate when they disclosed positive rather than negative self-disclosure. There was no effect of intimacy or the interaction of intimacy and valence on attraction. The authors concluded that disclosure of negative information to a stranger violates social norms of appropriateness and does not lead to attraction.

Wheeless (1976) also elaborated multiple dimensions of self-disclosure. A 32-item measure of various dimensions of self-disclosure was factor-analyzed and Wheeless concluded that there are at least five dimensions of self-disclosure: intention of self-disclosure, amount of self-disclosure (including frequency and duration), valence of self-disclosure (positive or negative), honesty-accuracy of self-disclosure (e.g., "I am not always honest in my self-disclosure"), and control of general depth or intimacy of self-disclosure. All of these dimensions of self-disclosure—intention, amount, valence, honesty-accuracy, intimacy of self-disclosure—may be consciously and intentionally altered in order to selectively present self.

The issue of selective self-disclosure is an issue for both FtF and online contexts. Self-disclosure is a vehicle of self-presentation, and self-disclosure is not always open and honest but may involve a conscious and intentional decision to reveal positive rather than negative aspects of self in order to be perceived as attractive and rewarding. This is no different from FtF self-disclosure. For instance, people can lie about their age, marital status, and even about their weight and height to a certain degree; however, lying online is easier because of the lack of or limited warranting information (Walther & Parks, 2002).

The focus of this chapter is on self-disclosure rather than self-presentation although we will talk about how self-disclosure can be modified to positive presentation of self. In the following section, we discuss how online self-disclosure differs from FtF self-disclosure and examine the variables that influence online self-disclosure.

Self-Disclosure on the Web

Certain attributes of CMC such as reduced nonverbal cues, asynchronity, and anonymity may influence self-disclosure (McKenna & Bargh, 2000; Walther, 1996) in addition to all the factors found to affect self-disclosure (motivation, setting, target, etc.). Indeed, the web offers unprecedented opportunities for people to disclose themselves that are unavailable offline. Countless people have taken advantage of this new setting to disclose information about self in the form of personal web pages, blogs, social networking sites, dating sites, etc.

Self-disclosure online differs in fundamental ways from FtF self-disclosure: Online self-disclosure typically is to multiple people whereas FtF self-disclosure typically occurs in dyads and small groups. However, FtF self-disclosure can occur in public settings to multiple people, such as an individual disclosing at a meeting, or in a classroom. But in these cases, the audience is physically present and the discloser can see the audience. Also, in FtF self-disclosure, whether to an individual or multiple people, we try to present our ideal selves in general. Thus, in this sense, self-disclosure in web pages does not differ from self-disclosure in public FtF settings.

On the web, communicators do not know who is on the other side of the screen. The authors can only imagine who their audience could be based on their assumptions about the audience when producing their messages. In some instances they imagine their audience as strangers, in other instances they imagine their audience predominantly as friends and relatives. As authors create a web page, self-presentation motivations may affect their self-disclosure. Constructing an ideal self reflects the authors' desires of how they want to be perceived by the invisible audience. Control of their self-disclosure, what they disclose and do not disclose about themselves, is one way authors construct their ideal self.

Most online self-disclosure research has been conducted in largely text-based and asynchronous online channels. With the emergence of social networking sites such as MySpace and Facebook, online self-presentations are no longer limited to text-based descriptions. The photograph is now a central component of online self-disclosure. On web pages and social networking sites, and even dating sites, pictures are prime means of conveying information about self. By posting pictures of themselves and sharing pictures of their family and friends, authors choose to reveal their ideal self. The viewers make assumptions about how the authors look and who they are based on these pictures.

In this chapter, we review studies that looked at CMC channels that potentially facilitate self-disclosure, ranging from asynchronous text-based channels to synchronous media-rich channels. We examine how self-disclosure may occur in such CMC channels including personal web pages, blogs, social networking sites, instant messaging, online dating sites (e.g., Match.com and eHarmony.com), and interactive online games (e.g., MOOs and MUDs). We ultimately explore what and how people reveal about themselves and factors that prompt people to reveal themselves in these online settings.

A Review of Research on Online Self-Disclosure

For this review, we compiled a sample of studies by computer searches of the following databases: *Psychological Abstracts, Social Sciences Index, Communication & Mass Media Complete, Communication Abstracts, Dissertation Abstracts*, and *Google Scholar*. The key search terms include *self-disclosure, self-presentation,* CMC, *online, Internet, computer, web*, and *cyber*. We also scanned the reference lists from the articles identified through the search. Of the articles collected, we selected empirical articles in peer-reviewed journals, conference papers, and theses/dissertations for inclusion in the review of findings

Self-Disclosure Online versus FtF

Early views of CMC argued that communication in online environments is less personal than communication in FtF contexts because of reduced nonverbal communication cues (e.g., facial expressions and tone of voice) in online text-based settings (Culnan & Markus, 1987).

Social information processing (SIP) theory (Walther, 1992) rejected the view that the absence of nonverbal communication restricts communicators' ability to engage in personal communication. Walther argued that communicators are just as motivated to reduce interpersonal uncertainty, form impressions, and develop affinity in online settings as in offline. When denied nonverbal communication, communicators substitute other cues to engage in impression formation and relational messages, such as content (including self-disclosure), style, and timing of verbal messages. Due to the limited nonverbal cues in online settings, people may use more verbal messages (including self-disclosure) to compensate for these limited cues. According to SIP, the rate of information exchange is slower online but, given enough time, communication conducted through CMC can be just as personal as FtF communication.

More recently, a third view emerged in which online communication, because of the characteristics of CMC, is often hyperpersonal (more personal than FtF communication including higher levels of self-disclosure) (Walther, 1996, 2007; Walther & Parks, 2002). Some have argued that Internet users come to know one another more quickly and intimately than in FtF relationships because the features of CMC may make self-disclosure easier online versus FtF. Individuals in CMC often are anonymous and the psychological comfort that comes from such anonymity may lead them to reveal more information about themselves (Wallace, 1999).

Walther (1996) specifically argued that CMC is hyperpersonal because of sender, receiver, message, channel, and feedback effects. Senders exploit CMC's absence of nonverbal cues for the purpose of selective self-presentation, presenting a positive and idealized image of self. Because most types of online interactions occur asynchronously, senders may have more opportunity to review and edit information about themselves. Receivers initially engage in stereotypically positive and idealized attributions of online partners. The channel facilitates goal-enhancing messages by allowing sources greater control over message construction. The process of feedback creates self-fulfilling prophecies among senders and receivers. Due to these features of CMC, online users have greater opportunity for "selective self-presentation" (Walther, 2007).

In sum, there are different claims on whether online communication is more or less personal than FtF communication. Thus, we conducted a meta-analysis, a quantitative method of summarizing the results from multiple studies, of self-disclosure in online versus FtF settings to provide empirical evidence to support one of these claims. For each study, we calculated the effect size for available dimensions of self-disclosure. Particularly, we focused on the amount and the intimacy of self-disclosure. The average effect size across studies was calculated. The effect size *d* is statistic that indicates how large an effect is. Cohen (1969) offered the guidelines for interpreting *d*: $d = .20$ is small; $d = .50$ is moderate; and $d = .80$ is large. To examine whether an effect size varies more than would be expected by chance across the group of studies, homogeneity of effect sizes is tested. The sample size in our study is small, only 11 studies involving 2,887 participants (see Table 1), thus any conclusions must be interpreted with caution.

Two of the 11 studies found online self-disclosure to be more intimate than FtF self-disclosure (Joinson, 2001; Schouten, Valkenburg, & Peter, 2007a). In Joinson's study, individuals were paired with strangers and assigned to a FtF or a CMC (online chat) condition. Subjects were asked to a make decision as to which five people in the world should be given place in shelter in event of nuclear war. They were stopped after 45 minutes if they had not already made a decision. Coders rated conversations for self-disclosure. Dyads in the CMC condition disclosed significantly more than dyads in the FtF condition using two measures of self-disclosure (number of self-disclosures and proportion of words coded as self-disclosure). This study appears to be an outlier (effect sizes were 1.33 and 1.90, which are unusually large effect sizes and are out of line and in opposite direction of almost all the other studies). This may be due to the nature of the task. As noted by the researcher, the amount of self-disclosure was extremely low in both conditions (there were only 38 instances of self-disclosure in total across the two conditions), which may be due to how self-disclosure was measured (only spontaneous self-disclosure was coded; self-disclosure in response to a partner's question was not coded as self-disclosure) or it may be due to the task, which obviously did not facilitate self-disclosure. Thus, one would not want to draw conclusions on the effect of CMC on self-disclosure based on this study.

The only other study that examined actual self-disclosure in FtF versus online conversations was conducted by Tidwell and Walther (2002). In this study, strangers met FtF versus online and were instructed to either get to know one another or work on a solution to a problem. Different time periods were allotted for CMC (up to 60 minutes) versus FtF (up to 15 minutes) conversations based on the argument that time in a FtF setting is not equal to time in an online setting (typing alone takes longer in an online setting). Self-disclosure was measured as the proportion of comments that contained self-disclosure. The results indicated that FtF partners disclosed more peripheral and intermediate self-disclosure than CMC interactants. There were no significant differences in core (intimate) self-disclosure.

The second study that found that self-disclosure was greater in CMC than FtF was conducted by Schouten et al. (2007a). In this study, 81 cross-sex dyads were randomly assigned to one of three experimental conditions (text-only CMC, visual CMC [i.e., webcam], and FtF communication) and engaged in a get-acquainted exercise for 24 minutes in CMC dyads and 12 minutes in FtF dyads. Unfortunately, a content analysis of these conversations was not conducted. Instead, participants provided self-report measure of how much they told their partner during the conversation on six topics, relationships, love, how you feel about your physical appearance, sex, secrets, and dating. Participants in both the text-based and visual-based CMC conditions reported that they disclosed more than participants in the FtF condition and these were large effect sizes (see Table 1).

The rest of the studies (see Table 1) are similar to Schouten et al. (2007a) in that they are self-report and other-report (perceptions of the partner's self-disclosure) studies of self-disclosure. Some are ratings of self-disclosure after conversation has taken place. Others are more general perceptions of self-disclosure. All of these studies found no differences between FtF and online self-disclosure (Parks & Roberts, 1998; Shaw, 2004) or that FtF self-disclosure is greater in amount or intimacy than online self-disclosure (Cho, 2007; Green, 2006; Mallen, Day, & Green, 2006; Schouten, Valkenburg, & Peter, 2007b; Stritzke, Nguyen, & Durkin, 2004; Wang & Andersen, 2007).

The results of the meta-analysis of these studies indicate an average effect size of $d = -.07$ ($k = 11$, $N = 2887$). This effect size is not significantly different from zero [$t(10) = -.378$, $p = .72$]. The results are not homogeneous across studies [$\chi^2(1, N = 11) = 50.41$, $p < .001$]. There is a great variation in the effects sizes ranging from large negative effect sizes, to no differences, to large positive effect sizes. However, we did not test for moderators given the small number of studies ($k = 11$). Eyeballing the data leads to no obvious differences in study design or measure of self-disclosure as the moderating variable. Nonetheless, there is very little evidence for hyperpersonal communication with only two of 11 studies finding online self-disclosure greater than FtF self-disclosure. There is more evidence for SIP (CMC self-disclosure is equal to FtF self-disclosure if given enough time to communicate in CMC) given that the overall effect size is not significantly different from zero. But there is also evidence that online self-disclosure may be less than FtF self-disclosure given the percentage of studies that found FtF greater than online self-disclosure. However, no definite conclusions can be drawn from this meta-analysis given the small number of studies, especially the small number of studies in which actual, rather than self-report, self-disclosure is examined. Future research needs to be conducted comparing FtF versus online self-disclosure using content analysis of self-disclosure in order to determine whether online self-disclosure is less personal, equally personal, or hyperpersonal than FtF self-disclosure.

Table 1. Effect Sizes for Differences in CMC vs. FtF Self-Disclosure

Authors	# of Participants	Method	Measure of Self-Disclosure	Effect Size (d)*
Cho (2007)	260	Survey of online chatting vs. FtF communication	Self-report: Amount and depth of self-disclosure	Ave. $d = -.464$ $-.583$ (amount, males) $-.596$ (amount, females) $-.486$ (depth, males) $-.192$ (depth, females)
Green (2006)	131	2 (CMC vs. FtF) × 2 (anonymous vs. non-anonymous) design: Ss completed demographic questionnaire which was exchanged with partner in non-anonymous condition, task-oriented discussion	Content analysis: # of self-disclosures per minute and depth of self-disclosure	Ave. $d = -.204$ # of self-disclosures per minute $d = -.407$ Intimacy: *ns* and no direction indicated = .0
Joinson (2001)	20 dyads	Online vs. FtF conversations between strangers	Content analysis: # of self-disclosures; proportion of words coded as self-disclosure	Ave. $d = 1.615$ # of self-disclosures $d = 1.33$; Proportion $d = 1.90$
Mallen, Day, & Green (2003; study 3)	32 dyads	Engaged in Internet chat or FtF conversation with stranger	Self-report of self-disclosure; Other-report (perception of partner's self-disclosure)	Ave. $d = -.449$ Self-report $d = -.454$; Other-report $d = -.444$
Parks & Roberts (1998)	235 current users of MOOs	MOO relationships vs. offline relationships	Self-report: Depth of self-disclosure	$-.037$

Continued on next page

Table 1. Effect Sizes for Differences in CMC vs. FtF Self-Disclosure *(continued)*

Authors	# of Participants	Method	Measure of Self-Disclosure	Effect Size (*d*)*
Schouten, et al. (2007a)	81 cross-sex dyads	Participants randomly assigned to one of 3 experimental conditions (text-only CMC, visual CMC, or FtF), get-acquainted exercise, 24 min for CMC dyads; 12 min for FtF dyads	Self-report: Self-disclosure to partner during conversation on six topics	Ave. d = .569 Text only d = .583 Webcam d = .554
Schouten et al. (2007b)	1203 adolescents	Survey of online (IM) vs. FtF same- and cross-sex friend self-disclosure	Self-report: Intimacy of self-disclosure for 7 topics (Personal feelings, the things I am worried about, secrets, being in love, sex, moments in my life I am ashamed of, and moments in my life I feel guilty about)	Ave. d = −.522 Cross-sex interaction d = −.344 Same-sex interaction d = −.699
Shaw (2004)	280 (148 CMC; 132 FtF)	Online vs. FtF conversations with strangers	Self-report of self-disclosure; Other-report (perception of partner's self-disclosure)	−.117
Stritzke et al. (2004)	134	Survey of everyday CMC and FtF interactions	Self-report	−.280
Tidwell & Walther (2002)	158 dyads	Online vs. FtF conversations between strangers	Content analysis: Proportion of utterances coded as peripheral, intermediate, and core self-disclosure	Ave. d = −.432 Peripheral d = −.541 Intermediate d = −.626 Core d = −.128
Wang & Andersen (2007)	353	Survey of online vs. FtF communication with long-distance friend	Self-report: Breadth, depth, and reciprocity aspects of self-disclosure	−.398

Note. +d = CMC self-disclosure greater than FtF self-disclosure

Sex Differences in FtF versus Online Self-Disclosure

Some researchers (e.g., Merkle & Richardson, 2000) have speculated that sex differences in self-disclosure would be less evident in online versus FtF communication. That is because in CMC, the anonymity of the Internet may permit users to step outside of constricting gender roles of communication. For instance, this may allow men to disclose more, making men's and women's self-disclosure equal.

However, anonymity and CMC are not synonymous. Indeed, Green (2006) manipulated technology (CMC/FtF) and anonymity (anonymous/non-anonymous) orthogonally. Green tested whether there was a difference in the amount and intimacy of self-disclosure observed in dyads that communicated FtF versus through CMC, anonymously versus non-anonymously, and whether there is an interaction between the effects of technology and anonymity on self-disclosure. One hundred and thirty-one participants completed a demographic questionnaire. For the non-anonymous conditions, the demographic questionnaire was exchanged with the conversational partner prior to interaction; the demographic questionnaire was not exchanged with the conversational partner prior to interaction in the anonymous conditions. Then partners communicated either FtF or through CMC. Subjects who interacted in the FtF condition met in a room with their conversational partner. Subjects in the CMC condition worked with their partner (who was located in a different room) over an instant messaging system. All subjects were given a social dilemma to solve and were told that they needed to reach agreement on the dilemma. Subjects in the FtF conditions were given 10 minutes to reach consensus; subjects in the CMC conditions were given 20 minutes to reach consensus.

The results of the study were that individuals who communicated FtF had a significantly greater number of self-disclosures than those who communicated via CMC. There was no significant difference for the main effect of anonymity or the interaction effect of technology and anonymity on the amount of self-disclosure. There were no significant differences for intimacy of self-disclosure. However, the results of this study should be interpreted with caution especially with respect to anonymity. The manipulation of anonymity is weak; in the non-anonymous condition subjects viewed demographic information about their partner; in the anonymous condition subjects received no information about their partner. However, the result is that anonymity and technology are not synonymous. A person can interact anonymously FtF and non-anonymously online.

Nonetheless, to determine whether sex differences in self-disclosure are less pronounced online than FtF, a search of the scholarly literature was conducted. Only one study could be located that compared sex differences FtF versus online. Wang and Andersen (2007) conducted a self-report survey of FtF versus online self-disclosure with a long-distance friend and indicated that women disclosed more than men in both FtF and CMC contexts. The effect size was larger for FtF contexts (η^2= .06) than CMC contexts (η^2 = .02) but no test of significance was conducted comparing the effect size for FtF versus CMC contexts.

A number of studies have examined sex differences in self-disclosure exclusively in online environments. We conducted a meta-analysis of sex differences in self-disclosure in CMC settings. Fourteen studies of sex differences in self-disclosure were included in the meta-analysis (one study includes two independent samples) (see Table 2). The results of the meta-analysis indicate that there are no sex differences in online self-disclosure (d = .048, k = 15, N = 5077, *ns*). This effect size was not homogeneous [$\chi^2(1, N = 15) = 36.79, p < .001$]; however, we did not test for moderators given the small number of studies included in the meta-analysis (k = 15).

The mean effect size for sex differences in self-disclosure in these studies can be compared to the results of a meta-analysis conducted of sex-differences in self-disclosure in FtF settings (Dindia & Allen, 1992) in which the mean effect size for 205 studies of sex differences in self-disclosure was d = .18. These results indicate that for FtF self-disclosure, women disclose more than men but the effect size is small (Cohen [1969], classified an effect size of d = .20 as small). The results were not homogenous and an analysis of moderator variables indicated that sex differences in self-disclosure in close relationships (friend, romantic partner, family member, etc.), whether measured through self-report data or observational data, are small, but there were no sex differences in self-disclosure between strangers using self-report or observational measures.

Dindia and Allen (1992) found overall effect size of d =.18 for sex differences in FtF condition; however, they found no sex differences when target was a stranger, which was the case in most studies investigating sex differences in self-disclosure in CMC conditions. Thus, the results for the meta-analysis on sex differences in online self-disclosure are similar to the results for Dindia and Allen's meta-analysis of sex differences in FtF self-disclosure. Two exceptions are Schouten et al. (2007b), who examined online self-disclosure to same- and opposite-sex friends, and Wang and Andersen (2007), who examined online self-disclosure to long-distance friends, both of whom found small effect sizes. Overall, there is no evidence that sex differences in self-disclosure are muted online in comparison to FtF, but that may be because sex differences in FtF self-disclosure are small to begin with.

Culture is a factor that affects self-disclosure. Culture regulates the way people communicate by defining what is appropriate and what is not. Some cross-cultural studies have investigated cultural differences in self-disclosure and found that people from the West tend to disclose more than people from the East. Cultural values influence the breadth and the depth of the information that people disclose in FtF contexts. A direct communication style is expected in individualistic cultures, whereas indirectness and restraint in self-disclosure are desirable in collectivistic cultures (Gudykunst & Nishida, 1986).

Table 2. Effect Sizes for Sex Differences in Online Self-Disclosure

Authors	# of Participants	Type of CMC	Measure of Self-Disclosure	Effect Size (*d*)*
Barak & Gluck-Ofri (2007)	480 (214 male; 266 female)	Online forums	Total self-disclosure, no self-disclosure, little self-disclosure, high self-disclosure	All *ns* (direction not indicated) $d = 0$
Cho (2007)	260 adolescents (132 male; 128 female)	Online chats	Self-report: Amount and depth of self-disclosure (also measured intention, valence, and honesty of self-disclosure but not included in meta-analysis)	Ave. $d = .014$ Amount $d = -.054$ Depth $d = .068$
Dominick (1999)	298 (258 male; 40 female)	Personal homepages at Yahoo	Content analysis: Descriptive biography; job description; resume; family information; friends information; introspective biography; spouse/partner information; # personal topics mentioned; likes and dislikes in music, art, TV, movies, books, travel, hobbies, entertainers, other areas; sports; personal views/opinions about politics, religion, social issues, others; philosophy	Ave. $d = .155$ Respectively: .235; .117; .205; .730; .117; .886; .324; .322; (*ns*); −.383; (*ns*); 1.022
Huffaker & Calvert (2005)	70 teenagers: (35 boys; 35 girls)	Blogs	Content analysis: First name, full name, age, birth date, contact information, email address, Instant messenger user name, location, homepage URL, relationships, homosexuality, self-references	Ave. $d = -.009$ Everything except self-references $d = -.031$ Self-references $d = .239$
Kane (2008)	293 (147 male; 146 female)	Social networking site: MySpace profiles	Content analysis: Amount of text written in "About me" and "Interests"; TV; heroes; general; music; movies; books	Ave. $d = .117$ .257;.323;.242; *ns*; *ns*; *ns*; *ns*

Continued on next page

Table 2. Effect Sizes for Sex Differences in Online Self-Disclosure *(continued)*

Kim (2007)	189 (81 male; 108 female)	Personal web pages and social networking sites	Self-report: Demographic self-disclosure; Emotional self-disclosure; Visual self-disclosure (e.g., pictures of self); Evaluative self-disclosure (opinions)	Ave. $d = -.027$ Demographics $d = -.70$; Emotions $d = .239$; Photos $d = .361$; Opinions $d = -.006$
Kim & Dindia (2008)	100 Americans (50 male; 50 female)	Social networking site: MySpace	Content analysis: 38 self-disclosure items including closed- and open-ended items (e.g., interests, about me, photos, blog entries, etc.)	Ave. $d = .072$ Closed-ended $d = -.05$ Open-ended $d = .143$
	100 Koreans (50 male; 50 female)	Social networking site: Cyworld	Content analysis: 13 self-disclosure items including closed- and open-ended items (e.g., interests, about me, photos, blog entries, etc.)	Ave. $d = -.398$ Closed-ended $d = -.19$ Open-ended $d = -.641$
Ma & Leung (2006)	591 (201 male; 390 female)	ICQ	Self-report: Depth, honesty/accuracy, valence, amount of self-disclosure	All *ns* (no direction reported) $d = 0$
Schouten et al. (2007b)	1203 Dutch adolescents (51% boys; 49% girls)	IM Online same and cross-sex friends	Self-report: Intimacy of self-disclosure for 7 topics (Personal feelings, the things I am worried about, secrets, being in love, sex, moments in my life I am ashamed of, and moments in my life I feel guilty about)	$d = .387$ (converted r to d) Cross-sex: 35% of boys vs. 28% of girls disclosed more online; same-sex: 21% of boys vs. 23% of girls disclosed more online

Continued on next page

Table 2. Effect Sizes for Sex Differences in Online Self-Disclosure *(continued)*

Stern (2004)	233 adolescents (163 male; 70 female)	Personal homepages	Content analysis: Self-expression/description: descriptive biography, photos of author, original poetry, diary or journal, original essays, favorite quotes, plans for future, appearance; Intimate topics: sex, religion/God, drugs/alcohol, self-destructive behaviors, depression, loneliness; Relationship variables: discussion, pictures, or links to family, friends, and romantic partners; Interests: discussed or linked to hobbies, various media, school, music or music sites, retail product/service sites, video games, sports, news, social causes	Overall ave. d = .076 Ave. d for intimate topics = .177 Ave. d for relationship variables = .125 Ave. d for interests = −.074
Trammell, Tarkowski, Hofmokl, & Sapp (2006)	295 (76 male; 219 female)	Polish blogs	Content analysis: Record of day; memory; feelings/thoughts; hobby/interest; family/friends; current projects/work; intimate details about life; political opinions	Ave d = .063 .299;.256; .285;−.561; .136;.197; −.06; −.048
Tufekci (2008)	512 (243 male; 269 female)	Social networking sites: MySpace and Facebook	Content analysis: Real name, favorite music, book, movie, political view, romantic status, sexual orientation, relationship, phone #, classes, and address	.024
Wang & Andersen (2007)	353 (106 male; 247 female)	Online communication with a long-distance friend	Self-report: Breadth, depth, and reciprocity aspects of self-disclosure	.267
Williams & Merton (2008)	100 adolescents (50 male; 50 female)	Social networking site	Content analysis: Demographic content, images, family issues, school issues, social issues, risk behaviors, sexual content, identity vulnerability	−.031

Note: +d = Women self-disclosure more than men

This principle may be replicated in CMC settings. Since there are not enough studies to conduct a meta-analysis (five studies), we provide a brief summary of cross-cultural CMC studies on self-disclosure (see Table 3). Kim and Papacharissi (2003) analyzed and compared the contents of 98 Korean and U.S. personal web pages from Yahoo! Korea and Yahoo! U.S. ccoders recorded the presence or absence of demographic information such as name, gender, age, occupation, and residence. The results show that Americans are more likely to provide information about their origin, present residence, and self-ascribed identity than Koreans. No significant difference was found in revealing name, gender, role-status, occupation, family information, religion, and physical description between Korean and American personal web pages. Kim and Papacharissi also indicated that Americans reveal themselves more directly than Koreans. For example, Americans presented themselves with still pictures, while Koreans were more likely to use manipulated graphics.

The finding in Kim and Papacharissi (2003) was replicated in a study of social networking sites by Kim and Dindia (2008). Kim and Dindia conducted a content analysis of Americans' self-disclosure in MySpace and Koreans' self-disclosure in Cyworld, the most popular social networking site in Korea. Some self-disclosure categories were suggested only by one of the two sites' general layout and therefore not suited for direct comparison. For example, Diary was present in Cyworld, but not in MySpace; General Interests, Music, Movies, TV, Books, and Heroes were present in MySpace, but not in Cyworld. However, the main effect of culture was examined for five open-ended self-disclosure items present both in MySpace and Cyworld (Headline, About Me, the number of self-references in About Me, number of photos, and number of blog entries). The results indicate that Americans in MySpace self-disclosed more in About Me and used more self-references in About Me than Koreans in Cyworld. Koreans in Cyworld had more photos and blog entries than Americans in MySpace. No difference was found in the number of words in Headline.

Yum and Hara's (2005) studied cultural differences in online self-disclosure. One hundred and twenty-six Japanese, 112 American, and 123 Korean college students completed a survey that consisted of the self-disclosure scales developed by Parks and Floyd (1996). The participants were asked to indicate the breadth and depth that they would disclose in CMC. In this study, Americans reported greater breadth and depth of self-disclosure in CMC than Japanese and Koreans ($d = 1.392$). These results are consistent with the results of extant cross-cultural researchin FtF contexts (Barnlund, 1975; Chen, 1995; Hofstede, 1998; Wheeless, Erickson, & Behrens, 1986) and the results of other cross-cultural CMC research (Kim & Papacharissi, 2003; Kim & Dindia, 2008).

One exception to this pattern is suggested by Choi, Kim, Sung, and Sohn (2008). In their study, Koreans showed a greater depth of self-disclosure than Americans ($d = -.883$). This stands in contrast to Yum and Hara (2005). Both studies employed college students and used the same self-disclosure measure by Parks and Floyd (1996). The different results might derive from differences in targets of self-disclosure. In Yum and Hara's study, participants were asked to think about their self-disclosure to a particular online partner regardless of type of CMC while Choi et al. asked participants to indicate how intimate their self-disclosure was in their social networking sites in general. It might be that any cultural difference in self-disclosure is moderated by the target of self-disclosure.

The four studies summarized above compared Americans and East Asians in their self-disclosure in CMC. One study compared Americans and Germans in their self-disclosure in MySpace profile pages (Banczyk, Kramer, & Senokozlieva, 2008). They conducted a content analysis of closed-ended self-disclosure items in MySpace profile pages and found that Americans self-disclose more than Germans (d = .344).

Although the amount of cross-cultural research on self-disclosure in CMC is small, the research findings provide consistent results that web authors reveal themselves in CMC consistent with their own cultural norms with the exception of Choi et al. (2008). This tendency seems to spill over into online environments. Americans disclose more than East Asians in FtF settings when the traditional definition of self-disclosure (verbal description of self) is used. However, online self-disclosure includes photos of self as well as verbal description of self. This review of research in CMC shows that Americans verbally reveal themselves more than East Asians in CMC while Koreans present more photos of themselves than Americans. This may indicate direct versus indirect expressions of cultural norms/behaviors. Future research needs to examine cultural differences in photographic self-disclosure. Also, cultural differences in self-disclosure need to be examined in various CMC environments to provide more generalizable knowledge about cultural differences on online self-disclosure.

Table 3. Effect Sizes for Cultural Differences in Online Self-Disclosure Americans vs. Non-Americans

Authors	# of Participants	Type of CMC	Measure of Self-Disclosure	Effect Size (*d*)*
Banczyk et al. (2008)	107 (53 Americans; 54 Germans)	Social networking site: MySpace	Content analysis: Relationship status, here for, sexual orientation, hometown, body type, ethnicity, religion, smoke/drink, children, education, occupation, income, schools, and companies	.344
Choi et al. (2008)	589 (349 Americans; 240 Koreans)	Social networking site: Not specified	Self-report: Depth and Breadth Measures by Parks & Floyd (1996). 5 depth items were used to compute the effect size	−.883
Kim & Dindia (2008)	176 (92 Americans; 84 Koreans)	Social networking sites: MySpace and Cyworld	Content analysis: Self-disclosure units in About me	.984

Authors	# of Participants	Type of CMC	Measure of Self-Disclosure	Effect Size (*d*)*
Kim & Papacharissi (2003)	98 (49 Americans; 49 Koreans)	Personal homepages	Content analysis: 10 categorical items: Name, gender, role-status, occupation, family information, self-ascribed identity, origin, religion, resident information, and physical description	.217
Yum & Hara (2005)	361(112 Americans; 123 Koreans; 126 Japanese)	Online relationship: Not specified	Self-report: Depth and Breadth Measures by Parks & Floyd (1996). Only depth measures were included to compute the effect size	Koreans $d = 1.392$ Japanese $d = .881$

Note. $+d$ = American self-disclosure higher than non-American self-disclosure

Other Predictors of Online Self-Disclosure

Besides the aforementioned variables (CMC versus FtF, sex, and culture), we were able to identify some other variables that may affect online self-disclosure (age, motivations, and website features); however, the number of studies that examine those variables relative to online self-disclosure is small and insufficient to conduct a meta-analysis. Thus, we provide a narrative review of the research for each variable.

Age. A direct test of the effect of age on self-disclosure is rare. CMC researchers tend to adopt a certain age group (e.g., adolescent or adult) and/or use age as a covariate to test the effect of variables of interest. However, age itself can be a predictor of online self-disclosure. The demographics of social networking clearly indicate that creation of profile pages on social networking sites is evolving faster among young people than older people. Internet usage overall is much higher among young adults (92% of those 18–29 years old) than older adults (37% of those 67 years or older); 55% of teens are using social networking sites compared to 16% of all U.S. adults (Pew Internet and American Life Project, 2008). And, instant messaging has become as popular as, if not more popular, than email for many people. Nearly 70% of teens and 24% of adults send more instant messages than email messages (AP-AOL Instant-Messaging Trends Survey, 2007). These patterns suggest that young people are more familiar with and comfortable with as well as more adept at using CMC for relationship building. The implication is that teens may be more inclined towards openness and self-disclosure in these media than their older peers.

Extant studies examining sex differences in online self-disclosure show that age can also be a predictor of online self-disclosure. Previous research has not found sex differences in adolescents' self-disclosure in online chatting rooms (Cho, 2007) and in blogs (Huffaker & Calvert, 2005) while other studies of adult netizens have shown sex differences in online self-disclosure (Dominick, 1999; Trammell et al., 2006). These results indicate that the gap between sexes is smaller in younger generations. Recently, Kim, Klautke, and Serota (2009) conducted a content-analysis of MySpace profile pages and found that younger users tend to disclose more than older users in open-ended self-descriptions (e.g., About me).

Motivations for using CMC *media.* Motivation is a key factor in media use. Individuals use media to fulfill their needs. Researchers have examined psychological and behavioral aspects of Internet users to identify a set of common underlying dimensions for Internet usage motivations (Lin, 1999). Papacharissi and Rubin (2000) examined Internet users' motivations and identified five factors: interpersonal utility, pastime, information seeking, convenience, and entertainment. Motivations may explain how different individuals use CMC channels with different results. Because the web provides an ideal setting for selective self-disclosure, individuals may strategically control self-disclosure based on individual goals. What information individuals reveal about themselves on their personal web pages, social networking websites, dating websites, etc., reflects their needs.

Some researchers have looked at motivations for using CMC media as a predictor of self-disclosure in CMC (Cho, 2007; Papacharissi, 2002; Kim, 2007). Conducting a survey and a content analysis, Papacharissi (2002) found that motives for using personal homepages affect the look of personal homepages. Papacharissi identifies six personal homepage motives: passing time, entertainment, information, self-expression, professional advancement, and communication with friends and family. Among these motives, the motives of professional advancement and self-expression are positively related to disclosure of personal information.

Kim's (2007) study provides further evidence of the relationship between motivation and online self-disclosure. Conducting an online survey of Korean personal web page and social networking site users, Kim found that motivations for using these CMC media influence the topics that the authors disclose in their web pages. In this study, those who were motivated to communicate with family and friends posted more demographic information to the web pages than those who did not have this motivation. People who used personal web pages or social networking sites to escape from real life engaged in more emotional self-disclosure than those who did not have this motivation. People who used these media for entertainment, communication, and self expression purposes posted more photos of themselves and others than those who did not have these motivations. Similarly, people who used personal web pages or social networking sites for self-expression, communication, and information purposes presented more opinions than those who did not have these motivations.

The effect of motivation on online self-disclosure has also been examined in an online chatting environment. Cho (2007) conducted a survey to Korean high school students about their online chatting experience with strangers. Three types of online chatting motives were identified: entertainment (to have fun, relax, kill time), information (to learn something, get information about one's environment, find economic, political, and cultural information), and initiation and development of interpersonal relationships (to make new friends, make a friend of the opposite sex, communicate with members in an online community). The results show that honesty-accuracy of self-disclosure did not vary across the three motives. The motivation to initiate and develop interpersonal relationships was associated with more amount of self-disclosure than the other two motives; however, ironically, the relationship motivation did not lead to more intimacy of self-disclosure. The information motivation was associated with greater intimacy and intent of self-disclosure, and more negative-positive self-disclosure the other motives.

Previous research has identified a number of motivational influences on self-disclosure. Among these motivations, interpersonal relationship motivation was considered as one single variable; however, each individual has a different level of interpersonal needs depending on which relational phase the person is in. Thus, examining different types of relational motivation is essential to understand the effects on online self-disclosure. In fact, relational motivations can range from initiating relationships (among strangers) to maintaining relationships (among friends, family, and long-distance partners), and this certainly will impact online self-disclosure.

Recently, some researchers have examined different kinds of relational motivations for authoring a social networking site. Kim, Klautke, and Serota (2009) content-analyzed MySpace users' profile pages and categorized them into two groups, those with and those without a romantic motive (dating or/and serious relationships versus friendship or/and networking only). They found that MySpace users with romantic relational motives provide more self-descriptions and relational interests on their profile pages and are more likely to disclose information on children and income, but less likely to disclose their sexual orientation, than those without romantic relational motives.

Park, Jin, and Jin (2009) examined two different relational motivations for using social networking sites: Relationship development and relationship maintenance. They conducted an online survey of Facebook users and found that the motivation for relationship development facilitated more self-disclosure than the motivation for relationship maintenance did. The motivation for relational development was negatively correlated with honesty of self-disclosure, whereas the motivation for relationship maintenance was not associated with it.

Website features. Some website features or structures may affect online self-disclosure. Some social networking sites solicit more self-disclosure than other social networking sites and this will affect the amount of self-disclosure. For example, MySpace has more close-ended self-disclosure items than Facebook (e.g., body type, ethnicity, income, etc.); thus, people are likely to disclose more personal information on MySpace than on Facebook. Also, convenience for posting and sharing photos will affect online self-disclosure. An individual may post more photos on Facebook than on MySpace because Facebook has more convenient function for posting and sharing photos (e.g., upload photos by e-mail or mobile phones, tagging, etc.) on than MySpace.

Online Self-Disclosure and Relationship Development

Online self-disclosure is a means to develop relationships. However, because there is less warranting information in CMC than in FtF settings, there is no guarantee that information about one's self is always true. Warranting information refers to information about one's self that the communicator cannot manipulate. An example of warranting information is postings by others on the wall in your Facebook page. You do not have control over the content of postings that are posted by others. People rely more heavily on warranting information than on self-descriptions because self-descriptions can be deceptive.

In a survey of online dating users, over 80% of participants registered concern that others misrepresent themselves (Gibbs, Ellison, & Heino, 2006). Several empirical studies suggest that, at least in contexts where future interaction is expected, such as online dating, deception tends to be subtle rather than extreme, and self-enhancing rather than overtly malicious (Ellison, Gibbs, & Heino, 2006). For instance, male online daters typically add a couple of inches to their height, whereas female daters subtract a few pounds from their weight to appear more attractive to the opposite sex (Toma, Hancock, & Ellison, 2008). This may not be different from deception that takes place in FtF settings. The difference is that there is more warranting in FtF settings. We can see the other person's body so they can only lie so much about age, weight, height, etc. In CMC settings, you can post a picture of yourself when you were thinner and younger or a picture that is not even a picture of yourself. Furthermore, people can even lie about their education, profession, or economic status both FtF and online.

This phenomenon leads people to develop a system by which they can detect deceptive messages including self-disclosure in online environments. This system is probably not that much different from detecting deception FtF. Research shows that perceptions of individuals who created online profiles in social networking systems are influenced by the comments and attributes of postings that others leave on those profiles, which is warranting information. According to Walther, Van Der Heide, Kim, Westerman, and Tong (2008), the content of friends' postings on profile owners' walls in Facebook affects perceptions of profile owners' credibility and attractiveness. The physical appearance of one's friends, as shown in those wall postings, affects the perceived physical appearance of the profile owner as well. Other research shows that when there is a discrepancy between a Facebook profile owner's self-generated information and others' comments, other people's comments override the profile owner's claims (Walther, Van Der Heide, Hamel, & Shulman, 2009).

Several researchers have talked about the potentially negative effect of hyperpersonal communication on relationship development. Cooper and Sportolari (1997) refer to this as the "boom and bust" phenomenon of online self-disclosure:

> When people reveal more about themselves earlier than they would in FtF interactions, relationships get intense very quickly. Such an accelerated process of revelation

> may increase the chance that the relationship will feel exhilarating at first, and become quickly eroticized, but then not be able to be sustained because the underlying trust and true knowledge of the other are not there to support it. (p. 12)

Cooper and Sportolari (1997) highlight media accounts of people who are certain they have found their "soul mate" and leave an established relationship, travelling across the country, to meet people who don't turn out to be who they seemed.

Others have noted that Internet romantic relationships progress through an inverted developmental sequence (Merkle & Richardson, 2000). In real life we meet people, then get to know them; in online we get to know someone and then choose to meet them (Rheingold, 1993). Some say this makes for an unstable relationship (Levine, 2000). However, the opposite is plausible. CMC may be characterized by a higher degree of personal investment of time and self-disclosure than is typical in FtF relationships. This greater investment may result in a stronger relationship (Merkle & Richardson, 2000).

Again, there is no empirical evidence to support the boom and bust phenomenon. Nonetheless, some have cautioned that when developing relationships online, you should move from virtual to FtF in a short period of time before unrealistic expectations have time to build up (Levine, 2000).

Conclusion

In this chapter, we provided a review of online self-disclosure research examining the effects of key variables in online self-disclosure across various CMC channels. Some CMC environments that are gaining more popularity, such as audio and video-conferencing, are not asynchronous. Similarly, audio environments include vocal cues and video-environments (video-conferencing such as Skype) include a number of nonverbal cues such as visual appearance, clothing and artifacts, etc. The only difference between video-conferencing and FtF communication would be the lack of touch and smell. These CMC environments are becoming more and more popular. Technology and the uses of technology (for instance, the relatively recent use of web cameras) change so quickly that any generalizations regarding self-disclosure on the Internet are problematic. Thus, this review of research on self-disclosure in CMC will likely be out-of-date before it is published.

References

AP-AOL Instant-Messaging Trends Survey (2007). *AP-AOL instant messaging trends survey reveals popularity of mobile instant messaging.* Retrieved July 20, 2009, from AOL: corporate web site http://corp.aol.com/press-releases/2007/11/ap-aol-instant-messaging-trends-survey-reveals-popularity-mobile-instant-mess

Banczyk, B., Krämer, N. C., & Senokozlieva, M. N. (2008, May). *"The wurst" meets "fatless" in MySpace. The relationship between self-esteem, personality and self-presentation in an online-community.* Paper presented at the annual conference of the International Communication Association, Montreal, Canada.

Barak, A., & Gluck-Ofri, O. (2007). Degree and reciprocity of self-disclosure in online forums. *CyberPsychology & Behavior, 10,* 407–417.

Barnlund, D. C. (1975). *Public and private self in Japan and the United States.* Tokyo: Simul.

Blau, P.M. (1964). *Exchange and power in social life.* New York: John Wiley.

Chen, G. M. (1995). Differences in self-disclosure patterns among Americans versus Chinese. *Journal of Cross-Cultural Psychology, 26,* 84–91.

Cho, S. H. (2007). Effects of motivations and gender on adolescents' self-disclosure in online chatting. *CyberPsychology & Behavior, 10,* 339–345.

Choi, W. M., Kim, Y., Sung, Y., & Sohn, D. (2008, May). *Motivations and social relationships: A comparative study of social network sites in the U.S. and Korea.* Paper presented at the annual conference of the International Communication Association, Montreal, Canada.

Cohen, J. (1969). *Statistical power analysis for the behavioral sciences.* San Diego, CA: Academic.

Collins, N. L., & Miller, L. C. (1994). Self-disclosure and liking: A meta-analytic review. *Psychological Bulletin, 116,* 457–475.

Cooper, A., & Sportolari, L. (1997). Romance in cyberspace: Understanding online attraction. *Journal of Sex Education and Therapy, 22,* 7–14.

Cozby, P. C. (1973). Self-disclosure: A literature review. *Psychological Bulletin, 79,* 73–91.

Culnan, M. J., & Markus, M. L. (1987). Information technologies. In F. Jablin, L. L. Putnam, K. Roberts, & L. Porter (Eds.), *Handbook of organizational communication* (pp. 420–443). Newbury Park, CA: Sage.

Derlega, V. J., Metts, S., Petronio, S., & Margulis, S. T. (1993). *Self-disclosure.* Newbury Park, CA: Sage.

Dindia, K. (2000). Self-disclosure, identity, and relationship development: A dialectical perspective. In K. Dindia & S. W. Duck (Eds.), *Communication and personal relationships* (pp. 147–162). Chichester, UK: Wiley.

Dindia, K. (2002). Self-disclosure research: Knowledge through meta-analysis. In M, Allen, R. W., Preiss, B. M., Gayle, & N. Burrell (Eds.), *Interpersonal communication: Advances through meta-analysis* (pp. 169–186). Mahwah, NJ: Lawrence Erlbaum.

Dindia, K., & Allen, M. (1992). Sex differences in self-disclosure: A meta-analysis. *Psychological Bulletin, 112,* 106–124.

Dominick, J. R. (1999). Who do you think you are? Personal home pages and self-representation on the World Wide Web. *Journalism and Mass Communication Quarterly, 76,* 646–658.

Ellison, N., Heino, R., & Gibbs, J. (2006). Managing impressions online: Self-presentation processes in the online dating environment. *Journal of Computer-Mediated Communication, 11*(2), article 2. Retrieved July 20, 2009, from http://jcmc.indiana.edu/vol11/issue2/ellison.html

Gibbs, J. L., Ellison, N. B., & Heino, R. D. (2006). Self-presentation in online personals: The role of anticipated future interaction, self-disclosure, and perceived success in Internet dating. *Communication Research, 33,* 1–26.

Gilbert, S. J. (1972).*A study of the effects of self-disclosure on interpersonal attraction and trust as a function of situational appropriateness and the self-esteem of the recipient.* Unpublished doctoral dissertation. University of Kansas, Lawrence, KS.

Gilbert, S. J., & Horenstein, D., (1975). The communication of self-disclosure: Level versus valence. *Human Communication Research, 1,* 316–322.

Goffman, E. (1959). *Presentation of self in everyday life.* New York: Doubleday.

Green, S. C. (2006). *Computer-mediated communication in society: A study on self-disclosure.* Unpublished doctoral dissertation, Indiana State University, Terre Haute, IN.

Gudykunst, W. B., & Nishida, T. (1986). Attributional confidence in low- and high- context cultures. *Human Communication Research, 12*, 525–549.

Hofstede, G. (1998). I, we, and they. In J. N. Martin, T. K. Nakayama, & L. A. Flores (Eds.), *Readings in cultural contexts* (pp. 345–356). Mountain View, CA: Mayfield.

Huffaker, D. A., & Calvert, S. L. (2005). Gender, identity, and language use in teenage blogs. *Journal of Computer-Mediated Communication, 10*(2), article 1. Retrieved July 20, 2009, from http://jcmc.indiana.edu/vol10/issue2/huffaker.html

Joinson, A. N. (2001). Self-disclosure in computer-mediated communication: The role of self-awareness and visual anonymity, *European Journal of Social Psychology, 31*, 177–192.

Jourard, S. M. (1971). *Self-disclosure: An experimental analysis of the transparent self.* New York: Wiley Interscience.

Jourard, S. M., & Jaffe, P .E. (1970). Influence of an interviewer's disclosure on the self-disclosing behavior of interviewees. *Journal of Counseling Psychology, 17*, 252–257.

Kane, C. M. (2008). *I'll see you on MySpace: Self-presentation in a social network website.* Unpublished master's thesis. Cleveland State University, Cleveland, OH.

Kim, J. (2007, November). *Gender, motivation and self-disclosure online: A study of Korean social networking websites.* Paper presented at the annual conference of the National Communication Association, Chicago, IL.

Kim, J., & Dindia, K. (2008, May). *Gender, culture, and self-disclosure in cyberspace: A study of Korean and American social network websites.* Paper presented at the annual conference of the International Communication Association, Montreal, Canada.

Kim, J., Klautke, H., & Serota, K. (2009, May). *Effects of relational motivation and age on online self-disclosure: A content analysis of MySpace profile pages.* Paper presented at the annual conference of the International Communication Association, Chicago, IL.

Kim, H., & Papacharissi, Z. (2003). Cross-cultural differences in online self-presentation: A content analysis of personal Korean and US homepages. *Asian Journal of Communication, 13*, 100–119.

Levine, D. (2000). Virtual attraction: What rocks your boat? *CyberPsychology and Behavior, 3* (4), 565–574.

Lin, C. A. (1999). Uses and gratifications. In S. G. Singletary & V. P. Richmond (Eds.), *Clarifying communication theories: A hands-on approach* (pp. 199–208). Ames: Iowa State University Press.

Ma, M. L., & Leung, L. (2006). Unwillingness-to-communicate, perceptions of the Internet and self-disclosure in ICQ. *Telematics and Informatics, 23*, 22–37.

Mallen, M. J., Day, S. X., & Green, M. A. (2003). Online versus face-to-face conversation: An examination of relational and discourse variables. *Psychotherapy: Theory, Research, Practice, Training, 40* (1–2), 155–163.

McKenna, K. Y. A., & Bargh, J. A. (2000). Plan 9 from cyberspace: The implications of the Internet for personality and social psychology. *Personality and Social Psychology Review, 4*, 57–75.

Merkle, E. R., & Richardson, R. A. (2000). Digital dating and virtual relating: Conceptualizing computer-mediated romantic relationships. *Family Relations, 49*, 187–192.

Papacharissi, Z. (2002). The self online: The utility of personal home pages, *Journal of Broadcasting & Electronic Media, 46*, 346–368.

Papacharissi, Z., & Rubin, A. M. (2000). Predictors of Internet use. *Journal of Broadcasting and Electronic Media, 44*, 175–196.

Park, N., Jin, B., & Jin, S. A. (2009, May). *Motivations, impression management, and self-disclosure in social network sites.* Paper presented at the annual conference of the International Communication Association, Chicago, IL.

Parks, M. R., & Floyd, K. (1996). Making friends in cyberspace. *Journal of Communication, 46,* 80–97.

Parks, M. R., & Roberts, L. D. (1998). Making MOOsic: The development of personal relationships online and a comparison to their offline counterparts. *Journal of Social and Personal Relationships, 15,* 517–537.

Pearce, W. B., & Sharp, S. M. (1973). Self-disclosing communication. *Journal of Communication, 23,* 409–425.

Pew Internet & American Life Project. (2008). Latest trends. Retrieved July 20, 2009, from http://www.pewinternet.org/trends.asp

Rheingold, H. (1993). *The virtual community: Homesteading on the electronic frontier.* Reading, MA: Addison-Wesley.

Schlenker, B. R. (1986). Self-identification: Toward an integration of the private and public self. In R. F. Baumeister (Ed.), *Public self and private self* (pp. 21–62). New York: Springer-Verlag.

Schouten, A. P., Valkenburg, P. M., & Peter, J. (2007a, May). *An experimental test of processes underlying self-disclosure in computer-mediated communication.* Paper presented at the annual meeting of the International Communication Association, San Francisco, CA.

Schouten, A. P., Valkenburg, P. M., & Peter, J. (2007b). Precursors and underlying processes of adolescents' online self-disclosure: Developing and testing an "Internet-attribute-perception" model. *Media Psychology, 10,* 292–315.

Shaw, L. H. (2004). *Liking and self-disclosure in computer-mediated and face-to-face interactions.* Unpublished master's thesis, University of California, Berkeley, CA.

Stern, S. R. (2004). Expressions of identity online: Prominent features and gender differences in adolescents' world wide web home page. *Journal of Broadcasting & Electronic Media, 48,* 218–243.

Stritzke, W., G. K., Nguyen, A., & Durkin, K. (2004). Shyness and computer-mediated communication: A self-presentational theory perspective. *Media Psychology, 6,* 1–22.

Tidwell, L. C., & Walther, J. B. (2002). Computer-mediated communication effects on disclosure, impressions, and interpersonal evaluations: Getting to know one another a bit at a time. *Human Communication Research, 28,* 317–348.

Toma, C., Hancock, J., & Ellison, N. (2008). Separating fact from fiction: An examination of deceptive self-presentation in online dating profiles. *Personality and Social Psychology Bulletin, 34,* 1023–1036.

Trammell, K. D., Tarkowski, A., Hofmokl, J., & Sapp, A. M. (2006). Rzeczpospolita blogów [Republic of Blog]: Examining Polish bloggers through content analysis. *Journal of Computer-Mediated Communication, 11*(3), article 2. Retrieved July 20, 2009, from http://jcmc.indiana. edu/vol11/issue3/trammell.html

Tufekci, Z. (2008). Can you see me now? Audience and disclosure regulation in online social network sites. *Bulletin of Science, Technology & Society, 28,* 20–36.

Wallace, P. (1999). *The Psychology of the Internet.* Cambridge, UK: Cambridge University Press.

Walther, J. B. (1992). Interpersonal effects in computer-mediated interaction: A relational perspective. *Communication Research, 19,* 52–90.

Walther, J. B. (1996). Computer-mediated communication: Impersonal, interpersonal and hyperpersonal interaction. *Communication Research, 23,* 3–43.

Walther, J. B. (2007). Selective self-presentation in computer-mediated communication: Hyperpersonal dimensions of technology, language, and cognition. *Computers in Human Behavior, 23,* 2538–2557.

Walther, J. B., & Parks, M. R. (2002). Cues filtered out, cues filtered in: Computer-mediated communication and relationships. In M. L. Knapp & J. A. Daly (Eds.), *Handbook of Interpersonal Communication* (3rd ed., pp. 529–563). Thousand Oaks, CA: Sage.

Walther, J. B., Van Der Heide, B., Hamel, L., & Shulman, H. (2009). Self-generated versus other-generated statements and impressions in computer-mediated communication: A test of warranting theory using Facebook. *Communication Research, 36*, 229–253.

Walther, J. B., Van Der Heide, B., Kim, S. Y., Westerman, D., & Tong, S. T. (2008). The role of friends' behavior on evaluations of individuals' Facebook profiles: Are we known by the company we keep? *Human Communication Research, 34*, 28–49.

Wang, H., & Andersen, P. A. (2007, May). *Computer-mediated communication in relationship maintenance: An examination of self-disclosure in long-distance friendships.* Paper presented at the annual meeting of the International Communication Association, San Francisco, CA.

Wheeless, L. R. (1976). Self-disclosure and interpersonal solidarity: Measurement, validation, and relationships. *Human Communication Research, 3*, 47–61.

Wheeless, L. R., Erickson, K. V., & Behrens, J. S. (1986). Cultural difference in disclosiveness as a function of locus of control. *Communication Monographs, 23*, 36–46.

Williams, A. L., & Merten, M. J. (2008). A review of online social networking profiles by adolescents: Implications for future research and intervention. *Adolescence, 43*(170), 253–274.

Worthy, M., Gary, A..L., & Kahn, G. M. (1969). Self-disclosure and exchange process. *Journal of Personality and Social Psychology. 13*, 59–63.

Yum, Y.-O., & Hara, K. (2005). Computer-mediated relationship development: A cross-cultural comparison. *Journal of Computer-Mediated Communication, 11*(1), article 7. Retrieved July 20, 2009, from http://jcmc.indiana.edu/vol11/issue1/yum.html

CHAPTER TEN

Multicommunicating and Episodic Presence: Developing New Constructs for Studying New Phenomena

Jeanine Warisse Turner

N. Lamar Reinsch, Jr.

> *"I can't go into a meeting unless I have this [Blackberry]. I am able to double and triple book myself because people know I can be reached by texting me. People just have to understand that that is my situation. If I couldn't do that in their meeting, I couldn't attend their meeting."* (Susan)
>
> *"A friend called me at work to tell me about a personal problem that he had. During the call, a co-worker IM'ed me on my computer to discuss lunch plans for the day. I was able to correspond on IM to finalize plans as well as chat about sports while at the same time listening to my friend. IM allowed me to feel like I was in my office rather than completely, emotionally immersed in the phone call."* (Steve)
>
> *"I often hold multiple conversations at once especially when using technology with slow response times like IM or text messaging. You can use the waiting time to answer the other parties you are conversing with–you can have multiple IM conversations while text messaging."* (Janet)

In what sense does Susan "attend" two or three meetings at the same time? What did Steve mean by "feel[ing] like I was in my office rather than ... in the phone call"? Do current communication theories help us to understand how "slow response times" are related to Janet's decision to engage in multiple conversations? Is there a name for what Susan, Steve, and Janet are describing?

These three excerpts taken from interviews with business executives discuss an emerging behavior (Cameron & Webster, 2005) that researchers have named multicommunicating (Reinsch, Turner, & Tinsley, 2008). Multicommunicating can involve any combination of media and takes place when an individual participates in two distinct conversations at once. It is defined as the practice of participating in two or more "speech events" (Hymes, 1972), using nearly synchronous media such as face-to-face, telephone, video conferencing, chat, and email (Reinsch et al., 2008). Examples can include talking on the phone while answering an email, instant messaging with multiple individuals, or text messaging during a business meeting. The above interviews reveal many of the tensions associated with this emerging practice including managing the goals of multiple conversations and audiences, pushing the limits of efficiency in the allocation of processing time, and regulating emotions associated with different conversations. The increase in virtual communication and online collaboration has created an intense pressure to be constantly available. In fact, the worldwide market for "presence-based" telecommunications services (which include instant messaging and push-to-talk) was expected to exceed $16 billion in 2009 as consumers continued to look for opportunities to enhance their availability (MarketWatch, 2008). This chapter will discuss this emerging phenomenon and review the available research. This chapter's authors have been working on the development of this construct for several years. We will review our research and trace our understanding of this behavior as it has emerged over time.

What Is a Construct?

The goal of science is theory. According to Fred Kerlinger (1986) a theory consists of "a set of interrelated constructs (concepts), definitions, and propositions that present a systematic view of phenomena by specifying relationships among variables, with the purpose of explaining and predicting the phenomena" (p. 9). Thus the goal of scientific study is to develop explanations that allow us to understand and, if possible, to control what we are studying.

In this chapter we're going to emphasize the parts of theory that Kerlinger describes as constructs and definitions. Research on multicommunicating and presence began when scholars observed people like Susan, Steve, and Janet engaging in some new or perhaps simply previously unnoticed behaviors. Those behaviors seemed to require not only new theories but something even more basic, new constructs. The research that we review has therefore focused for the most part on developing the construct multicommunicating and the construct episodic presence, and defining them clearly.

Before beginning to review the research, however, we should offer a brief comment on Kerlinger's perhaps puzzling expression "constructs (concepts)." Both the word construct and the word concept can be used to refer to the building blocks of theory (Wallace, 1983, pp. 45–69), that is, notions that are described with nouns and can be linked together with propositions. For example, the construct/concept "feedback" and the construct/concept "emotional intelligence" may be included in a theory of interpersonal communication perhaps, for example, in the specific claim that persons who receive both favorable and unfavorable feedback while growing up are likely, when mature, to demonstrate high levels of emotional intelligence.

Kerlinger's view–which we will adopt–is that both constructs and concepts appear in theories. However, Kerlinger uses the word concept to describe "an abstraction formed by generalization from particulars" and the word construct to describe a concept that has been "deliberately and consciously invented or adopted for a special scientific purpose" (Kerlinger, 1986, pp. 26–27). Thus, "construct" is a sub category of "concept," consisting of those concepts that have been created or adapted for scientific use and we will refer to multicommunicating and episodic presence as constructs. However, not everyone distinguishes between construct and concept in the same way that Kerlinger does (cf. Chinn & Kramer, 2004, pp. 26–27). Thus various fields refer to the building blocks of their theories as either "concepts" (e.g., Morse, Mitcham, Hupcey, & Tason, 1996) or "constructs" (e.g., Churchill, 1979; Lewis, Templeton, & Byrd, 2005).

What Is Multicommunicating?

We define multicommunicating as the practice of engaging in multiple conversations at any one time (Reinsch et al., 2008). With the development of new ways of communicating that include mobile devices with both synchronous and asynchronous features, the number of messages that individuals receive is compounded by the number or relationships in which they are a part. And these messages can be received at any time or place. Additionally, Internet and social networking applications that allow individuals to post status updates provide the opportunity for one individual to simultaneously communicate with several individuals at once, generating even more messages within the span of a day. Instant messaging applications or chat technology is often used to create a persistent connection so that dyads are open to each other's messages for hours throughout the day, even if they only send messages occasionally (Nardi, Whittaker, & Bradner, 2000). Through this persistent connection, individuals are able to create episodes of presence where they are interpersonally connected to colleagues or friends for spans of time. These episodes can occur simultaneously. These opportunities lead to the practice of multicommunicating.

When we first began conceiving of the idea of multicommunicating, we connected the behavior to multitasking. A concept that describes an orientation towards multitasking is polychronicity. Therefore we originally developed the term polychronic communication (Turner & Tinsley, 2002) to describe the behavior of communicating with several individuals at any one time. While individuals have been engaging in this behavior for centuries (for example, two individuals conferring during a presentation), we saw that many current technologies seemed to facilitate the behavior, so we decided to write about it. The initial theory that supported multicommunicating was Hall's (1959) notion of polychronicity.

Hall suggested that different cultures exhibit two opposing orientations to time: monochronic and polychronic. A monochronic orientation supports the preference to accomplish one activity at a time and the polychronic orientation supports the preference to accomplish more than one activity at any one time. Bluedorn and his colleagues moved this concept of time orientation from a cultural variable to an organizational variable (Bluedorn, 2000; Slocombe & Bluedorn, 1999) and an individual variable (Bluedorn, Kallaith, Strube, & Martin, 1999). At the individual level, a polychronicity orientation suggests an individual prefers to complete multiple tasks at the same time and believes that this is the best way to work. Empirical research has suggested that polychronicity is negatively associated with deadlines, punctuality, and routine (Kaufman, Lane, & Lindquist, 1991).

However, as we further explored the polychronicity concept, we recognized several ways that polychronic orientations did not necessarily match up with multicommunicating behavior. Specifically, Bluedorn's work on polychronicity was focused on the managing of multiple tasks at any one time (Bluedorn, Kaufman, & Lane, 1992). While communication could be one of many tasks being accomplished, communication is a fundamentally different activity than individually oriented tasks (like working on a spreadsheet while eating lunch). Communication involves the complex activity associated with the management of dialogue and the strategic presentation of messages with other individuals in mind (we will discuss more of these characteristics later). Additionally, a polychronic orientation does not have a focus on multitasking for the end goal of efficiency or getting more things accomplished within a specific time period. Many of our early interviews that described multicommunicating behavior reflected an underlying concern about accomplishing tasks within a specific time period. While multitasking can be a preference for individuals, multicommunicating is often a response to someone else's needs (perhaps, for example, one's boss). For example, a person might multicommunicate often without actually initiating much of the conversation. They might multicommunicate because they feel compelled to respond to a certain number of messages within a time frame rather than because that is a preference for communication. A polychronic orientation suggests a tendency or preference for doing multiple things at once (Bluedorn et al., 1992). Therefore multicommunicating is a behavior that spans both polychronic and monochronic orientations or preferences.

First, let's examine the communication process to better understand how multicommunicating occurs. A person participating in one-on-one conversation engages in (a) interpretation, (b) goal generation, (c) planning, (d) enactments, (e) monitoring, and recycling (processes b through e) (Burleson & Planalp, 2000). During this process, in our everyday face-to-face interactions, conversations are governed by "unspoken rules" (Taylor & Van Every, 2000) that specify that individuals do not allow gaps of silence and do not overlap in speaking. These rules require that individuals are attentive to the conversation so that they are aware of speaking turns. Group interactions can involve multiple communicators and complex speaking patterns. However, because individuals share a common space (either physical or virtual), interaction norms encourage members to coordinate their behaviors so as to share messages successfully. In sequential conversations, conversations that closely follow each other (for example, conversations of a telephone receptionist or a desk staff person), messages are punctuated by explicit openings or closings that might involve phrases like, "I will be right with you" or "Please hold." Parallel conversations can occur in a group setting where two individuals in a group are talking together while the rest of the group is meeting (Reinsch et al., 2008). These interaction norms govern the presence episode–when individuals feel they are connected to one another. Multicommunicating occurs when any one individual in a conversational pair begins participating in two or more simultaneous conversations.

Multicommunicating is possible from a human standpoint because humans can think more rapidly than they speak or type (Greene, 2000). Ideally, with this extra time, we are processing the content of the conversation that we are in and creating strategies for response. However, now, we can open a new conversation with that extra time. Two specific characteristics that contribute to this ability are compartmentalization and flexibility of tempo (Reinsch et al., 2008; Turner & Reinsch, 2009).

Compartmentalization describes the cross-conversational availability of cues. Specifically, media with this characteristic provide the opportunity to hide activity in one conversation from a conversational partner in a second conversation. For example, chat technology allows someone to communicate with several communication partners without any of the partners knowing how many chat windows are open at one time. Two overlapping chat interactions are fully compartmentalized, while two face-to-face conversations provide no compartmentalization.

Flexibility of tempo describes the extent to which a participant may delay a response (allow a gap of silence) without giving offense or disrupting the interaction. Within two overlapping chat conversations, the individuals involved do not have visual cues of the participants to know who is at his or her desk and who is not. Therefore the speed of response can be governed not only by a person's typing skill, but the possibility that he or she has been delayed in receiving the message (perhaps absent from his or her desk) or distracted by other tasks (Reinsch et al., 2008). Thus chat is socially constructed to allow for delays in response as being acceptable. Additionally, messages without an enduring form (Herring, 2003) such as a live conversation, do not allow for flexibility of tempo. Email, on the other hand, does allow for flexibility of tempo.

What Does Multicommunicating Look Like?

The two characteristics of compartmentalization and flexibility of tempo provide a general description of how multicommunicating episodes take place, but to get a better picture of the technologies that are often paired, we used a critical-incident technique to explore successful and unsuccessful incidents of multicommunicating from the perspective of 201 MBA students (Turner & Reinsch, 2009). We found a frequent pairing of the telephone (providing compartmentalization) and email (providing compartmentalization and flexibility of tempo) as well as pairing of email and instant messaging (providing compartmentalization and flexibility of tempo). Several themes emerged from our analysis. For example, in most of the successful episodes, participants engaged in multicommunicating around a similar topic. We referred to this practice as quantitative multicommunicating where all conversations were devoted to the same topic, providing the participants with more clarity on a specific problem. An example might involve a salesperson who is talking to a customer about a product while emailing her colleagues about the specifics of the product or price issues so that she can have a more comprehensive discussion with the client.

Many unsuccessful episodes described a breaking point where the individual tried to do too many things at once. Most of these episodes were what we call qualitative multicommunicating, that is, occasions when the stacking of conversations required a participant to switch between topics and, frequently, between social roles. Here an individual might be discussing a personal issue with a friend, while emailing a client about a proposal, and while scheduling a conference call with a virtual team. At a certain point, the individual reaches a breaking point where he or she is confused about the messages being sent or the roles he or she is trying to manage (Turner & Reinsch, 2009). Our point is not that qualitative multicommunicating never succeeds but, rather, that it is more intense and more difficult and therefore less often successful.

As many of our interviews would describe a breaking point or a time in which multicommunicating became too stressful or overwhelming, we decided to identify specific characteristics that lead to the intensity of the multicommunicating experience (Reinsch et al., 2008). We argued that multicommunicating varies in intensity based on four distinct factors: the number of open conversations, the pace of each conversation, the integration of the social roles involved, and the number of topics being discussed. Number of open conversations refers to the number of conversations that have been opened but not exited from. We suggest that the number of open conversations or episodes provides an index as to how an individual has divided her "presence" among speech events. In fact, we developed the term presence allocator to describe the person who is multicommunicating (Turner & Reinsch, 2007). We suggest that all things being equal, conducting three overlapping conversations is a more intense form of multicommunicating than conducting two (Reinsch et al., 2008). Pace describes the expectation for feedback, often facilitated by the technological features associated with a specific medium. Texting is limited by the skill and typing ability of the texters, while a phone conversation requires a quicker response on the part of the communicators. As the pace at which an individual is communicating across several conversations increases, the intensity of the experience also increases. Integration of social roles describes the specific social roles evoked by various conversations. Transitions from one role to the next (for example, supervisor to subordinate to parent to friend) can be difficult when those roles are not well integrated. As role transitions may create role conflict, the more distinct roles involved (especially when those roles are not integrated) may increase the intensity of the multicommunicating episode. Finally, number of topics involved can contribute to the intensity of the experience, especially when the topics are qualitatively different from one another (Reinsch et al., 2008). Now that we have described the multicommunicating construct, we will connect this behavior to some of the research done within the field of computer-mediated communication (CMC).

Linking Multicommunicating to CMC Research

Much of the research exploring CMC has examined message exchange associated with one conversation at a time. Whether the question involved media choice, media use, or mediated messages, the focus was on one conversation at a time and one medium at a time. However, many of the variables of this CMC research have interesting implications for multicommunicating.

Task Equivocality and Status

Daft and Lengel's (1986) media richness model suggests that individuals choose a medium based on a match between the equivocality of the message and the richness offered by a particular medium. Richness is defined by the extent to which a medium offers the opportunity for rapid feedback, natural language, number of types of cues, and personalization of messages. The richer the medium, the more capacity the medium has for carrying an equivocal message. While media richness theory has been criticized for focusing too narrowly on the objective attributes of channels and the assumption that communicators choose a medium on a purely rational basis (Carlson & Zmud, 1999; Fulk, Schmitz, & Steinfield, 1990), this theory remains relevant because of its parsimonious explanation of how people make media choices. This explanation not only resonates as people discuss their media choice strategies, it is also evident in individual discussions of their multicommunicating practices.

In a study that explored the factors associated with multicommunicating, we found that individuals often use complexity of the topic as a criteria for choosing whether or not to multicommunicate (Turner & Reinsch, 2007). For example, one interviewee commented:

> If it's a complicated issue or a complex issue . . . I'm going to focus on it like one-on-one kind of thing, one in sequence. If it's just somebody asking me "hey, are you going to this meeting or where's this meeting located" or something really insignificant then I could have multiple [interactions] going on. (Turner & Reinsch, 2007, p. 44)

The other issue that individuals considered important when making the decision to multicommunicate was the status of the receiver. Specifically, when an individual was communicating with their boss, it made sense to focus just on that person. When a boss multicommunicated, the behavior was often forgiven or explained away by his or her important role within the organization. Said one participant:

> If it's somebody that works for me and I go talk to them, then you know, to be honest, I'm their boss I have priority and you [they] need to pay attention to that. I just say, "I need to talk about this" and then they stop what they're doing and we talk. If

it's an equal . . . then I'm pretty tolerant of [multicommunicating]. (Turner & Reinsch, 2007, p. 45)

Social Norms

Social norms have also been a powerful influence on the choices that individuals make regarding communication technology (Fulk, 1993; Fulk, Steinfield, Schmitz, & Power,1987; Fulk et al., 1990). Research has found that the extent to which an individual's co-workers use a specific technology influences the likelihood that that individual will use that technology, especially when he or she cares about the work group (Fulk et al., 1987). Social norms and specifically organizational norms have also influenced decisions to multicommunicate. In a study of a high-tech organization instrumental in the development of instant messaging technology, we found a "dominant media norm" in the form of instant messaging technology specifically and as a result, multicommunicating behavior (Turner, Grube, Tinsley, Lee, & Opell, 2006). Returning to the discussion of polychronicity, we found that the strong norm for communicating via instant messaging (IM) influenced employee use of IM, and even more so when individuals had a strong polychronic orientation (preference for doing multiple activities at once). Supervisor use of IM and email also influenced employee use. And those employees who followed the organizational norms by using IM and email were awarded higher performance ratings by their supervisors (Turner et al., 2006).

Interviews with employees found that the practice of multicommunicating often accompanied the use of IM and email within the organization. For example, one employee noted, "Since I started working here, IM has become like the backbone of what we do here and how we communicate." Said another, "Being online is like breathing, you have to be there" (Turner et al., 2006, p. 18). Combining IM with multicommunicating, one participant offered, "I can talk to five different people at once and pay attention while doing like five different IM . . . it gets really hectic when you are trying to do your busy work on top of an Excel or whatever document you are working in and on top of that switch it back and forth to an Excel and browser so you can IM people but it really doesn't bother me that much. I've gotten used to it so it's a norm (Turner et al., 2006, p. 19).

Implications of Multicommunicating

The trend towards the development of technologies that support and encourage multicommunicating seems to be accelerating. The development of social media applications that provide status updates with special feeds that direct these messages into an email program or viewer means that the number of open conversations that an individual participates in on a daily basis can seem infinite. When we initially thought about multicommunicating, we considered that episodes would be delineated by the number of simultaneous conversations occurring that had been opened with a "hello" but never really closed with a "goodbye." However, a status update, communicated through technologies like Facebook or Twitter create a constant presence with a group of people (friends or followers). Clive Thompson of *Wired* magazine argues that these technologies "create a social sixth sense" (Thompson, 2009). Thompson writes:

> Twitter and other constant-contact media create social proprioception. They give a group of people a sense of itself, making possible weird, fascinating feats of coordination. For example, when I meet Misha for lunch after not have seen her for a month, I already know the wireframe outline of her life: She was nervous about last week's big presentation, got stuck in a rare spring snowstorm, and became addicted to salt bagels.

This awareness provides an ongoing and constant sense of what the key individuals in someone's life are doing. Are these ongoing conversations? If I look at my Blackberry and read status updates during a conference call with colleagues, am I in an ongoing conversation with the senders of those status updates?

The practice of multicommunicating has begun to influence the norm that participating in more than one conversation at a time is rude or disrespectful (Reinsch et al., 2008; Turner & Reinsch, 2004, 2007). This changing norm, coupled with the relative ease with which an individual can process a Twitter update (messages are limited to 140 characters) may lure an individual into believing that he or she is "truly paying attention" to the communicators involved. One person suggested that it only took a minute to glance down at his Blackberry during a conversation with his colleague about an upcoming client presentation, and notice that his cousin was engaged and that his best friend from grade school was ordering a value meal from a fast food restaurant. This person argued that he could keep up with the presentation while also remaining better informed about his interpersonal network. An interview response that we often receive when we ask individuals how they can multicommunicate is, "I listen when it pertains to me." Our question is, how do they know if it pertains if they are not listening? When we think back to the characteristics that lead to intensity of multicommunicating, the opportunity for misunderstandings or unsuccessful conversations seems probable.

Channel expansion theory (Carlson & Zmud, 1999) argues that as individuals gain more experience with a medium, the medium becomes richer for them and provides more opportunities for dialogue. The richness is not only in the match between message and medium characteristics but also involves the skill and experience of the communicator. Maybe this idea is also relevant to multicommunicating. Are those individuals of a generation that grew up texting and multicommunicating better at this practice than those who are from a generation that tries to hide the behavior for fear of showing disrespect?

Productivity implications. Many people that we have interviewed argue that they "have to multicommunicate to get everything done." "If I didn't multicommunicate, I couldn't do my job." It seems that many individuals connect efficiency with multicommunicating when, in fact, research regarding cognitive switching argues the opposite. Considerable research has explored the executive control processes associated with switching from one task to another (for review of experiments, see Rubinstein, Meyer, & Evans, 2001). Research is inconclusive as to the precise relationship associated with task switching. However, the extent to which the rules associated with a specific task move from complex to familiar or familiar to complex seem important. Much of this research has been conducted with relatively routine tasks associated with multiplication problems or the identification of patterns or shapes. These tasks, while varying in their routine and complexity, do not resemble the complexity associated with changing message, recipient, and medium. Additionally, when answering a text or email, other than the subject line, the communicator often has no warning of the context associated with the task. Therefore an individual could be texting about a sick friend while talking on the phone about a complicated project. Each message requires different emotional energy and mental concentration. We suggest that the feeling of productivity may come from an addiction to the accomplishment associated with sending a message or responding to a request, rather than the absolute efficiency associated with multicommunicating.

Presence implications. Davenport (2001) argues that in our economy, capital, labor, information, and knowledge are in plentiful supply; what is in short supply is attention. He refers to attention as the current currency of business and the ability to manage attention is the key to business success. He explores this by discussing the strategies used by marketers and web designers to hold the eyes and attention of their customers. However, in many ways, our ability to allocate presence successfully and create episodes of presence constructively may determine our success in both our relationships and workplaces.

Conclusions

As shown in this literature review, the study of multicommunicating and the study of episodic presence are just beginning. We believe that multicommunicating affects workplace productivity–enhancing productivity at low intensities and reducing productivity at higher intensities (Reinsch et al., 2008). However, so far, the available data consist of self-reports (which aren't always accurate) and anecdotes (which aren't always representative).

We believe also that episodic presence will prove to be an important conceptual tool for understanding the efforts of a multicommunicating person (for example, Susan who was quoted at the beginning of this chapter) to "be" in more than one place at a time, and for understanding where the multicommunicating person (for example, Steve who was quoted at the beginning of this chapter) and his or her conversation partners "are" (psychologically speaking).

We are confident that managers will need to become increasingly smart about the deployment of technologies and increasingly aware of the advantages and disadvantages of both multicommunication and of dividing or allocating one's episodic presence.

Preparing this chapter has also stimulated us to ponder the speed with which technologically assisted communication continues to change. When–only a few years ago–we began to study multicommunication, we found that some of our colleagues (and some of the individuals who reviewed our manuscripts) reacted with skepticism. They seemed to feel that multicommunication was a temporary aberration or a local anomaly confined to the one workplace where we first observed it, or even something that should be stamped out as quickly as possible. Today, we don't get that reaction as often (though there are still people who view multicommunication with alarm–as do we at times).

On the other hand, we also note that technology and the frontier of innovative technology-assisted communication behavior have continued to move. With Twitter, the distinction between interpersonal communication and mass communication seems to have been completely obliterated. Episodic presence may be allocated not only among a couple of conversations but into the ongoing tweetlogs of multiple friends and celebrities. Will the constructs of multicommunication and of episodic presence–which are rooted in our understanding of interpersonal communication–prove adaptable enough and sturdy enough to stimulate and guide research in the age of Twitter? Only time will tell–and we look forward to finding out.

References

Bluedorn, A. C. (2000). *Polychronicity, change orientation, and organizational attractiveness*. Paper presented at the Annual Meeting of the Society for Industrial and Organizational Psychology, New Orleans, LA.

Bluedorn, A. C., Kallaith, T. J., Strube, M. J., & Martin, G. D. (1999). Polychronicity and the inventory of polychronic values (IVP): The development of an instrument to measure a fundamental dimension of organizational culture. *Journal of Managerial Psychology, 14*(3/4), 205–230.

Bluedorn, A. C., Kaufman, C. F., & Lane, P. M. (1992). How many things do you like to do at once? An introduction to monochronic and polychronic time. *The Academy of Management Executive, 6*, 17–26.

Burleson, B. R., & Planalp, S. (2000). Producing emotion(al) messages. *Communication Theory, 10*, 221–250.

Cameron, A. F., & Webster, J. (2005). Unintended consequences of emerging communication technologies: Instant messaging in the workplace. *Computers in Human Behavior, 21*, 85–103.

Carlson, J., & Zmud, R. (1999). Channel expansion theory and the experiential nature of media richness perceptions. *Academy of Management Journal, 42*(2), 153–170.

Chinn, P. L., & Kramer, M. K. (2004). *Integrated Knowledge Development in Nursing*. St. Louis, MO: Mosby.

Churchill, G. A. J. (1979). A paradigm for developing better measures of marketing constructs. *Journal of Marketing Research, 16*, 64–73.

Daft, R., & Lengel, R. (1986). Organizational information requirements, media richness, and structural design. *Management Science, 32*(5), 554–571.

Davenport, T. H., & Beck, J. C. (2001). *The Attention Economy: Understanding the New Currency of Business*. Boston, MA: Harvard Business School Press.

Fulk, J. (1993). Social construction of communication technology. *Academy of Management Journal, 36*, 921–950.

Fulk, J., Schmitz, J., & Steinfield, C. (1990). A social influence model of technology use. In J. Fulk & C. Steinfeld (Eds.), *Organizations and communication technology* (pp. 71–94). Newbury Park, CA: Sage.

Fulk, J., Steinfield, C., Schmitz, J., & Power, J. G. (1987). A social information processing model of media use in organizations. *Communication Research, 14*(5), 529–552.

Greene, J. (2000). Evanescent mentation: An ameliorative conceptual foundation for research and theory on message production. *Communication Theory, 10*(2), 139–155.

Hall, E. T. (1959). *The silent language*. New York: Doubleday.

Herring, S. C. (Ed.). (2003). *Computer-mediated discourse*. Malden, MA: Blackwell.

Hymes, D. (1972). Models of the interaction of language and social life. In J. Gumperz & D. Hymes (Eds.), *Psychological perspectives on the self, 1* (pp. 231-262). Hillsdale, NJ: Lawrence Erlbaum.

Kaufman, C. F., Lane, P. M., & Lindquist, J. D. (1991). Time congruity in the organization: A proposed quality of life framework. *Journal of Business and Psychology*, 6, 79–106.

Kerlinger, F. N. (1986). *Foundations of behavioral research* (3rd ed.). New York: Holt, Rinehart and Winston.

Lewis, B. R., Templeton, G. F., & Byrd, T. A. (2005). A methodology for construct development in MIS research. *European Journal of Information Systems, 14*, 388–400.

MarketWatch. (2008). Push to talk and instant messaging revenue will top 16 billion in 2009, says Insight Research Corporation. Retrieved July 29, 2009 from http://www.send2press.com/ newswire/2008-12-1204-001.shtml

Morse, J., Mitcham, C., Hupcey, J., & Tason, M. (1996). Criteria for concept evaluation. *Journal of Advanced Nursing*, 24, 385–390.

Nardi, B., Whittaker, S., & Bradner, E. (2000). *Interaction and outeraction: Instant messaging in action*. Paper presented at the Computer Supported Cooperative Work, New York.

Reinsch, L. N., Jr., Turner, J. W., & Tinsley, C. (2008). Multicommunicating: A practice whose time has come? *Academy of Management Review, 33(2)*, 391–403.

Rubinstein, J. S., Meyer, D. E., & Evans, J. E. (2001). Executive control of cognitive processes in task switching. *Journal of Experimental Psychology: Human Perception and Performance*, 27(4), 763–797.

Slocombe, T. E., & Bluedorn, A. C. (1999). Organizational behavior implications of the congruence between preferred polychronicity and experienced work-unit polychronicity. *Journal of Organizational Behavior*, 20, 75–99.

Taylor, J., & Van Every, E. (2000). *The emergent organization: Communication at its site and surface*. Mahwah, NJ: Lawrence Erlbaum.

Thompson, C. (2009). Clive Thompson on how Twitter creates a social sixth sense, *Wired*, Vol. 15.07. Retrieved January 9, 2010 from http://www.wired.com/techbiz/media/magazine/15-07/st_thompson

Turner, J. W., & Reinsch, L. N., Jr. (2004). *Except when it's my boss: Predictors of intent to communicate polychronically*. Paper presented at the Academy of Management, New Orleans, LA.

Turner, J. W., & Reinsch, L. N., Jr. (2007). The business communicator as presence allocator: Multicommunicating, equivocality, and status at work. *Journal of Business Communication*, 44(1), 36–58.

Turner, J. W., & Reinsch, L. N., Jr. (2009). Successful and unsuccessful multicommunication episodes: Engaging in dialogue or juggling messages? *Information Systems Frontiers*, 9, in press.

Turner, J. W., Grube, J., Tinsley, C., Lee, C., & Opell, C. (2006). Exploring the dominant media: Does media use reflect organizational norms and impact performance? *Journal of Business Communication, 43*(3), 1–31.

Turner, J. W., & Tinsley, C. (2002). *Polychronic Communication: Managing Multiple Conversations at Once*. Paper presented at the Academy of Management, Denver, CO.

Wallace, W. A. (1983). *From a realist point of view: Essays on the philosophy of science* (2nd ed.). Lanham, MD: University Press of America.

CHAPTER ELEVEN

The More Things Change, the More They Stay the Same: The Role of ICTs in Work and Family Connections

Paige P. Edley

Renée Houston

The ubiquitous presence of information communication technology (ICT) offers working families opportunities to reassign, privilege, and attend to the intersections of work and family (Bryant & Bryant, 2006). Amidst increasing connectedness and perpetual development, ICTs allow men and women to engage in paid work from home, run their own businesses from a home office, and work at the office while keeping close ties with latchkey kids via mobile phones. Mobile ICTs bring work into the home and home into work. Wireless laptops, mobile network cards, flash drives, email, and smart phones (previously referred to as PDAs), like BlackBerry and iPhone, that combine mobile phones with Internet and web access for downloading applications (apps) for music, global positioning systems (GPS), calendars, email, and social networking (Facebook, Twitter, Linkedin, etc.) constitute what Weiser (1995; cited in Quesenberry & Trauth, 2005) referred to as "ubiquitous computing"—discreet computers that are so naturally interspersed into our everyday lives that they are deemed invisible and normal. They are considered an extension of the self (Lemish & Cohen, 2005), a fashion accessory (Gant & Kiesler, 2001; Green, 2003; Ling, 2003, 2004; Hulme & Peters, 2001; Katz et al., 2003; Lobet-Maris, 2003; Skog, 2002; all cited in Campbell, 2007; Wei & Lo, 2006), a second skin (Katz, 2003; Katz & Aakhus, 2002), and a fifth limb (Ellwood-Clayton, 2006).

While on one hand ICTs appeal to working parents as a means to make life easier and work more flexible, ICTs also can create longer work hours at home (Kivimaki, 2004; cited in Gregg, 2008; Wajcman et al., 2008) and on the road since they provide both temporal and spatial extensions of work into the home during family and leisure time (Araújo, 2008), as well as travel and commute times (Totten et al., 2007), during weekends, evenings, and early morning hours (Araújo, 2008; Edley, 2001, Edley & Houston, 2006; Hilbrecht et al., 2008). Treating these extended work sessions as normal and necessary means that we ignore the negotiated tensions and contradictions that parents engage in as they try to manage the multiple and changing roles, identities, and social constraints produced and reproduced in the gendered discourses of work and family (Araújo, 2008; Edley & Houston, 2006; Golden, 2009b).

Here we offer an interdisciplinary perspective focusing on how ICTs intersect with family communication. We recognize the enormity of summarizing a complex social phenomenon and thus focus on selecting cross-disciplinary literature that connects work-life studies with the adoption and uses of ICTs to reveal how family relationships and responsibilities are managed. The review begins by examining assumptions concerning gender that constrain how women participate with technology and in technology-based careers. We posit that cultural meanings and patriarchal structures shape the meanings of ICTs and the ways that people interact with and relate to new media (Kelan, 2007; Lee, 2006; Wajcman, 1991). Technology[1] is not only a socially constructed system of knowledge but also a historical, cultural, and politically situated communication artifact that becomes gendered, raced, and classed within dominant systems of knowledge and inequitable power relations. These meanings thereby produce and reproduce stereotypical ways in which women and men use ICTs for family communication.

Next, we turn to how varied literatures reveal gendered implications of work and family management through the uses and innovations of new media. Each new technology is examined to inform social and political implications of work and parental identity construction and the presentation of self via new media and social networking processes. ICT discourses and practices are both empowering and constraining. While we view gender constructions as constantly in flux, we critique how these forms of ubiquitous computing also make the lives of working parents different and more complex, while also producing and reproducing gender stereotypes and the corporate colonization (Deetz, 1992) of relationships and work practices. We conclude by offering potential contributions, critiques, and future research directions.

Review of Literature

Gender and Technology

A swelling tide of research focuses on the challenges, opportunities, and impact of women's opportunities to work and play with ICTs. Scholars who draw on the cultural construction of both women and technology investigate how gendered constructions constrain women's opportunities for utilizing technology outside of gendered norms. Within the field of technology studies, several scholars examine how technology is imbued with a masculine identity. In terms of gender, technology is constructed as masculine, which means that men are naturally more at ease, proficient, and in control (Kelan, 2007, 2008; Lee, 2006; Wajcman, 1991; Youngs, 2001). Further, this construction casts men as creators of technology and women as users of men's creations (Consalvo, 2006). As noncreators, women are not practically constrained in their ability to create or use technology, yet they are socially constrained to use technology along stereotypical gender lines. When examining the networks of expertise among library and information technology professionals, Houston and Ricigliano (2003) found that women equally recognize both men and women as technical experts, whereas men only recognize men's expertise. Technology's construction as masculine parallels the traditional, mechanistic ideology of work[2] where rationality is privileged over emotionality. Men's work is technical and rational while women's work is caring and emotional (Hochschild, 1989; Kelan, 2008). Interestingly enough, however, many scholars have argued that women are good ICT workers because of their strong social skills (CDI, 2002; Donato, 1990; Funken, 1998; Panteli et al., 2001; Schelhowe, 1997; Woodfield, 2000; cited in Kelan, 2008). Since contemporary ICT work is service and teamwork oriented, ICT workers need to be able to interact with both customers and colleagues. They need to be "hybrids" with both strong technological skills and strong social skills (Woodfield, 2000; cited in Kelan, 2008).

Cautioning against essentializing gender within a binary, Kelan (2007) suggests that men and women may engage with ICTs in gendered ways, not in hierarchically different ways, but in shifting and fluid ways (see also Butler, 1989; Wajcman, 2007, 2008; West & Zimmerman, 1987). Kelan (2007) interrogates gendered differences by examining how women and men relate to technologies. Women view technology as tools, while men view them as toys. She argues that these relationships with technology perpetuate the gendered and stereotypical constructions of technology use and the gendered binaries within society. Kelan (2007) advocates for considering gender as multiple and shifting in technology use and design and also argues for viewing one's relationship with technology as also shifting one's subject positions. Hence, we argue that these shifting subject positions include the multiple roles that working parents enact in the course of managing work and family relationships and responsibilities via their relationships with ICTs.

Managing Work and Family Relationships and Responsibilities

Given the vast literature on work-life balance, this review begins with a discussion of some underlying assumptions of work-life balance and then focuses on technology's intersection with women's paid work whether at the worksite or home office. The language of *balancing home and work* is problematic in that it suggests the balancing act is between two separate gendered spheres, the public sphere of work and masculinity and the private sphere of home, family, and femininity (Kirby, Golden, Medved, Jorgenson, & Buzzanell, 2003). The division of public and private spheres into separate, mutually exclusive realms reproduces the "myth of separate worlds" (Kanter, 1977, 1989) or "false dichotomy," (Martin, 1990), a gendered work/family binary in which men and women traverse the boundaries of public and private in different ways and with different values and privileges attached to their movement (Hochschild, 1989, 1997; Nippert-Eng, 1995; Rakow & Navarro, 1993). Clark (2000) refers to border crossing as the movement both spatially and psychologically between work and home. Alternatively Medved and Graham (2006) suggest thinking dialectically about the tension of work-life spheres may best capture its complex nature. Further, Medved et al. (2006) found that women more often than men were steered towards careers that would allow them to manage family responsibilities. In particular, for employed mothers[3] the boundaries are blurred as work and family responsibilities spill over into both areas of their lives. Women engage in paid work and domestic labor simultaneously, thus scholars also refer to this phenomenon as "work-family conflict," "work-family spillover," and more recently have adopted the term "work-life balance."[4]

ICTs blur the boundaries of work and home (Avery & Baker, 2000; Edley, 2001, 2004; Edley et al., 2004: Ellison, 2005; Golden, 2009a, 2009b; Hochschild, 1997; Jackson, 2002, 2008; Mallia & Ferris, 2000; Perrons, 2003; Rakow & Navarro, 1993; Townsend & Batchelor, 2005; Wajcman et al., 2008). The literature is mixed as to whether the ICT impact is helpful, harmful, or both, to employed parents and their families. For example while online communication technologies may support increases in family communication and interaction, some technologies may create more isolation (Bryant & Bryant, 2006). As Sproull and Kielser (1991) outlined nearly 20 years ago, a contingency perspective on technology recognizes the fragile balance between claiming the beneficent or deleterious impact of technology. ICTs then present a paradox that may both help and hinder (Edley, 2001; Edley & Houston, 2006; Golden, 2009a, 2009b). Sullivan and Lewis (2001) and Jackson (2002) argue that employed parents are physically present in one realm while being virtually present in another through their ICT use. This usage creates a "dual presence" (Jackson, 2002, p. 28). Jackson (2008) also states:

> A boundaryless world means that coming home doesn't signal the end of the workday anymore than being on vacation is a time for pure relaxation or being under one roof marks the beginning of unadulterated family time. The physical and virtual worlds are always with us, singing a siren song of connection, distraction, and options. (p. 63)

Edley (2001) argued this as well; "The public and private spheres are blurred and blended, simultaneous and parallel, remote and onsite. There is no distinction" (p. 33). The blurring of boundaries can be seen as an intertwining of needs that mutually affect and co-construct needs and practices of employers and families (Edley et al., 2004; Martin, 1990, 1994). Moreover, Presser (2003) stipulates that with the U.S. society moving toward a "24/7 economy," 2/5 of the U.S. workforce work predominantly at "nonstandard times—in the evening, at night, on a rotating shift, or during the weekend" (p. 1). The result of this scheduling is that workers' family life and their health are both being negatively impacted. Hence, the blurring of boundaries has both positive and negative impacts on working parents. ICTs enable workers to extend the work day and place, but not reduce the workload—a paradox that leads to a complex negotiation of tensions between contradictory discourses.

Two contradictory discourses characterize research on the role of technology in negotiating work and family: exploitation and empowerment (Edley, 2001; Edley & Houston, 2006; Kirby et al., 2003). In the discourse of exploitation, ICTs allow employers to intrude into employees' personal lives to gain additional work time and effort—a managerial bias of extended control. Alternatively, in the discourse of empowerment, ICTs enable workers to: (a) control when and where they do their work, (b) connect with family/friends, and (c) manage work and family responsibilities (Kirby et al., 2003).

More specifically, Perron's (2003) research on boundaries demonstrates that, "new working patterns have eroded the boundaries and collective rhythms of working life and the concept and reality of a fixed working day have declined for many people" (p. 69). Similarly, Townsend and Batchelor (2005) express concern about "the way technology, including mobile phones and the Internet, blurs the boundaries between work and non-work time allowing work to spill over into home life and leisure time" (p. 261). Chesley (2005) notes that previous studies found ICT use may blur work/family boundaries with negative consequences for employees. Despite being located at home, individuals work longer hours, sleep fewer hours, and develop health problems, which result in more work-family conflict. Even the market for PDAs attempted to reify public/private boundaries by introducing "Audrey" a PDA-type "appliance" designed as "the family nerve center" (Rodino-Colcino, 2006) that feminized the home computer by locating it in the kitchen. Since Audrey was not portable it failed to reduce work for mothers; the device actually translated into more work by requiring mothers to input and sync all family members' schedules into the appliance. Rodino-Colocino (2006) demonstrated how domesticated technology created more work for mother, consistent with Cowan's (1983) and Schor's (1991) earlier work. Mobile phones allow mothers true mobility to traverse the public/private boundary, "No matter how far they roam, mothers drag the domestic sphere with them...and [with] cell phones the domestic sphere moves with mothers" (Rodino-Colocino, 2006, p. 386). Since the mobile phone is the only ICT that actually moves with a busy parent as s/he navigates a hectic schedule, smart phones that actually support family communication would include work-life management "apps" that allow easier and less time-consuming ways to sync family schedules with work schedules.

Mobility is not the only work-life management use of ICTs: another is perpetual connection. Scholars (Haddon & Silverstone, 2000; Lewis & Cooper, 1999; Valcour & Hunter, 2005) argue that ICT use allows for constant access and what Katz and Aakhus (2002) refer to as "perpetual contact." While perpetual contact is positive for achieving efficiency of work (Golden, 2009b) and positive for family access (Chesley, 2005; Christensen, 2009), for the busy parent this constant access can be extremely stressful. If one is always connected and always accessible, the individual feels like s/he is unable to turn off the access (Edley & Houston, 2006). There is no escape—no relief from the perpetual "on call" state that is paradoxically a relief and a stress at the same time. Perpetual contact thus becomes normative, natural, and expected through ubiquitous computing.

ICTs and Ubiquitous Computing

Ubiquitous computing is defined as the pervasiveness of discrete computers (e.g., smart phones and wireless laptops, etc.). Weiser (1995; cited in Quesenberry & Trauth, 2005) first used the phrase to explain "situations where multiple computers are invisible, indistinguishable, and available to an individual throughout a physical environment, and thus woven into the fabric of everyday life" (Quesenberry & Trauth, 2005, p. 45). This omnipresent availability of computing has created expansive new alternatives for when, where, and how work and family occurs (Hill et al., 2003; Useem, 2000; cited in Quesenberry & Trauth, 2005). According to Quesenberry and Trauth (2005), although men and women both report high levels of work-life imbalance (Duxbury & Higgins, 2003), women often do most of the domestic work (Hochschild, 1989, 1997; Perlow, 1998), and women and men exhibit different behaviors regarding work-life balance (Mennino & Brayfield, 2002; cited in Quesenberry & Trauth, 2005).[5] Of course men may choose to stay at home with children, which may impact work-life management; however, recent trends are inconsistent. For example a 2009 Sloan Work and Family Research reports that in 2007 165,000 dads chose to stay at home while in 2008 year the number of stay-at-home dads dropped to 140,000. In addition, women in the IT profession tend to experience more stress due to imbalances with work and family than do their male counterparts (Duxberry et al., 1992; Gallivan, 2003; Igbaria et al., 1997; cited in Quesenberry & Trauth, 2005).

Telework

Worldatwork.com reports that the number of Americans who worked from home or remotely at least one day a month increased from 12.4 million in 2006 to 17.2 million in 2008. This breaks down to 3.7 million Americans telecommute at least once a month 24.2 million telecommute at least one day a week, and 13.5 million telecommute almost every day. In addition, most U.S. telecommuters are typically upper middle class, 40-year-old male college graduates.

Organizational communication scholars have explored how telework aids working parents in the balancing of work and family responsibilities (Edley, 2001; Hylmö, 2004; 2006; Hylmö & Buzzanell, 2002; Mallia & Ferris, 2000; Tremblay, 2003). Studies of telecommuters found that job autonomy significantly reduced work-family conflict because they could choose their own work schedule and take breaks to tend to family concerns, like attending children's ball games (Goldstein, 2003; Pratt, 1999; cited in Ahuja et al., 2007). However, research on telework and its effects on employees and families is mixed. Mirchandani (1998) found teleworkers need to protect distinctions between work and home. Wajcman et al. (2008) argue that ICTs used for telework, especially cell phones, don't make work-life balance any easier; rather, it adds to "uncontrollable work-family spillover" (p. 648). Not only has virtual work been linked to stress and exhaustion, work-family conflict is a strong predictor of stress and exhaustion (Salaff, 2002; cited in Ahuja et al., 2007).

Research on working families in Canada offers a different understanding of the role of telework in managing family relationships. Tremblay, Paquet, and Najem (2006) found work obligations rather than family obligations led Canadians to work an average of six hours a week from home without pay. These extra hours amount to almost a full day's work. They also found that (presumably male) managers and professionals were more likely to be given the necessary technology to work from home than women, and by virtue of their privileged status, found it easier to telework (Tremblay et al., 2006). Further, only a small percentage of participants indicated that they engaged in telework due to family obligations; the predominant reason cited was employer requirements.

As employers demand telework, employees need to weigh the pros and cons of working from home for free. Golden (2009b) argues, using ICTs to manage work and family responsibilities can disadvantage both women and children. According to Golden (2009b), technology,

> demands of us heightened reflexivity and greater attentiveness to those who depend on our attention. Using technology to manage work and family can disadvantage women by reproducing traditionally gendered divisions of responsibility, and disadvantage children by extending the boundaries of work into the home; however, used mindfully, technology can also extend men and women's abilities to engage fully with work and family. (pp. 352–53)

Golden also argues that how people use technology is influenced by "particular families' ways of interacting, and by broad cultural values, such as the 'spirit of capitalism' that Weber observed, as it is by affordances of any technology in and of itself" (2009, p. 352).

Working parents who telecommute experience both costs and benefits of negotiating family and work responsibilities (Edley, 2001, 2004; Hilbrecht et al., 2008). Hylmö (2004) found the gendered downside to telecommuting is that telework allows women more flexibility, but their "work-life balance is likely to be achieved at the cost of advancement" (p. 67). Social interaction is imperative to achieve advancement at work (Davies, 1998; Seibert et al., 2001; cited in Hylmö, 2004). According to Hylmö, telecommuting women, who wish to be considered for advancement, need to discuss work options and promotion expectations with their supervisors. If they fail to set expectations and recognition of distributed work then teleworkers are less likely than their co-located counterparts to be considered for promotion. While telecommuting is a structure that clearly places public work in the center of the private sphere, the widespread adoption of mobile phones keep parents in touch with both family and work.

Mobile Telephones

The proliferation of mobile phones highlights changing communication patterns among family, friends, and employers. Palen et al. (2001) define the pervasiveness of mobile phones as "social artifacts...that support coordination with others" (p. 1). In early 2008 more than a quarter billion people in the United States were mobile phone subscribers, who used more than a trillion minutes of calling time in the first half of 2007 (Rice & Katz, 2008). Current statistical reports on mobile phone providers by the International Telecommunications Union suggest that 87% of the U.S. population owns a mobile phone. (http://www.itu.int/ITU-D/ict/statistics/). The impact of cell phones on relationships highlights the variety of experiences mobile phone users have as they maintain family and work relationships. Blinkoff and Blinkoff's (2002) study of 160 mobile phone users in six countries revealed that mobile phones were predominantly used as a relationship tool. In addition, Katz and Rice (2002) found a large population of mobile phone users thought their cell phones led to improved relationships and reported feeling more accessible to their social network. Furthermore, they described the content of their mobile phone conversations as being "affect-oriented and psychologically pleasurable" (cited in Katz, Rice, Acord, Dasgupta, & David, 2004, p. 341). Palen et al. (2001) found that while some people thought of their mobile phone as a leash that kept them accessible all the time, others interpreted the mobility of the technology as giving them freedom.

Whether a leash or freedom, mobile phones can provide constant access and perpetual contact (Katz & Aakhus, 2002). Leung and Wei (2000) found two significant gratifications tied to cell phone use: mobility and immediate access. Hence the mobile phone is an exemplar of instant gratification by providing instant access to family, friends, and work. Furthermore, Leung and Wei found instrumental gratifications applied to mobile access to co-workers and business partners, whereas mobility and affection gratifications were associated with communication with family members. Wei and Lo (2006) confirmed Leung and Wei's findings and stipulated that the mobile phone strengthens and enhances family ties. As Hampton and Gupta (2008) argue, mobile phones and public wi-fi create a trend towards "public privatism" in which engaging with mobile communication creates "a private sphere of interaction within public spaces" (p. 835).

The gendered aspects of mobile technology, as articulated by Ling (2004), suggest that the mobile phone "makes women accessible as caregivers and thus can enforce and extend the private sphere" (p. 63). Lemish and Cohen (2005) argue that mobile phone use in Israel has blurred the gender boundaries and has actually become feminized because of its use in maintaining social networks. Edley (2001) found working mothers using their mobile phones as electronic babysitters, and Rakow and Navarro (1993) coined the term "remote mothering" to describe working mothers' using their cell phones to bridge the gap between home and work as they checked on their children after school. Wajcman, Bittman, and

Brown (2008) argue, "communication is central to contemporary practices around intimacy" and suggest that women's use of mobile phones could "signify their responsibility for emotional work" (p. 648).

Ling and Yttri (2002) show how mobile phone use has changed the way people organize their everyday lives. The authors refer to frequent mobile phone calls among family members as "micro-coordination," which they argue is closely associated with "nuanced instrumental coordination" (p. 139; cited in Christensen, 2009, p. 436). Wei and Lo (2006) found that the mobile phone "enhances the ties between family members, with women in particular relying on extensive use of cell phones to show affection to their families while on the move" (p. 68). Baym, Zhang, and Lin (2004) also found mobile phones, Internet, and email used in conjunction with face-to-face communication to maintain intimate relationships. Based on this research, we also can envision how mobile phones allow mothers to manage the "third shift" (Bolton, 2000) or the anxiety women feel about how they handle conflicting obligations.

Christensen argues that ICT use changes "intra-familial communication by integrating mediated communication as an inseparable part" of family relationships. Yet face-to-face family time, or "physical co-presence," is still the "bedrock on which close relationships are built" (2009, p. 445). In addition to mediated communication, face-to-face activities such as the family meal are still very important, as are morning rituals, family holidays, and weekend activities. These activities "form the basis for vivid and close family relationships" (p. 445). Starting with Licoppe's (2004) theory of "connected presence," Christensen (2009) describes the mobile phone's phenomenological implications within the family setting. Licoppe argues that each of the mediated interactions "reactivates, reaffirms, and reconfigures the relationship" (2004, p. 138). Christensen further argues that with each new text message or phone call a sense of closeness solidifies the relationship despite the fact that the individuals are physically separate. Wajcman, Bittman, and Brown (2008) also confirmed that mobile phones are used by working parents primarily for contacting family members rather than being used predominantly for work purposes. They found that mobile phones significantly increase people's capacity to maintain intimate relationships from a distance and thus allow them to document this intimacy multiple times a day.

This type of communication connection may be particularly important for nonresidential parents, the majority of whom are fathers (Braithwaite, Schrodt, & Baxter, 2006). An important implication of the use of mobile phones is that it allows for divorced parents to communicate with their noncustodial children without the custodial parent acting as gatekeeper. Noncustodial parents give their children cell phones so they can talk without surveillance by the custodial parent and so they can call and text to say "good night" and "I love you" (Livingstone, 2002; cited in Kaare et al., 2007).

Similar to previous research on mobile phones (Licoppe, 2004; Christensen, 2009), Wellman et al. (2008) of the Pew Internet and the American Project found that mobile phones and the Internet are bringing closeness to family members rather than pulling them apart. The study found family members using ICTs to accomplish everyday practices, such as email for work, education, and personal purposes. Teens and tweens (10–12 year olds) use them to keep up with their social network as well as doing school work. Smart phones allow students to access textbooks online and connect with libraries and databases to research term papers, while parents use ICTs for work and for staying connected with children and spouses or significant others. Smart phones with GPS allow parents to track the whereabouts of their children, while video technology in the phones allow parents to view children's accomplishments via smart phones when they cannot attend a school play, dance recital, or sporting event.

Pew research suggests that households depicting a traditional nuclear family with married parents and children are more likely than other family configurations to have mobile phones, home computers, and broadband Internet access on a regular basis (Wellman et al., 2008). More than half of these families owned multiple mobile phones and multiple computers. Of course, this finding has political, socio-economic, and cultural implications in that traditional family configurations generally fit into higher socio-economic brackets than single parent households and those of same sex partner families.

As Pew research has followed technology and family configurations, they also examine how ICTs allow for new types of family connectedness through mobile phone use and communal Internet experiences. Pew scholars describe these new family connections as "hey, look at this" moments. Working parents are highly connected via ICTs and can coordinate their busy schedules with each other to determine who is picking up the kids or who will cook dinner. On the other hand, high-tech families are less likely to have dinner together and less likely to express satisfaction with their leisure time. Their dinners are often fast food eaten in the car on the way to soccer practice or dance class. As for the lack of satisfaction with leisure time, they have no leisure time; ubiquitous computing makes more work not less work for family members, especially employed mothers (Rakow, 2007; Rodino, 2003; Rodino-Colocino, 2006). Ubiquitous computing only allows family members to manage responsibilities by connecting with each other to talk about where they're going, what they're doing, and when they'll see one another again.

Teens' and Tweens' Use of Mobile Phones

Family context is an important part of teenagers' use of mobile phones (Ishii, 2006). Teens use mobile phones to maintain social networks with friends without being closely monitored by their parents. Interestingly, this is a common reason cross culturally and seen as a way to achieve freedom from family surveillance (Kim, 2002; Oksman & Turtiainen, 2004; cited in Ishii, 2006). Also, ICTs give children new ways of expressing both content and meaning that were not easily communicated before (Kaare et al., 2007). Kaare et al.'s survey revealed that by age 10, Norwegian children are proficient users of mobile phones and Internet. By 10–13 half of them have their own mobile phones and send an average of 1.5 messages a day. By age 16, 100% of Norwegian teens have their own mobile phones (Kaare et al., 2007).

Bachen (2001) argues that for teenagers, the mobile phone is important for being connected to their friends, but also, to many teens' chagrin, the freedom to connect also means that parents can check on them more easily. In addition, Palen et al. (2001) utilized a similar metaphor to describe the mobile phone as an "umbilical cord" that allows parents and children "to retain a permanent channel of communication in times of spatial distance" (Geser, 2004, p. 12). Nafus and Tracey (2002) argue that mobile phones "enable parents and children to check in with each other as the children explore new spaces" (p. 212). An extension of the parent-child relationship is found in parents' distinguishing differences in needs based on age. Parents in Nafus and Tracey's (2002) study in the United Kingdom felt there was an appropriate age for acquiring cell phones and that was when s/he attained a driver's license at age 16.

In addition, Roos (2001) argues that mobile phones enable more flexible negotiation and a new form of interaction between children and parents. Similarly, Katz (1999) argues that being connected via ICTs makes people more accountable to others. One consequence of wireless communication is that contact makes us more responsible for our actions and responsible to the people with whom we are connecting. It makes us "more subject to social control" (p.17).

Blended Uses of Talking, Texting, and Instant Messaging (IM)

Text messaging is used predominantly by working parents to communicate with friends and family, rather than for business purposes (Castells et al., 2007; Christensen, 2009; Kaare et al., 2007; Wajcman et al., 2008). Castells et al. (2007) argue, "Unlike voice calls, which are generally point-to-point and engrossing, messaging can be a way of maintaining ongoing background awareness of others, and of keeping multiple channels of communication open" (p. 105; cited in Manghani, 2009). Green (2003) and Kaare, Brandzæg, Heim, and Endestad (2007) both found that kids tend to use mobile phones to call their parents, but prefer to text their friends. Similarly, half of online teens interact with siblings via IM (Bryant & Bryant, 2006). In addition, Christensen (2009) found that children would use both talk and text functions on their mobile phones to communicate with their parents; however, parents always used the talk function when calling their kids, as well as when they called each other. Ishii (2006) suggests that teenagers' shyness and immature social skills lead to their preference for text messaging rather than calling. Furthermore, Wajcman, Bittman, and Brown's (2008) study of families in Australia found that perpetual contact through text messaging creates new levels and forms of intimacy. The ability to form, deepen, and even dissolve relationships via text messaging enhances families' ability to be "communicatively present while being physically absent" (p. 648).

Time delay and duration of messages are also important parts of texting for families. Texting is convenient when calling is inconvenient or inappropriate (Ito & Okabe, 2006; cited in Hampton & Gupta, 2008). It is a quick connection, especially if one is trying to reach a teenage child who is in class or a partner who may be in a meeting. They cannot answer a call but nevertheless need to receive the message to call back as soon as possible. Texting is also convenient when a connected call would take more time and emotional energy than one currently has due to a hectic schedule or lack of sleep.

Social Networking

Social networking sites (SNSs) include a variety of websites that offer personal connections among users. These SNSs operate around a central focus on a person's profile, which includes a name or nickname and information about her/himself. SNSs require people to indicate their relationships, or friendships, with other participants on the networking site (boyd, 2006). The most commonly accessed SNSs are Facebook and Myspace, both of which began as networking sites for college students and have expanded to include the general public. Research studies (Gross et al., 2005; Lampe et al., 2006; Stutzman, 2006; all cited in Tufekci, 2008) show that 80–90% of college students use SNS and have a profile on at least one. Academic researchers are only just beginning to realize the impact of online social networking. Acar (2008) examined the differences in users' online networks versus real life networks and discovered that "gender and extroversion are the major predictors of both online social network size and time spent online for social networking" (p. 75). In particular, women have larger networks and spend more time communicating online than men do.

The amount of interactivity available to children and their parents is astounding with more than 250 million users on Facebook (http://www.facebook.com/press/info.php?statistics), and a membership that has increased 228% in the last year. Although not specific to U.S. families, one example of the potential change in communication among families using SNSs emerged from a specialized SNS site for working mothers in Hong Kong named "Happy Land." Chan (2008) conducted a virtual ethnography and participated in the Happy Land parenting community, a network site in which the members ask for advice on parenting and balancing work and family responsibilities. Ironically, most of the interactions are conducted online during the work day when the boss is not watching. Chan found virtual relationships constructed online often turned into offline relationships for not only the employed mothers but their families as well. Members would gather for play dates and find that they built relationships among the adults too. Thus, the women started including spouses and entire family groups would meet for picnics and other family activities. Similarly, in her study of entrepreneurial mothers Edley (2004) found that women with home-based businesses utilize online communities as resources for advice on professional and family issues and how to blend work and family in the same space. Hlebec, Manfreda, and Vehovar (2006) also found a social support system among Internet users in Slovenia. Their network analysis found Internet users providing support to each other within social networks that existed both online and offline, supplementing previous research by Haythornthwaite (2000; cited in Hlebec et al., 2006) that online communication complements other communication media.

An additional social interaction tool, Twitter, which is a 140-character microblogging website emerged in 2006 to engage people in the question: what are you doing? According to Nielsen Online, *twitter.com* has increased its number of tweeters by 1382% in a year to reach 7 million in February 2009. However, contrary to popular belief, the largest group of tweeters consists of people aged 35–49 not college students. Nielson Online found a majority of people visit *twitter.com* from their workplace (http://blog.nielsen.com/nielsenwire/online_mobile/twitters-tweet-smell-of-success/).

Crane and Morales (2009) argue that Twitter is being used more often by working parents, especially working fathers, to stay connected with their families. Twitter's ease of not having to create a profile or update a webpage allows working fathers to connect quickly, easily, and discreetly from their mobile phones during breaks and, if necessary, in meetings, without disrupting work flow.

The Dark Side of Social Networking

Young people's use of SNSs suggests there may be a dark side of providing personal information on a public site. Recent inquiries into how SNSs are used by employers demonstrate what is at risk for teens or young adults who are just "trying on" new identities (see Livingstone, 2008, for strategies of presenting the self online) or sharing transient moments with friends. According to Roberts and Roach (2009), human resource (HR) professionals are using SNSs to either screen out potential job candidates or to network with potential employees. Potential employers can rule them out as job candidates based on what the employer might see as poor decision-making and unprofessional presentation of self.

The presentation of teen selves on SNSs means that parents may need to help reign in a space that has provided an opportunity for teens to interact with peers. Risks to college students who post pictures of themselves and friends in various states of inebriation and stages of undress should be explained. While many teens set their profile to the private option (Lenhart & Madden, 2007), they "may disclose personal information with up to several hundred people known only casually" (Livingstone, 2008, p. 404). Livingstone (2008) also notes that teens may disclose "personal information that previous generations often have regarded as private (notably age, politics, income, religion, sexual preference)" (p. 404). While it's not clear how the courts would rule if information that was intended to be private were obtained without permission by a potential employer (Brandenberg, 2008), it suggests a new coaching role for parents that could help teens establish appropriate boundaries for displaying personal information.

Contributions and Implications for Future Research

At a fundamental level as we reflect on the assumptions of technological use, we realize that ICT adoption is often equated with change and that adaptation patterns in family communication would be the anticipated outcome. Still, the review of gendered notions of technology reveals a pro-masculinity bias of technology, which may constrain women's exposure to and use of ICTs. Despite the proliferation of technology-based communication patterns and rhythms of family, coordination and communication continue to be maintained by women. Mom is communication central.

Still some trends in changing family communication dynamics are clear. First, lives are saturated with technology. Embracing any number of ICTs means that the time family members can devote to interpersonal communication may shift from moment to moment. In spite of changes in attention demanded by ICTs, scholars found face-to-face communication is still an essential part of relationships (Baym et al., 2004; Christensen, 2009; Ishii, 2006). Both face-to-face communication and online and mobile communication are necessary for close relationships among family members and close friends. Both online and offline communication enhance relationship maintenance and stability. For example, several scholars (Christensen, 2009; Golden, 2009b; Wellman et al., 2008) found that family dinners are still an important component of family relationships.

ICTs and Work-Life Balance: The Three Illusions

ICTs are often promoted as tools that can help families, in particular parents, manage work-life balance. In our review, this *illusion of empowerment* raises the question, how are we empowered if we are actually working longer days by choosing when and where we work? Araújo (2008) argues that the time spatial component erases our ability to judge how much time something takes to do because it is broken into parts and accomplished late at night or early morning when everyone else is sleeping. ICTs also elevate the standards of what we should be doing. For instance, feminist critiques (Cowan, 1983; Schor, 1991) of technology in the home suggest that when vacuum cleaners and washing machines were invented, our standards of cleanliness and our workload were increased. The same phenomenon has occurred as ICTs increase demands for more availability and productivity at both home and work.

In a similar vein, organizations encourage us to base our everyday decisions whether purchasing a new version of Word or the latest iPhone on what is best for the organization, or its bottom line. Hence we are colonized by organizations for which we work and by organizations that produce the products and services we consume. However, this *illusion of control*, would suggest that we are able to extend this control to child-rearing. ICT connections mean that children may be increasingly accountable for their whereabouts, which leads to a feeling of security among parents. Since children may be required to check in by text or by calling, working parents may feel they have better control and access to their children via the ICT umbilical cord. Still, the connection may be fragile since phones may lose a signal or batteries may need to be charged. Setting aside the technical aspects of connectivity, mobile devices still require active participation of users. The first author had a recent experience in which her son texted her that he would not be home by curfew. Not knowing that her phone's battery had died, he assumed she received the text. She called him after curfew and he didn't pick up. Later he explained that he felt a need to protect his friend and her siblings from their drunken father, so he stayed until the father fell asleep. Despite putting himself in potential danger of drunken violence, the decision demonstrated a sense of responsibility and maturity; however, he should not have assumed that his mother received the text since she didn't respond. Of course the parent had no control in this situation. Similarly, the second author was with a friend whose teen daughter went out for a smoothie with a classmate. When time came to touch base with the daughter, she ignored her mother's calls and texts. This meant they had to search for the daughter who was not where she was supposed to be. Clearly both scenarios address situations which demonstrate the illusory control ICTs offer parents who attempt to provide opportunities for children to demonstrate responsible behavior and are left feeling helpless when faced with a lack of communication.

Illusions of choice show up in three ways: the push to purchase and the class- and race-based assumptions of ICT use. In the *illusion of choice to purchase*, marketers colonize our lives with gadgets that must be acquired in order to be successful, productive, or to make our lives easier. For example, Gregg (2008) argues that ad campaigns have "enshrined the laptop and mobile phone as *compulsory companions* (emphasis added) of the successful businesswoman finally realizing her dream of equality in the workplace, asserting her independence through challenging, well compensated work" (p. 290). Gregg's description illustrates that success is equated to connectivity.

A second aspect of the illusion of choice is the class-based nature of ICTs. According to Pew (2004), only 44% of parents earning less than $30, 000 a year have internet access. Comparatively, 89% of all households whose income is over $75, have internet access. Parents employed in jobs that don't pay living wages are not afforded the same opportunities to connect work and family realms via telework or flex time. Many of these jobs are production or service-oriented and require onsite work focused on machinery or clients. As Gregg (2008) notes, the closest immigrant women in Silicon Valley get to technology is to work in processing plants. Hossfeld (2001) highlights the class-based nature of those who benefit from technology versus those who are "indentured" to it:

> When they aren't indentured to the strict schedules of computer assembly lines, their experience of "working from home" is to produce the labels we creatives wear in performing our cool fashion credentials via an increasingly aestheticised information workstyle. To meet the demanding orders for the international brand names servicing the West, these women often enlist the help of children and other extended family members in what amount to home sweatshops. (As cited in Gregg, 2008, p. 294)

Race, the third illusion of choice, is invisible in most of the new media and work-life balance studies. Young (2005; cited in Rice & Katz, 2008) found that African Americans and Latino/as are no longer lagging behind in their use of mobile technology. Both groups are now early adopters and innovators of wireless technology. Still, when it comes to the issue of choice, race[6] may play a factor. We see an imperative need for new media and work-life balance studies to include race in more cultural terms than simply discussing the digital divide, which is important, but as scholars we need to push the interrogation of race, ICTs, and work-life balance to address specifically the issues of power, race, gender, and class and how these constructions impact everyday lives of people at the margins of society.

The New Public/Private Split

Previous research (e.g., Golden, Kirby, & Jorgenson, 2006; Kirby et al., 2003) describes a shift in the boundaries of public and private time and place. In previous workplace configurations, the public was the domain of work, which did not overlap with the private arena of the home, yet as we have noted ICTs have erased this public/private boundary. While this removal will remain a notable trace of ICTs' footprint on working families' lives, there are other public/private divides that are shifting or switching. As previously mentioned, Hampton & Gupta (2008) note the trend towards "public privatism," which is when private communication occurs in public spaces. In this instance, family members may be conducting personal affairs in public places. Moreover, social networking has introduced a new split wherein the personal communication of private messages appears to mass audiences. What used to be "family" business is now available to a multitude of Facebook friends that may turn viral and introduce new family conflicts. Parenting and family arrangements may come under scrutiny by friends, thereby increasing the potential for family conflict as children experience normative pressures from peers. Further, a generational divide appears when teens spend any number of hours divulging details of their lives online to a multitude of friends within the confines of their home but would not talk to their parents about personal issues, such as dating or sex. While at times teens' priorities for information resources shift, Hetsroni (2008) found that after peers the Internet is a teen's preferred source of information on sex.

Gendered Critiques of Technological Control

Rakow argues that gaining control over our personal and professional lives is not better or even helpful if we don't stop to question "that these choices occur in conditions not of our own making" (2007, p. 404). Building on Rakow's (2007) call for more critical voices in new media studies, we consider how technology is a gendered paradox (as argued in Edley, 2001; Edley, Hylmo, & Newsom, 2004; Golden, 2009b). ICT use is socialized by the needs of the people using it, as well as being constructed by the marketing campaigns that convince consumers to buy them. Although ICTs can be viewed as connection to family (Rodino, 2003), as a way of establishing and maintaining connected presence (Christensen, 2009; Licoppe, 2004), and allow women to balance work and family, we still need to question the social systems that created this need—Americans' ostentatious lifestyle of consumption that requires both parents to work to keep up with this lifestyle. Also, divorce and transitive lifestyles take nuclear families away from extended family and friendship networks that serve as support systems. Hence, we create the need for technologies to help us manage family and work and to maintain a presence with both family members and co-workers. As Bryant and Bryant (2006) suggest, new forms of communication technologies may offer distributed families opportunities to stay connected.

This creation of technological needs is also influenced by corporate discourses that suggest what products are necessary and which ones will improve lives. Thus, the choice to manage work and family technologically may not be a choice at all but rather a form of corporate colonization (Deetz, 1992), not only by employers but also marketing campaigns to seduce consumers into buying the next great technological solution to complex lives. Since the cycle of technology production exists in perpetuity new ICTs must be purchased on cycle. And the more we work the more technology we need to buy in order to manage our work and family lives and maintain our connected presence so that we can continue working in order to continue consuming. This vicious cycle constitutes a big influence in the 2009 economic recession in the United States. Americans want to buy cheap goods, thus American manufacturers are forced to outsource and utilize sweatshops in China and the developing world, so that customers will buy (and Wal-Mart will sell at the lowest price) and the vicious cycle of consumption, layoffs, unemployment, and the need for balancing work and family lives via technological advancements continues.

A Note on Theory, Method, Subjects, and Future Research in Family Communication

In an attempt to focus on the intersection of technology and the shifting patterns of communication among family members, we purposefully have limited our discussion of the role of theory. Since recent scholarship on technology/new media has already offered observations of both history of the discipline and potential future directions in communication and new media research (Lievrouw, 2009;

Scott, 2009), we offer some limited remarks concerning theories used by scholars examining the intersection of technology and family communication.

Most work-life balance literature exists outside the communication discipline in management and sociology. A great deal of literature stems from Australia and the United Kingdom with the United States lagging far behind in both academic research and corporate policies. Kirby et al.'s (2003) review of literature in *Communication Yearbook 27* demonstrates the majority of the research in this area approaches work-family issues as static concepts rather than changing, dynamic discursive constructs. Another valuable review of literature is Golden et al.'s (2006) integration of work-life research literatures from family communication and organizational communication perspectives in *Communication Yearbook 30*. Some notable communication technology-focused exceptions can be found in the 2000 special issue of *The Electronic Journal of Communication* on Communication Perspectives on Work and Family, edited by Annis Golden (Avery & Baker, 2000; Farley-Lucas, 2000; Golden, 2000; Jorgenson, 2000; Kirby, 2000; Mallia & Ferris, 2000) and the 2006 special issue of *The Electronic Journal of Communication* on Communication and the Accomplishment of Personal and Professional life, edited by Erika Kirby. This issue includes essays that address issues of technology and telework options (Gill, 2006; Golden & Geisler, 2006; Shuler, 2006). Over the past 15 years, only a few communication-focused research studies on telework and working from home have been published (Edley, 2001, 2004; Edley et al., 2004; Hylmö, 2004, 2006; Hylmo & Buzzanell, 2002; Jackson, Leonardi, & Nelson, 2003; Jorgenson, 1995, 2000; Shumate & Fulk, 2004; Tremblay, 2003; Tremblay et al., 2006).

Studies of ICTs bridge many disciplines and sub-disciplines, from information theory to mass communication to interpersonal and family communication. The theoretical perspectives employed are as rich and varied as the traditions within which they fall. Also, the methodologies range from quantitative theory-based predictions to the thick descriptions of ethnographic convention. Our interest here is not to exhaustively recount all theories, but to indicate a few trends and present exemplars that offer potential directions for family communication scholars interested in exploring the use of ICTs.

First, the wide variety of ICTs available for study means that it can be difficult to unify the literature in any meaningful way to assemble results that may offer holistic insights into changing family communication dynamics. For example, while some scholars (Holloway, 2007; Hylmö, 2004, 2006; Hylmö & Buzzanell, 2002; Kivimaki, 2004, cited in Gregg, 2008; Mirchandani, 1998; Tremblay, 2003; Tremblay et al., 2006) focus on the use of Internet technology in the home, others focus on the use of mobile phones (Bachen, 2001; Christensen, 2009; Geser, 2004; Lemish & Cohen, 2005; Ling, 2003; cited in Ling, 2004; Ling & Yttri, 2002; Rice & Katz, 2003; Wajcman et al., 2008; Wei & Lo, 2006). While specific, focused research offers important insights into how and why people use a particular ICT in a particular way or for a specific reason, it is challenging to understand how the use of multiple technologies may impact the use of any one technology. While not comprehensive in its approach to understanding ICTs, one notable exception is Christensen's (2009) study of families' use of ICTs to connect with each other. He aptly identifies the different preferences teens and parents have for texting versus talking on a mobile or landline phone with particular audiences. His interview research offers greater insights into the dynamic use of multiple functions within one ICT that interlocutors use for different audiences.

The second issue concerns the concept of boundaries. If scholarly interest centers on working adults (Edley & Houston, 2006; 2007; Golden, 2009a, 2009b) or teens (Kaare, Brandtzæg, Heim, & Endestad, 2007; Valkenburg & Peter, 2009) then communication scholars have done an excellent job of understanding particular subgroups of users. Few scholars have investigated families as a unit of analysis (again, a notable exception is Christensen, 2009) where we might discover patterns particular to children's age, income, and uses of particular technologies. It is easy to envision, for example, how parents may use mobile phones when children are younger to perform third shift/parallel shift work (Edley, 2001; Rakow & Navarro, 1993), yet when children are old enough to employ technologies themselves the conversation shifts to text messaging, Twitter, or IM. Family communication scholars may wish to consider how to approach understanding whole family dynamics. This call is consistent with Bryant and Bryant (2006) who suggest a multitheoretical, multilevel (MTML) approach to integrate "multiple theories and social relationships" to account for the complexities of studying ICTs in families.

The diversity of literature examining the intersection of ICTs and family communication stems from a variety of traditions and is characterized by a wide range of theories. Four strands of theoretical perspectives characterize much of the research on ICTs and family communication: mass communication, social interaction, critical and new media approaches. Some research draws on the rich traditions of mass communication to examine individual technology uses (e.g., Wei & Lo, 2006) while other scholars seek to examine consequences via relational quality and interaction (e.g., Baym, Zhang, Kunkel, Ledbetter, & Lin, 2007) or social support (Rice & Katz, 2008). A third critical strand of research seeks to address the inequalities that appear in the construction or uses of technology whether based on gender, social class[7]or race. This scholarly work addresses how technology coupled with cultural constructions may create and impact social groups in a different way. The fourth research strand, new media, examines the impact of the variety of new media technologies and is reported on by scholars of communication, business, sociology, and information systems. Some of the most promising research stems from the idea of how people attend to multiple messages. While what counts as "new media" is time- and context-dependent, some of the research in this area has begun to examine potential legal challenges or changes as well as changing patterns of communication connectedness on social networking sites and Twitter.

Social presence and media richness theories have inspired new communication theories to explain the complexity of digital connectedness. Turner and Reinsch, Jr., describe a theory of multicommunicating as a "form of polychronic behavior and multitasking" (2007, p. 36). Using the ideas of message equivocality and interlocutor status, Turner and Reinsch (2007) examine how people allocate their personal presence across multiple interactions. While their research examined workplace instant-messaging interactions, one could easily imagine how family communication scholars might investigate the tensions they experience as they must choose to privilege work over family or children over spouses. Similarly, Christensen (2009) builds on Licoppe's theory of connected presence and offers opportunities to consider how the number or content of messages may reinforce strong bonds within families. Christensen argues that the uptick in the number of messages that have emerged as a result of mobile phones work to "reaffirm the close relations between family members and contributes to the continuous conversation that is the foundation of close and meaningful relationships" (2009, p. 446). Family communication scholars might also note that Christensen focuses on social interaction among family members via interviews with parents and children about the content and medium of ICT communication (i.e., texting, calling on mobile or landline).

Furthermore, it is necessary to focus on the materiality of the ICT-mediated relationships not just the discursive constructions of the new media and its uses. The gendered enactments of everyday life and the management of family and work relationships and responsibilities are complex constructions that are discursive as well as material. The material bodies of working parents, their children, and their aging parents interact with each other through mobile technologies. The mundane accomplishments of work and family management by gendered bodies are the epitome of materiality. As Lee (2006) argues, women's relations to new ICTs are both materialist and discursive. They are also economic and cultural, and gendered and political. Materialist perspectives also need to address the consumption of ICTs and the fact that not all socioeconomic classes of people worldwide are equally able to access or engage ICTs.

Of additional note, our review found class and race notably absent in the discussions of ICT uses in the management of work and family relationships and responsibilities. International and intercultural studies gave us insight into cultural differences and similarities in the uses of ICTs, but not one discussed the social implications of race or class in the balancing acts of working parents and their children. This lack of focus on race and class is problematic, just as ignoring gender is unacceptable when discussing a social phenomenon that is pervasive. As scholars we need to attend to the nuances of all social and political inequities of family communication via mobile technologies.

Finally, we note the importance for U.S. scholars and corporate policymakers to begin to engage in serious consideration of telework and telecommunications policies to produce more family-friendly work/life policies and thus more case studies for scholars to investigate. The studies reviewed here were predominantly from outside the United States and included Australia and Canada, both of which had nation wide scholarly studies of ICT use and management of work and family relationships and responsibilities. Studies also emanated from northern Europe, especially the United Kingdom and Norway. The common denominator among these countries is their socialist approach to work-family balance issues that offer a family-friendly approach to work and government policies. Government sponsored programs and policies offer parents opportunities to manage work and family through telework and through their more highly developed and marketed ICT products that give parents and children more reliable network services, software apps and hardware innovations. In order to remain competitive in a world market and meet the needs of working families, the United States needs to address its antiquated work-life practices and policies.

Notes

1 A note on assumptions concerning technology: Following Houston and Jackson (2003) we recognize the intersection of context and technology and start with an integrationist perspective that assumes technology and context are inseparable.

2 For an excellent review of the meaning and discourse surrounding STEM work; see Kisselburgh, Berkelaar, & Buzzannel (2009).

3 The terms "working mother" or "working parents" are contentious in that all mothers, whether "stay-at-home" mothers or mothers who do paid work outside the home, are working mothers and all parents are working parents. Hence in this essay we use the terms "employed mothers" and "employed parents" to mean those who do paid work either inside or outside the home. Thus, this terminology encompasses teleworkers and entrepreneurs as well.

4 The term "work-family balance" suggests a static, idealized state in which work and family responsibilities are equalized. The other much used term to describe this area of research is the term "work-life balance," which implies that life and work are two separate categories. We find this term problematic as well since it suggests that work is not life and life is not work. Our lived experience tells us that work is a large part of our lives and even labors of love, such as caring for children, spouses, and aging parents, are also forms of work albeit unpaid work.

Following Kirby et al.'s (2003) and Golden et al.'s (2006) synthesis of the literatures on work-life issues, we advocate the re-labeling of this area of research. In this manuscript we strive to use the terms "balancing" and "managing" of work and family relationships and responsibilities. The terms "balancing" and "managing" are both active verb forms and both suggest a dynamic process rather than a steady state, while the combined term "work-family" illustrates our specific focus. Moreover, our use of the term "relationships and responsibilities" shows our commitment to the relational perspective on which we focus. Kirby (2009) has changed over time her way of describing the phenomenon formerly known as work-family balance, which she argued was too exclusionary, to adopting the phrase work-life balance, which she also found problematic in that work is still a part of life, to using "personal-professional life" (Kirby, 2006), to adopting (specifically for a particular forum in 2009 Communication Monographs) to using the terms "working/institutional (WI)"and "personal/family life (PI)" (Kirby, 2009).

5 An exception to this line of research lies in a recent study by Edley and Houston (2007) in which working fathers were found to be putting in more time around the house and doing more than just "helping" with the children and housework.
6 We recognize that race and social class are often culturally intertwined.
7 Social class as we discuss would encompass research on the digital divide.

References

Acar, A. (2008). Antecedents and consequences of online social networking behavior: The case of Facebook. *Journal of Website Promotion, 3*(1 & 2), 62–83.

Ahuja, M. K., Chudoba, K. M., Kacmar, C. J., McKnight, D. H., & George, J. F. (2007). IT road warriors: Balancing work-family conflict, job autonomy, and work overload to mitigate turnover intentions. *MIS Quarterly, 31*(1), 1–17.

Araújo, (2008). Technology, gender and time: A contribution to the debate. *Gender, Work, and Organization, 15*(5), 477–503.

Avery, G. C., & Baker, E. (2000). Understanding technology use within Australian households. *The Electronic Journal of Communication/La Revue Electronique de Communication, 10*(3). Retrieved July 24, 2009, from http://www.cios.org/www/ejcrec2.htm

Bachen, C. (2001). *The family in the networked society: A summary of research on the American family.* Paper presented at the Center for Science Technology Society, Santa Clara University, Santa Clara, CA. Retrieved July 24, 2009, from http://www.sts.scu.edu/nexus/Issue1-1/Bachen_TheNetworkedFamily.asp

Baym, N.K., Zhang,Y. B., Kunkel, A., Ledbetter, A., & Lin, M. C. (2007). Relational quality and media use in interpersonal relationships. *New Media & Society, 9*(5), 735–752.

Baym, N. K., Zhang, Y. B., & Lin, M.-C. (2004). Social interactions across media: Interpersonal communication on the Internet, telephone, and face-to-face. *New Media & Society, 6*, 299–318.

Blinkoff, R., & Blinkoff, B. (2000). *Wireless Opportunities: A Global EthnographicStudy*, Context-Based Research Group ethnographic report, December.

Bolton, M. K. (2000) *The third shift: Managing hard choices in our careers, homes, and lives as women.* San Francisco, CA: Jossey-Bass.

boyd, d. (2006, December). Friends, friendsters, and top 8: Writing community into being on social network sites. *First Monday* [Online], *11*(12).

Braithwaite, D.O., Schrodt, P., & Baxter, L. A. (2006). Understudied and misunderstood: Communication in stepfamily relationships. In K. Floyd & M.T. Morman (Eds.), *Widening the family circle: New research on family communication* (pp. 153–169). Thousand Oaks, CA: Sage.

Brandenburg, C. (2008). The newest way to screen job applicants: A social networker's nightmare. *Federal Communications Law Journal, 60*(3), 597–626

Bryant, J. A., & Bryant, J. (2006). Implication of living in a wired family: New directions in family and media research. In L. H. Turner & R. West (Eds.), *The family communication sourcebook* (pp. 297–314). Thousand Oaks, CA: Sage.

Butler, J. (1989). *Gender trouble: Feminism and the subversion of identity.* New York: Routledge.

Campbell, S. W. (2007). A cross-cultural comparison of perceptions and uses of mobile telephony. *New Media & Society, 9*(2), 343–363.

Castells, M., Fernandez-Ardèvol, M., Qui, J. L., & Sey, A. (2007). *Mobile communication and society: A global perspective.* Cambridge, MA: MIT Press.

Chan, A. H. N. (2008). "Life in Happy Land": Using virtual space and doing motherhood in Hong Kong. *Gender, Place, and Culture, 15*(2), 169–188.

Chesley, N. (2005). Blurring boundaries? Linking technology use, spillover, individual distress, and family satisfaction, *Journal of Marriage and Family, 67*(5), 1237–1248.

Christensen, T. H. (2009). 'Connected presence' in distributed family life. *New Media & Society, 11*, 433–451.

Clark, S. C. (2000). Work/family border theory: A new theory of work/family balance. *Human Relations, 53*, 747–770.

Consalvo, M. (2006). Gender and new media. In B. J. Dow & J. T. Wood (Eds.), *The Sage handbook of gender and communication* (pp. 355–369). Thousand Oaks, CA: Sage.

Cowan, S R. (1983). *More work for mother: The ironies of household technology from the open hearth to the microwave*. New York: Basic.

Crane, S., & Morales, M. (2009). Working fathers: Maintaining roles, responsibilities, and communication for work-life balance. Unpublished manuscript.

Deetz, S. A. (1992). *Democracy in an age of corporate colonization: Developments in communication and the politics of everyday life*. Albany: State University of New York Press.

Edley, P. P. (2001).Technology, working mothers, and corporate colonization of the lifeworld: A gendered paradox of work and family balance. *Women and Language, 24*, 28–35.

Edley, P. P. (2004). Entrepreneurial mothers' balance of work and family: Discursive constructions of time, mothering, and identity. In P. Buzzanell, H. Sterk, & L. Turner (Eds.), *Gendered approaches to applied communication contexts* (pp. 255–274). Thousand Oaks, CA: Sage.

Edley, P. P., & Houston, R. (2006, November). *Working mothers' use of technology in work-family balance*. Paper presented at the annual meeting of the National Communication Association in San Antonio, TX.

Edley, P. P., & Houston, R. (2007, October). *What about Dad?: How working fathers balance work and family*. Paper presented at the annual meeting of the Organization for the Study of Communication, Language, and Gender. Omaha, NE.

Edley, P..P.., Hylmö, A., & Newsom, V. A. (2004). Alternative organizing communities: Non-traditional organizing practices. *Communication Yearbook 28*, (pp. 87–126). Thousand Oaks, CA: Sage.

Ellison, N. B. (2005). *Telework and social change: How technology is reshaping the boundaries between home and work*. Westport, CT: Praeger.

Ellwood-Clayton, B. (2006). All we need is love—and a mobile phone: Texting in the Philippines. *Cultural Space and Public Sphere in Asia*, 357–369.

Farley-Lucas, B. (2000). Communicating the (in)visibility of motherhood: Family talk and the ties to motherhood with/in the workplace. *The Electronic Journal of Communication/La Revue Electronique de Communication, 10* (3). Retrieved July 24, 2009, from http://www.cios.org/www/ejcrec2.htm

Geser, H. (2004). Towards a sociological theory of the mobile phone. Retrieved July 24, 2009, from http://socio.ch/mobile/t_geser1.pdf

Gill, R. (2006). The work-life relationship for "people with choices": Women entrepreneurs as crystallized selves? *Electronic Journal of Communication, 16*(3–4).

Golden, A. G. (2000). What we talk about when we talk about work and family: A discourse analysis of parental accounts. *The Electronic Journal of Communication/La Revue Electronique de Communication, 10* (3).Retrieved July 24, 2009, from http://www.cios.org/www/ejcrec2.htm

Golden, A. G. (2009a). Employee families and organizations as mutually enacted environments: A sensemaking approach to work life interrelationships. *Management Communication Quarterly, 22*, 385–415.

Golden, A.G. (2009b). A technologically gendered paradox of efficiency: Caring more about work while working in more care. In S. Kleinman (Ed.), *The culture of efficiency* (pp. 339-354). New York: Peter Lang.

Golden, A. G., & Geisler, C. (2006). Flexible work, time, and technology: Ideological dilemmas of managing work-life interrelationships using personal digital assistants. *The Electronic Journal of Communication /La Revue Electronique de Communication, 16*(3–4). Retrieved July 24, 2009, from http://www.cios.org/www/ejc/v16n34.htm

Golden, A. G., Kirby, E. L., & Jorgenson, J. (2006). Work-life research from both sides now: An integrative perspective for organizational and family communication. In C. Beck (Ed.), *Communication Yearbook, 30* (pp. 143–195). Thousand Oaks, CA: Sage.

Green, N. (2003). Outwardly mobile: Young people and mobile technologies. In J. Katz (Ed), *Machines that become us: The social context of personal communication technology*, (pp.201-19). New Brunswick, NJ: Transaction Publishers.

Gregg, M. (2008). The normalization of flexible female labour in the information economy. *Feminist Media Studies, 8*(3), 285-299.

Haddon, L., & Silverstone, R. (2000). Home information and communication technologies and the information society, in K. Ducatel, J. Webster,. & W. Herrmann (Eds.) *The information society in Europe: Work and life in an age of globalization*, (pp. 233-58). Lanham, MD: Rowman and Littlefield Inc.

Hampton, K. N., & Gupta, N. (2008). Community and social interaction in the wireless city: Wi-fi use in public and semi-public spaces. *New Media & Society, 10*(6), 831–850.

Hetsroni,A. (2008). Dependency and adolescents' perceived usefulness of information on sexuality: A cross-cultural comparison of interpersonal sources, professional sources and mass media. *Communication Reports, 21*(1), 14–32.

Hilbrecht, M., Shaw, S., Johnson, L. C., & Andrey, J. (2008). "I'm home for the kids": Contradictory implications for work-life balance of teleworking mothers. *Gender, Work, and Society, 15*(5), 444–476.

Hlebec, V., Manfreda, K. L., & Vehovar, V. (2006). The social support networks of Internet users. *New Media & Society, 8*(1), 9–32.

Hochschild, A. R. (1989). *The second shift*. New York: Avon.

Hochschild, A. R. (1997). *The time bind: When work becomes home and home becomes work*. New York: Metropolitan.

Holloway, D. (2007). Gender, telework and the reconfiguration of the Australian family home. *Continuum: Journal of Media & Cultural Studies, 21*(1), 33–44.

Hossfeld, K. (2001). "Their logic against them": Contradictions in sex, race, and class in Silicon Valley. In A. Nelson & T. L. Tu with A. H. Hines (Eds.), *Technicolor: Race, technology, and everyday life*, (pp. 34–63). New York: New York University Press.

Houston, R., & Jackson, M. (2003). Technology and context within research on international development programs: Positioning an integrationist perspective. *Communication Theory, 13*(1), 57–77.

Houston, R., & Ricigliano, L. (2003) *Pink collar workers in the digital age: Technology and gender in academic libraries*. Paper presented at the International Communication Association Annual Meeting, San Diego, CA.

Hylmö, A. (2004). Women, men, and changing organizations: An organizational culture examination of gendered experiences of telecommuting. In P. Buzzanell, H. Sterk, & L. Turner (Eds.), *Gendered approaches to applied communication contexts* (pp. 47–68). Thousand Oaks, CA: Sage.

Hylmö, A. (2006). Telecommuting and the contestability of choice: Employee strategies to legitimize personal decisions to work in a preferred location. *Management Communication Quarterly, 19*(4), 541–569.

Hylmö, A., & Buzzanell, P. (2002). Telecommuting as viewed through cultural lenses: An empirical investigation of the discourses of utopia, identity, and mystery. *Communication Monographs*, *69*(4), 329–356.

Ishii, K. (2006). Implications of mobility: The uses of personal communication media in everyday life. *Journal of Communication*, *56*(2), 346–365.

Jackson, M. (2002). *What's happening to home: Work, life, and refuge in the information age*. Notre Dame, IN: Sorin.

Jackson, M. (2008). *Distracted: The erosion of attention and the coming dark age*. Amherst, NY: Prometheus.

Jackson, M. H., Leonardi, P. M., & Nelson, N. A. (2003, May). *Technology and the construction of telework practices: The case of broadband*. Paper presented at the annual meeting of the International Communication Association, San Diego, CA.

Jorgenson, J. (1995). Marking the work-family boundary: Mother-child interaction and home-based work. In T. Socha & G. Stamp (Eds.), *Parents, children and communication: Frontiers of theory and research*. Mahwah, NJ: Lawrence Erlbaum.

Jorgenson, J. (2000). Interpreting the intersections of work and family: Frame conflicts in women's work. *The Electronic Journal of Communication/La Revue Electronique de Communication, 10*(3). Retrieved July 24, 2009, from http://www.cios.org/www/ejcrec2.htm

Kanter, R. M. (1977). *Men and women of the corporation*. New York: Basic.

Kanter R. M. (1989). *When giants learn to dance*. New York: Simon & Schuster.

Kaare, B. H., Brandzæg, P.B., Heim, J., & Endestad, T. (2007). In the borderland between family orientation and peer culture: The use of communication technologies among Norwegian tweens. *New Media & Society*, *9*(4), 603–624.

Katz, J. E. (1999): *Connections, social and cultural studies of the telephone in American life*. London: Transaction.

Katz, J. E. (Ed.) (2003). *Machines that become us: The social context of interpersonal communication technologies*. New Brunswick, NJ: Transaction Publishers.

Katz, J. E., & Aakhus, M. (2002). *Perpetual contact: Mobile communication, private talk, public performance*. Cambridge, UK: Cambridge University Press.

Katz, J. E., & Rice, R. E. (2002). *Social consequences of Internet use: Access, involvement and interaction*. Cambridge, MA: MIT Press.

Katz, J.E., Rice, R.E., Acord, S., Dasgupta, K., & David, K. (2004). Personal media communication and the concept of community in theory and practice. In P. J. Kalbfleisch (Ed.), *Communciation Yearbook*, *28*, 315–371.

Kelan, E. K. (2007). Tools and toys: Communicating gendered positions towards technology. *Information, Communication & Society*, *10*, 358–383.

Kelan, E.K. (2008). Emotions in a rational profession: The gendering of skills in ICT work. *Gender, Work and Organization*, *15*, 49–71.

Kirby, E. L. (2000). Should I do as you say or do as you do?: Mixed messages about work and family. *The Electronic Journal of Communication/La Revue Electronique de Communication, 10* (3). Retrieved July 25, 2009, from http://www.cios.org/www/ejcrec2.htm

Kirby, E. L. (2006). Communication as the accomplishment of personal and professional life: An introduction. *The Electronic Journal of Communication/La Revue Electronique de Communication, 16*(3–4). Retrieved July 25, 2009, from http://www.cios.org/www/ejc/v16n34.htm

Kirby, E. L. (2009). "Helping you make room in your life for your needs": When organizations appropriate family roles. *Communication Monographs,* 73(4), 474–480.

Kirby, E. L., Golden, A. G., Medved, C. E., Jorgenson, J., & Buzzanell, P. M. (2003). An organizational communication challenge to the discourse of work and family research: From problematics to empowerment. In P. J. Kalbfleisch (Ed.), *Communication Yearbook* 27 (pp. 1–44). Mahwah, NJ: Lawrence Earlbaum.

Kisselburgh, L. G., Berkelaar, B. L., & Buzzannel, P. M. (2009). Discourse, gender, and the meaning of work: Rearticulating science, technology, and engineering careers through communicative lenses. In C. Beck (Ed.), *Communication Yearbook 33* (pp. 259–299). Mahwah, NJ: Lawrence Earlbaum.

Lee, M. (2006). What's missing in feminist research? *Feminist Media Studies*, 6(2), 191–210.

Lemish, D., & Cohen, A. A. (2005). On the gendered nature of mobile phone culture in Israel. *Sex Roles*, *52*(7/8), 511–521.

Lenhart, A., & Madden, M. (2007). *Social networking websites and teens: An overview.* Retrieved July 25, 2009, from http://www.pewinternet.org/pdfs/PIP_SNS_Data_Memo_ Jan_2007.pdf

Leung, L., & Wei, R. (2000). More than just talk on the move: Uses and gratifications of cellular phone. *Journalism and Mass Communication Quarterly*, 77(2), 308-320.

Lewis, S., & Cooper, C. L. (1999). The work-family research agenda in changing contexts. *Journal of Occupational Health Psychology*, 4(4), 382-393.

Licoppe, C. (2004). "Connected presence": The emergence of a new repertoire for managing social relationships in a changing communication technospace. *Environment and Planning D: Society and Space*, *22*(1), 135–156.

Lievrouw, L. A. (2009). New media, mediation, and communication study. *Information, Communication & Society*, 12(3), 303–325.

Ling, R. (2004). *The mobile connection: The cell phone's impact on society.* Amsterdam: Morgan Kaufmann.

Ling, R., & Yttri, B. (2002). Hyper-coordination via mobile phones in Norway. In J. Katz & M. Aakus (Eds.) *Perpetual contact: Mobile communication, private talk, public performance*, (pp. 139-169). Cambridge: Cambridge University Press.

Ling, R., & Yttri, B. (2006). Control, emancipation, and status: The mobile telephone in teens' parental and peer relationships, In R. Kraut, M. Brynin, & S. Kiesler (Eds.), *Computers, phones, and the Internet: Domesticating information technology* (pp. 219–34). Oxford, UK: Oxford University Press.

Livingstone, S. (2008). Taking risky opportunities in youthful content creation: Teenagers' use of social networking sites for intimacy, privacy and self-expression. *New Media & Society*, *10*, 393–441.

Loscocco, K. A., & Roschelle, A. R. (1991). Influences on the quality of work and nonwork life: Two decades in review. *Journal of Vocational Behavior*, *39*, 182--225.

Mallia, K. L., & Ferris, P. S. (2000). Telework: A consideration of its impact on individuals and organizations. *The Electronic Journal of Communication/La Revue Electronique de Communication*, *10*(3). Retrieved July 25, 2009, from http://www.cios.org/www/ejcrec2.htm

Manghani, S. (2009). Love messaging: Mobile phone texting seen through the lens of tanka poetry. *Theory, Culture & Society*, *26* (2–3), 209–232.

Martin, J. (1990). Rethinking feminist organizations. *Gender and Society*, *4*, 182–206.

Martin, J. (1994). The organization of exclusion: Institutionalization of sex inequality, gendered faculty jobs and gendered knowledge in organizational theory and research. *Organization*, *1*, 401–431.

Medved, C., & Graham, E. (2006). Communicating contradictions: Reproducing dialectical tensions through work, family, and balance socialization messages. In L. H. Turner & R. West (Eds.), *The family communication sourcebook* (pp. 353–372). Thousand Oaks, CA: Sage.

Medved, C. E., Brogan, S., McClanahan, A. M., Morris, J. F., & Shepherd, G. J. (2006). Work and family socializing communication: Messages, gender, and power. *Journal of Family Communication, 6*, 161–180.

Mirchandani, K. (1998). Protecting the boundaries: Teleworker insights on the expansive concept of "work." *Gender & Society, 12*(2), 168–187.

Nafus, D., & Tracey, K. (2002). Mobile phone consumption and concepts of personhood. In J. E. Katz & M. Aakhus (Eds.), *Perpetual contact: Mobile communication, private talk, public performance* .(pp. 206–221). Cambridge, UK: Cambridge University Press.

Nippert-Eng, C. E. (1995). *Home and work: Negotiating boundaries through everyday life.* Chicago: University of Chicago Press.

Palen, L., Salzman, M., & Young, E. (2001). Going wireless: Behavior & practice of new mobile phone users. Retrieved July 24, 2009, from http://www.cs.colorado.edu/%7Epalen/Papers/cscwPalen.pdf

Perlow, L. (1998). Boundary control: The social ordering of work and family time in a high-tech corporation. *Administrative Science Quarterly, 43*(2), 328-357.

Pew Research Center (2004). Comments from the Pew Research Center's Internet & American Life Project in the Matter of: *Empowering Parents and Protecting Children in an Evolving Media Landscape.* Retrieved July 24, 2009 from http://www.pewinternet.org/Commentary/2010/February/~/media/Files/Reports/2010/Pew%20Internet%20Project_FCC%2009-194 % 2002-24-2010.pdf

Perrons, D. (2003). The new economy and the work-life balance: Conceptual explorations and a case study of new media. *Gender, Work, and Organization, 10*(1), 65–93.

Presser, H. B. (2004). The economy that never sleeps. *Contexts, 3*(2), 42–49.

Quesenberry, J. L., & Trauth, E. M. (2005). The role of ubiquitous computing in maintaining work-life balance: Perspectives from women in the information technology workforce. In *Designing ubiquitous information environments: Socio-technical issues and challenges* (pp. 43–55). Boston, MA: Springer. Retrieved July 24, 2009, from electronic database at SpringerLink. https://commerce.metapress.com/content/427q830n44q8n155/

Rakow, L. (2007). Follow the buzz: Questions about mobile communication industries and scholarly discourse. *Communication Monographs, 74*(3), 402–407.

Rakow, L. F., & Navarro, V. (1993). Remote mothering and the parallel shift: Women meet the cellular telephone. *Critical Studies in Mass Communication, 10* (2), 144–157.

Rice, R. E., & Katz, J. E. (2003). Comparing internet and mobile phone usage: Digital divides of usage, adoption, and dropouts. *Telecommunications Policy, 27*(8/9), 597-623.

Rice, R. E., & Katz, J. E. (2008). Assessing new cell phone text and video services. *Telecommunications Policy, 32*, 455–467.

Roberts, S. J., & Roach, T. (2009). Social networking web sites and human resource personnel: Suggestions for job searches. *Business Communication Quarterly, 12*, 110–114.

Rodino, M. (2003). Mobilizing mother. *Feminist Media Studies, 3*(3), 375–378.

Rodino-Colocino, M. (2006). Selling women on PDAs from "Simply palm" to "Audrey": How Moore's law met Parkinson's law in the kitchen. *Critical Studies in Media Communication, 23*(5), 375–390.

Roos, J. P. (2001, August). Postmodernity and mobile communications. Paper presented at ESA Helsinki Conference. Retrieved July 26, 2009, from http://www.valt.helsinki.fi/staff/jproos/mobilezation.htm

Schor, J. (1991). *The overworked American: The unexpected decline of leisure.* New York: Basic.

Scott, C. R. (2009). A whole-hearted effort to get it half right: Predicting the future of communication technology. *Journal of Computer-Mediated Communication, 14*(3) 753–757.

Shuler, S. (2006). Working at home as total institution: Maintaining and undermining the public/private dichotomy. *Electronic Journal of Communication, 16*(3–4).

Shumate, M., & Fulk, J. (2004). Boundaries and role conflict when work and family are co-located: A communication network and symbolic interaction approach. *Human Relations, 57*, 55–74.

Sloan Work & Family Research Network. (2009). Conversations with the experts: The daddy shift: Stay-at-home fathers. Retrieved July 25, 2009, from http://wfnetwork.bc.edu/The_Network_News/ 60/The_Network_News_Interview60.pdf

Sproull, L., & Kiesler, S. (1991). *Connections: New ways of working in the networked organization.* Cambridge, MA: MIT Press.

Sullivan, C., & Lewis, S. (2001). Home-based telework, gender, and the synchronization of work and family: Perspectives of teleworkers and their co-residents. *Gender, Work, and Organization, 8*(2), 123–145.

Totten, J., Schuldt, B.,Taylor, A., & Donald, D. (2007). Employment differences regarding the impact of family & technology issues on sales careers. *Proceedings of the Academy of Marketing Studies*, 12(1), 45–50.

Townsend, K., & Batchelor, L. (2005). Managing mobile phones: A work/nonwork collision in small business. *New Technology, Work and Employment, 20*(3), 260–267.

Tremblay, D.-G. (2003). Telework: A new mode of gendered segmentation? Results from a study in Canada. *Canadian Journal of Communication, 28*(4), 461–478.

Tremblay, D.-G., Paquet, R., & Najem, E. (2006). Telework: A way to balance work and family or an increase in work-family conflict? *Canadian Journal of Communication, 31*(3), 715–731.

Tufekci, Z. (2008). Grooming, gossip, Facebook and Myspace: What can we learn about these sites from those who won't assimilate? *Information, Communication & Society, 11*(4), 544–564.

Turner, J. W., & Reinsch, Jr., N. L (2007). The business communicator as presence allocator: Multicommunicating, equivocality, and status at work. *Journal of Business Communication, 44*(1), 36–58.

Valcour, P. M., & Hunter, L.W., (2005). Technology, organizations, and work-life integration. In E. E. Kossek & S. J. Lambert's (Eds.), *Work and life integration: Organizational, cultural, and individual perspectives* (pp. 61-84). Mahwah, NJ: Lawrence Earlbaum.

Valkenburg, P. M., & Peter, J. (2009). The effects of instant messaging on the quality of adolescents' existing friendships: A longitudinal Study. *Journal of Communication, 59*, 79–97.

Wajcman, J. (1991). *Feminism confronts technology*. Cambridge, UK: Polity.

Wajcman, J. (2007). From women and technology to gendered technoscience. *Information, Communication, and Society, 10*(3), 287–298.

Wajcman, J., Bittman, M., & Brown, J. E. (2008). Families without borders: Mobile phones, connectedness and work-home divisions. *Sociology, 42*, 635–652.

Wei, R., & Lo, V. H. (2006). Staying connected while on the move: Cell phone use and social connectedness. *New Media & Society, 8*(1), 53–72.

Wellman, B., Smith, A., Wells, A., & Kennedy, T. (2008). Families, mobile, new media ecology. Pew Internet & American Life Project, October 19, 2008, http://www.pewinternet.org/PPF/r/213/report_display.asp.

West, C., & Zimmerman, D. H. (1987). Doing gender. *Gender & Society, 1*(2), 125–151.

Youngs, G. (2001). The political economy of time in the Internet era: Feminist perspectives and challenges. *Information, Communication, and Society, 4*(1), 14–33.

PART 3

Influences of CMC on Relational Contexts

CHAPTER TWELVE

CMC and the Conceptualization of "Friendship": How Friendships Have Changed with the Advent of New Methods of Interpersonal Communication

Amy Janan Johnson

Jennifer A. H. Becker

With many new channels of communication available, the ways individuals engage in interpersonal relationships are changing. Scholars interested in interpersonal communication must consider whether and how such changes affect the way they conceptualize and explore certain relational variables. For example, the advent and popularity of new technologies has dramatically increased the possibilities and expectations for sustaining close connections despite geographic distance (Adams, 1998). College students in particular are availing themselves of new technologies that allow them to communicate with their long-distance friends more easily (Pew Internet, 2002a). Pew Internet (2002a) found that 72% of college students from the United States reported they used the Internet mainly to communicate with friends, most commonly with friends from high school (35%), followed by friends on campus (24%), and friends off campus (20%).

These changes in the ways that individuals communicate with their friends lead to a need to re-examine theoretical concepts related to friendship (Adams, 1998). In the past, friendship has theoretically been viewed as a potentially fragile relationship (Fehr, 1999; Wiseman, 1986). Specifically, these traditional theories suggest that the lack of face-to-face contact resulting from greater distance has high maintenance costs. Hence, commitment dissipates, and the friendship moves towards its end. However, such views assume the primacy of face-to-face communication in interpersonal relationships, irrespective of the increasingly diverse means of communication open to individuals.

Today, rather than conceptualizing friendships as fragile (Wiseman, 1986), many friendships may be more aptly described as *flexible* (Becker, Johnson, Craig, Gilchrist, Haigh, & Lane, 2009). Friendships today can be highly elastic, as one's commitment to a friendship varies over time, in response to individual, relational, and extra-relational turning points (Johnson, Becker, Craig, Gilchrist, & Haigh, 2009). The ease of computer-mediated communication potentially helped initiate change in the ways friendships are being conducted (Boneva, Kraut, & Frohlich, 2001; Johnson, Haigh, Becker, Craig, & Wigley, 2004; Johnson, Haigh, Becker, Craig, & Wigley, 2008). A research line by the authors has examined how college students engage in modern friendships and argues for this reconceptualization of friendship from a "fragile" relationship to a "flexible" one (Becker et al., 2009; Johnson, 2001; Johnson, Wittenberg, Villagran, Mazur, & Villagran, 2003; Johnson et al., 2004; Johnson, Wittenberg, Haigh, Wigley, Becker, Brown, & Craig, 2004; Johnson, Haigh, Craig, & Becker, 2009).

However, even though individuals can interact more frequently with friends using CMC channels, this increase in accessibility may not always lead to greater relational closeness. Adams (1998) discusses how technology may actually hasten the end of some long-distance friendships that were based mostly on past experiences. Also, increases in ability to contact long-distance friends may lead to the friendship ending if the individuals do not avail themselves of these new technologies (Adams, 1998). There is a need to examine how computer-mediated communication can both aid in maintaining friendships and also potentially hasten the ending of some friendships.

This chapter explores these changes in friendship. The research by the authors is reviewed, along with other research that explores how friendship is enacted using mediated channels such as computer-mediated communication. The authors discuss how friendship should be conceptualized in the interpersonal communication and computer-mediated communication literatures. In addition, promising future research directions in this area are presented.

Traditional Views of Friendship

How researchers conceptualize a type of relationship affects how they study and interpret research findings related to that relationship. Traditionally, friendship has been viewed as a voluntary, potentially fragile relationship (Wiseman, 1986) due to a lack of external pressures to continue the relationship (Cramer, 1998). Blieszner and Adams (1992) claim that friendships are vulnerable due to their lack of institutional ties and availability of alternative friendships. Prior research has assumed that frequent interaction is necessary for maintaining friendships (Fehr, 2000), while transitions, especially an increase in distance between the friends, are perceived as particularly disruptive for friendships (Fehr, 1999). Early research illustrated that an increase in geographic distance between the friends was one of the main reasons cited for friendships ending (Rose, 1984).

One reason that friendship has been perceived as fragile is a heavy focus on face-to-face contact exclusive to other possible communication channels. Therefore when researchers have considered long-distance friendships, they have often assumed that these relationships are rare and naturally less close than geographically close relationships (Stafford, 2005). Some researchers have even expressed doubt that close, long-distance friendships can be maintained (Kelley et al., 1983).

There are three reasons that long-distance friendships are perceived as more likely to erode than geographically close friendships: (1) individuals must invest more time and energy in long-distance friendships; (2) individuals cannot as easily engage in frequent talk; and (3) individuals cannot as readily provide emotional and instrumental support (Fehr, 1999). While all of these reasons may be relevant if one focuses only on relational maintenance that is enacted face-to-face, new channels such as CMC decrease the costs associated with long-distance relationships. For example, with mobile phones and instant messaging, one may still be able to engage in frequent talk and provide emotional support for a long-distance friend. Therefore, this assumption that relational maintenance is enacted primarily face-to-face is one reason that friendships, especially long-distance ones, have been seen as so vulnerable to dissolution.

In general therefore traditional views of relational maintenance have focused on maintenance as occurring mainly face-to-face and have not considered maintenance that might occur over other channels, such as CMC (Stafford, Kline, & Dimmick, 1999). Stafford et al. discuss how this view is now dated. This traditional view perceived geographic distance as adding costs to the relationship (Davis, 1973). Geographic distance does potentially decrease the number or ease of engaging in maintenance activities, and relational maintenance research has assumed that relationships will deteriorate without maintenance behaviors (Canary & Stafford, 1994). Even though technology may help to eliminate this gap in maintenance behaviors for geographically close and long-distance friendships, Johnson (2001) found that geographically close friends still reported more maintenance behaviors than long-distance friends. However, in that study and in several additional studies (Johnson, 1999, 2000; Guldner & Swensen, 1995), individuals in long-distance and geographically close relationships did not differ in reported closeness or relational satisfaction.

Another way of conceptualizing friendship that portrays friendship as fragile is a replacement view of friendship. Rose (1984) reported that individuals chose between their long-distance and geographically close friends, which often meant ending their long-distance friendships and replacing them with geographically close ones. However, Pew Internet (2002a) found that college students are currently more likely to keep in contact with their high school friends, leading them to have wider social networks than their parents. In addition, Johnson, Haigh, et al. (2009) suggest that long-distance and geographically close friends may provide unique benefits, providing incentives to maintain both.

A last aspect of interpersonal communication research that has resulted in a portrayal of friendship as fragile is a traditional view of relational development and maintenance as linear. Dindia (1994) claims that in the relational maintenance literature, relational development is viewed as progressing linearly to a high level, which individuals seek to maintain. These linear models of relational development have focused on series of stages of increasing intimacy that are reversed when the relationship deteriorates (e.g., Altman & Taylor, 1973; Knapp, 1984). Therefore decreases in such variables as relational closeness or relational commitment are perceived as relational deterioration pointing towards the dissolution of the relationship. Such traditional models would predict that long-distance relationships such as friendships should be fragile because if the increase in distance leads to a decrease in closeness, then this should lead to the ending of the relationship.

However, a nonlinear view of relational development based on dialectics theory potentially questions this belief of friendships as fragile. For instance, Baxter and Montgomery (1996) claim that one should use the term *relational change process* rather than relational development. One of the basic tenets of relational dialectics theory is that relational partners continually fluctuate between stability and change (Altman, Vinsel, & Brown, 1981). Deterioration is not seen as forecasting the end of the relationship but rather providing opportunities for future growth (Altman et al., 1981).

An alternative conceptualization of friendship as flexible rather than fragile fits better with this dialectical perspective. Friendships may go through a time when they are very stable and others when they are changing rapidly. Friendships, especially those separated by distance, may go in and out of periods of dormancy (Rawlins, 1994). In addition, friendships, especially among college students, may even alternate between being geographically close and long-distance as transitions occur in the individuals' lives (Becker et al., 2009). Baxter and Montgomery (1996) claim that nonlinear relational development implies that for many relationships, progression through relational stages is not orderly and can reoccur repeated times. Becker et al. (2009) provide evidence for this by illustrating that even though the majority of LD and GC friends reported a linear progression through the stages of casual, close, and best friends, others reported backtracking in relational stages before growing closer again. Long-distance friends were more likely than geographically close friends to report a relational sequence where they backtracked to casual friends before becoming closer once again.

Therefore whether a friendship continues depends not only on transitions that occur in the friendship, but also how individuals interpret and handle these transitions. In dialectics theory the element of totality states that both internal and external forces relate to whether a relationship develops or deteriorates (Baxter & Montgomery, 1996). The advent of computer-mediated communication has been an external force that has transformed how friendships are conducted. The next section will discuss research that has examined how individuals engage in modern friendships and how a view of friendship as flexible rather than fragile might benefit both interpersonal and computer-mediated communication researchers.

Friendships as Flexible

Individuals today have become increasingly mobile and this fact along with additional and new communication channels has increased people's abilities to form and maintain relationships over distance (Blieszner & Adams, 1992; Wood, 1995). Therefore rather than expecting relationships to end when distance is increased, people increasingly hold the expectation that these relationships will continue (Adams, 1998), although they may have different expectations for how they will be enacted (Johnson, 2001).

A line of research by the authors has examined how individuals engage in modern friendships. A traditional view of friendship as fragile would suggest that long-distance and geographically close friends should be very different, but Stafford (2005) claims that they are actually quite similar. The research line by Johnson, Becker, and colleagues illustrates what variables may explain how and why individuals maintain relationships over distance. These studies focus on three factors: (1) examining maintenance behaviors that individuals use to enact relationships face-to-face and over distance, including the use of CMC; (2) determining relational developmental patterns in friendship and whether they support a conceptualization of friendship as fragile or flexible; and (3) separating traditional interpersonal communication concepts from such an exclusive focus on face-to-face contact by examining relationships that are enacted over geographic distance. Therefore this line of research helps determine what is unique in interpersonal relationships to face-to-face contact versus what aspects of these relationships can be enacted through other mediated channels, such as CMC.

Maintenance of Geographically Close and Long-Distance Friendships through Face-to-Face and Computer-Mediated Means

Johnson (2001) examined how individuals maintained geographically close and long-distance friendships by having participants keep a diary for six weeks describing their maintenance activities with one geographically close and one long-distance friend. Individuals reported the maintenance activities of *social networks* and *joint activities* more with their geographically close friends, while long-distance friends reported the maintenance activities of *cards, letters, and calls* more often. Although individuals reported a greater number of maintenance activities with their geographically close friends, they did not report significantly different closeness levels, satisfaction levels, or expectation that the relationship would continue with geographically close friends as compared to long-distance friends.

Johnson et al. (2008) had 226 individuals collect their emails to interpersonal family, friends, and romantic partners for a week to examine the maintenance behaviors they enacted over this channel. They found that romantic partners and family were more likely than friends to use assurances of the relationship's importance, while family members were more likely than romantic partners to mention social networks. Few differences were found between long-distance and geographically close relationships, especially friendships, although long-distance partners (collapsing across relationship types) reported a lower percentage of assurances than geographically close partners. This finding may reflect long-distance relationships using email to keep up with day-to-day activities, reducing the proportion of the email that is used for assurances. This study suggests that when examining computer-mediated communication channels, rather than face-to-face contact, there are few differences in maintenance behaviors between long-distance and geographically close friends.

Interpreting Friendship Trajectories as Flexible

Johnson, Wittenberg, et al. (2003, 2004) examined friendships to see whether they illustrated a traditional linear view of relational development. Such a view would be shown by perceiving a relationship to develop to a high degree of closeness, with drops in closeness signaling deterioration and forecasting eventual dissolution. Johnson et al. (2003) found that 49% of the friends reported linear relational development trajectories, while Johnson, Wittenberg, et al. (2004) found that 51% of friendships that had ended reported their friendship had followed a linear trajectory. For Johnson et al. (2003) the most commonly recalled turning point that was indicative of a downturn in relational closeness was an increase in geographic distance between the friends, while the third most commonly reported turning point associated with an upturn in closeness was a decrease in geographic distance. Johnson, Wittenberg, et al. (2004) found that an increase in geographic distance was the fourth most commonly reported reason that the reported friendship ended (13% of individuals reported), after less affection (23% of friends), friend or self changed (21% of friends), and stop spending as much time together (15% of friends). Future research should examine why certain friendships can navigate this turning point of increased geographic distance, while for other friendships geographic distance ends the friendship. How individuals use or fail to use computer-mediated channels to continue communicating over distance may be an important factor.

Johnson et al. (2003) claim that relational development itself can be considered a dialectic with individuals vacillating between periods of development and deterioration. Such a view suggests that turbulence in friendship closeness is natural rather than problematic, supporting a view of friendship as flexible rather than fragile.

Reconsidering Interpersonal Communication Concepts in Light of Geographic Distance

Adams (1998) calls for the need to divorce how we conceive of relational variables from dependence on face-to-face contact. Becker et al. (2009) and Johnson, Becker, et al. (2009) examined friendship levels and commitment levels to explore whether enacting a relationship over distance related to how these variables changed over time. Becker et al. found that a linear sequence of friendship level change (from casual to close to best) was observed most often for both geographically close and long-distance friends. However, a nonlinear sequence which included a shift back to casual friend with later recovery to deeper levels of friendship was reported more often by long-distance friends than geographically close friends. This supports Stafford's (2005) claim that long-distance friends are more likely to go through a period of dormancy. Becker et al. claim that these results illustrate LD friendships "tremendous potential for resiliency" (p. 367) and support a view of friendship as flexible rather than fragile.

Johnson, Becker, et al. (2009) examined beliefs about relational commitment. Being able to continue a relationship over distance questions two views about this concept, that relationships are dependent on external barriers to dissolution (Johnson, 1991) and that commitment is dependent on rewards and costs (Fehr, 1999) that are most easily exchanged face-to-face (Davis, 1973). Overall, Johnson, Becker, et al. found that geographically close friends reported higher commitment levels. They also reported higher degrees of couple identity and relational primacy and higher social pressure to continue the friendship (Stanley & Markman, 1992). Men were more likely to report perceived linear trajectories in commitment changes for both geographically close and long-distance friends. Nineteen percent of long-distance friends said that their commitment levels were currently falling, while 4% of geographically close friends reported the same. Long-distance friends also reported more downturns in commitment than geographically close friends (suggesting greater turbulence in relationships over distance). However, long-distance and geographically close friends reported similar types of turning points. These findings support a view of friendship as flexible rather than fragile because the majority of both long-distance and geographically close friendships reported high, currently rising levels of commitment.

A third study by the authors examined how the concept of closeness can be divorced from a focus on face-to-face contact by comparing how long-distance and geographically close friends differ in how they define relational closeness (Johnson, Haigh, et al., 2009). There was an interaction between gender of friends and geographic distance between friends. For instance, geographically close female friends were more likely to list the definitions of "comfort and ease" and "frequency of interaction" than long-distance female friends. Long-distance female friends were more likely to list "understanding." Both geographically close and long-distance male friends reported "self-disclosure" as their most common definition of closeness, but approximately 75% of long-distance males reported this definition, while 50% of geographically close males did. Long-distance males were significantly more likely to list "trust" than geographically close males. Participants also differed in what they were likely to rank as prototypical of friendship intimacy (Fehr, 2004), with differences mainly focusing on geographically close friends being expected to provide more practical help, favors, physical support, and "being there" for their friend. There was no significant difference between geographically close and long-distance friends on self-reported closeness (Johnson, 2001). This study helps to illustrate what aspects of the relational closeness concept are tied inextricably to physical distance and which are not (Adams, 1998). It also suggests a need for a multidimensional conceptualization of relational closeness, where individuals who are separated by physical distance and those with greater amounts of face-to-face contact have similar overall perceived closeness levels but differ on individual components that comprise this overall assessment.

Therefore these studies all illustrate that friendships are more flexible than fragile and help to illustrate what aspects of interpersonal relating are tied tightly to face-to-face communication while others can be enacted through more varied means. Greater numbers of communication channels, such as those that are computer-mediated, have increased individuals' abilities to continue relationships that are conducted over distance. These studies illustrate how individuals maintain relationships over distance both through email (Johnson et al., 2008) and other channels (Johnson, 2001). Having greater opportunities to communicate in manners besides face-to-face supports a conceptualization of friendship as flexible rather than fragile and suggests that friendships are definitely not doomed to end due to geographic separation. Johnson, Wittenberg, et al. (2003, 2004) illustrate how friendships (both geographically close and long-distance) often exhibit nonlinear relational development patterns where an increase or decrease in geographic distance is a transition that the relationship must navigate. CMC channels may help individuals navigate these transitions.

In addition, the increase in potential communication channels means that interpersonal scholars need to separate their relational concepts from such a heavy reliance on frequent face-to-face contact (Adams, 1998). Becker et al. (2009), Johnson, Haigh, et al. (2009), and Johnson, Becker, et al. (2009) examine the concepts of friendship level, relational closeness, and relational commitment to determine factors about these concepts that are tied more closely to face-to-face contact. The next section will focus on different types of mediated communication and former research that has examined how people use these channels to communicate with friends.

Mediated Communication and Interpersonal Relationships

According to equity theory, people continue relationships that are equitable in terms of rewards and costs (Canary & Stafford, 1992; Walster, Berscheid, & Walster, 1973). Traditionally, geographic distance has been seen as negative for relationships because of the increased costs associated with a lack of face-to-face contact (Davis, 1973). However, new channels of communication, such as computer-mediated channels, have tipped this balance of rewards and costs for long-distance relationships. Today, individuals have even more channels to communicate with people who live long distance. These channels can provide opportunities to continue relationships but also can affect how the relationship is conducted. The channels of telephone, email, instant messaging, and social networking sites (especially Facebook) will be examined individually. Friends' use of multiple channels to maintain their relationships will then be discussed.

Telephone

The telephone is a long-standing and commonly used mediated channel of communication. One early study on long-distance female friends (Rohlfing, 1995) found that four sets of female LD friends communicated mostly by telephone, and their frequency of contact varied from once a month to once a year. They were frustrated with the telephone and felt that they could not continue their history of self-disclosure as effectively as they could previously face-to-face. However, they reported more frequent talk about the importance of their friendship, which was a common topic in their conversation. Therefore the channel of telephone actually shaped conversations among these long-distance friends.

The advent of the mobile phone has had particularly dramatic effects on long-distance friendships by facilitating frequent and inexpensive communication. Mobile phones tend to be used to communicate with strong social ties, such as family members and close friends (Kim, Kim, Park, & Rice, 2007). Utz (2007) suggests that the telephone is a richer medium than email, for example, and allows close friends to more easily convey emotion and communicate about highly personal topics. In addition to the telephone, there are several methods of using computer-mediated channels that are commonly used to communicate with friends. These methods will be discussed next.

Email

Boneva et al. (2001) claims that email can help counteract the greater costs of long-distance relationships. However, even though email may provide a way for long-distance individuals to stay in contact more frequently, it and CMC channels in general can also actually change how people communicate with their interpersonal partners.

Walther and Parks (2002) discuss how individuals can be more strategic when communicating over CMC channels as compared to face-to-face. Email also provides an opportunity to communicate asynchronously at one's own leisure (Boneva et al., 2001). Stafford et al. (1999) found that individuals perceived the lack of geographic limitation as one of the main advantages of using email along with its lower cost, convenience, and ability to be similar to conversation. Stafford (2005) suggests that Internet communication may be used differently by geographically close and long-distance partners, with geographically close individuals using the Internet for small talk and long-distance partners reflecting more intimate content. However, Johnson et al. (2008) found that geographically close partners actually reported a larger percentage of assurances in their email content than long-distance partners did.

Email can be used to enact those maintenance activities that individuals can also enact face-to-face. For instance, Johnson et al. (2008), Wright (2004), and Rabby (1997) all found that individuals used email or communication over the Internet to communicate positivity, openness, and assurances. Individuals may also be more likely to obtain support from long-distance family, friends, and romantic partners by using email. For instance, Tognoli (2003) talks about how staying in contact with one's family through email can combat homesickness for college students. The same could be true for staying in contact with one's high school friends.

However, email does not always have a positive impact on interpersonal relationships. Editing of communication in mediated channels may cause individuals to have distorted views of one another (Walther & Parks, 2002) and avoid conflict (O'Sullivan, 2000). Adams (1998) discusses how relationships may suffer with increased communication allowed by email if they were mainly based on pleasant memories and it becomes clear that the individuals have changed. Rabby and Walther (2003) wondered if individuals who currently communicate mostly online, even though they communicated face-to-face in the past, illustrate a hyperpersonal perspective in which they believe their levels of intimacy are deeper than they are and are based on idealized views of each other.

Instant Messaging

Instant messaging (IM) has quickly become a popular form of mediated communication, particularly among teenagers and young adults (Boneva, Quinn, Kraut, Kiesler, & Shklovski, 2006). Like email, IM is based on typed text transmitted over the Internet. However, whereas email is asynchronous, IM requires real-time "instant" messages between participants. IM is used predominantly for friendship maintenance, and can be an important signifier of friendship among teenagers. In a longitudinal study of Dutch teenagers, Valkenburg and Peter (2009) found that IM stimulates intimate self-disclosure among existing friends, which enhances the quality of friendship. Specifically, IM allows teenagers to share highly personal information that they would not otherwise easily share in a face-to-face setting. Moreover, teenagers view mutual self-disclosure as an important sign of close friendship.

Social Networking Sites

Social networking sites (also known at social network sites and SNSs) have become an increasingly popular way to initiate, develop, and maintain friendships online and to show one's social network of friends (boyd & Ellison, 2008). Although hundreds of SNSs exist, Facebook has emerged with the highest number of users. According to Facebook.com at the time of this writing (July, 2010) there are more than 400 million active users. Although Facebook was initially created for and remains heavily used by college students, more than two-thirds of current Facebook users are outside of college, and the fastest growing demographic is those 35 years old and older. Additionally, Facebook expands the traditional boundaries of CMC, as more than 100 million active users currently access Facebook through their mobile phones, and these users are twice as active on Facebook as non-mobile users.

As the term suggests, SNSs provide opportunities for individuals to engage in varied types of relationships, including business and romantic relationships, although most are friendships and acquaintanceships. Some SNSs truly focus on networking, or relationship initiation among strangers based on shared interests, activities, language, or identities (boyd & Ellison, 2008). However, boyd and Ellison contend that most SNS users "are not necessarily 'networking' or looking to meet new people; instead, they are primarily communicating with people who are already a part of their extended social network" (p. 211).

On Facebook, users who agree to include each other in their social networks are then identified as "friends" on their respective profile pages. Although posting on another friend's profile can compromise one's personal privacy, many Facebook users "disclose personally identifying information as a signaling tool to their peers" (Lange & Lampe, 2008, p. 1). Lange and Lampe also note that many Facebook users disclose surprising, even alarming, amounts and types of personal information on their profiles and their friends' profile pages.

Due to the ease of "friending," Facebook users often have far more friends than individuals in a traditional (i.e., offline) social network (Tong, Van Der Heide, Langwell, & Walther, 2008). Donath and boyd (2004) argue that SNS users may have more close relationships than those individuals who do not avail themselves of this technology, since SNSs allow friends to keep informed of each other's status updates and activities. However, given the tremendous size of some users' networks, not all Facebook friends are close (boyd & Ellison, 2008).

Some researchers have investigated the manner in which friendship is performed on SNSs. Tong et al. (2008) found that Facebook users with relatively high (e.g., 902) or relatively low (e.g., 102) numbers of friends cast off less favorable impressions of themselves. Moreover, Walther, Van Der Heide, Kim, Westerman, and Tong (2008) found that observers' perceptions of an individual user are related to the attractiveness of the user's network of friends. SNS users can manage their self-presentation through strategic inclusion of certain friends and their profile content (boyd & Ellison, 2008).

The loose connections between some Facebook friends should not be devalued. Research has demonstrated the effectiveness of SNSs in building social capital, particularly as SNS users navigate their friendships via online and offline channels (boyd & Ellison, 2008). In a longitudinal study of young adults, Steinfield, Ellison, and Lampe (2008) found that Facebook use generates *bridging social capital*, which is a large, heterogeneous network of weak ties between friends and acquaintances. The weak ties can be particularly helpful for young adults seeking employment or other opportunities. Steinfield et al. observed that Facebook users can browse their friends' profiles and thereby gain valuable information that can subsequently "lower the barriers to initiating communication" in face-to-face and online settings (p. 18). Thus, SNSs can facilitate smoother and more comfortable face-to-face interaction.

In a different study, Ellison, Steinfield, and Lampe (2007) found that Facebook use is also associated with *bonding social capital* (i.e., relationships with close family and friends) and *maintained social capital* (i.e., relationships from a long-distance network, such as high school friends). Ellison et al. also reported that Facebook use is especially beneficial for young adults with low self-esteem and low life satisfaction. After reflecting on the body of research about Facebook, Steinfield et al. (2008) argue that rather than dismissing large Facebook networks as shallow or superficial, researchers should recognize Facebook's increasingly central role in relationship initiation, development, and maintenance.

The "Web" of CMC Channels

Computer-mediated communication researchers often look at the channels of communication separately rather than considering how individuals use them together to enact their interpersonal relationships (Rabby, 1997). However, as Ledbetter (2008) points out, most friends rely on multiple channels of communication. Wang and Andersen (2007) found that individuals in more intimate relationships reported using a wider variety of channels to communicate with their long-distance friends. Therefore to fully understand modern friendships, one should focus on the whole pattern of channel use in a particular relationship.

A small body of research has investigated friends' use of multiple channels of communication. In a study of college students, Wang and Andersen (2007) examined how long-distance friends (who had established their friendships in previous face-to-face interactions) maintained their friendships. Telephone, email, and IM were the top three channels used, with face-to-face communication following as the fourth most commonly used channel. Less than 3% of participants reported that they did not communicate with their long-distance friends at all, providing further support for the robustness of the long-distance friendship. Friendship closeness was positively correlated with the number of channels used for communication as well as the frequency of communication via these channels. Finally, Wang and Anderson found that face-to-face self-disclosure was greater than CMC self-disclosure, although they suggest that self-disclosure via multiple channels contributes to friends' relationship quality.

In a related study, Baym, Zhang, Kunkel, Ledbetter, and Lin (2007) found that the proportion of face-to-face, telephone, and Internet communication was not associated with friends' relational quality (i.e., closeness and satisfaction); however, they focused on the proportion and not the frequency of channel use. The Baym et al. study was subsequently contradicted by Ledbetter's (2008) findings. Using a longitudinal panel data from 1987 to 2002, and investigating telephone, postal mail, and face-to-face communication among friends, Ledbetter found that telephone communication was most predictive of relational closeness among friends in 2002. He also found that face-to-face communication was *less* predictive of relational closeness among friends in 2002. Ledbetter argues, "It is plausible that the telephone is an ideal medium for maintaining relational closeness, offering intimacy that is relatively unfettered by the need for spatial proximity" (p. 560). Likewise, Utz (2007) found support for the primacy of telephone communication in maintaining relational closeness. In a comparison of telephone and email communication, she found that the closer the friendship, the greater use of email. However, closer friends were also more likely to communicate over the telephone. In analyzing the content of email and telephone communication, Utz found that the closer the friendship, the more intimate the telephone conversation. However, emails did not increase in intimacy, indicating that "personal matters are still discussed on the phone" (p. 694). The combined findings of Ledbetter (2008) and Utz (2007) support the notion that mediated channels of communication can facilitate the flexibility of friendships over time and across geographical distance.

Another of Ledbetter's studies provides further explanation for differences in friends' channel use. Ledbetter (2009) hypothesized that the relationship between family communication patterns and friendship closeness for young adults is mediated by the frequency of friendship maintenance behaviors via face-to-face and online communication. "Specifically, conversation orientation was positively associated with both face-to-face and online relational maintenance, and conformity orientation was inversely associated with face-to-face maintenance" (p. 142). Ledbetter argues that young adults from conversation-oriented families are better equipped to communicate flexibly and spontaneously in face-to-face environments, whereas young adults from conformity-oriented families prefer computer-mediated channels for the ability to carefully craft a message without time pressures. These findings provide explanation for variance in channel use, maintenance behaviors, and closeness of friends. Future research should continue to examine how individuals use multiple channels to develop and maintain their interpersonal relationships.

Future Directions in Friendship Research

There are some possible conclusions and future directions for those who study friendship as illustrated by the studies in this review. For instance, Johnson, Haigh, et al. (2009) claim that the expected rewards may differ for those friendships that are geographically close versus those that are long-distance. Individuals expect physical help, favors, and "being there" more from their geographically close friends (Adams, 1985-1986; Johnson, Haigh, et al., 2009), but when individuals describe closeness in long-distance friends, females focus more on understanding and males focus more on trust and self-disclosure than when describing closeness in their geographically close friendships. Research has consistently found that long-distance friendships are of longer duration than geographically close friendships (Adams, 1985-1986; Johnson, Becker, et al., 2009; Johnson, Haigh, et al., 2009), suggesting that individuals may have more invested in these relationships, providing incentive to continue these relationships even if the costs are higher (Rusbult, 1980). Indirect investments, such as shared memories, may be particularly important for long-distance friends. One needed direction for future research is to examine how individuals use their network of both geographically close and long-distance friends and both face-to-face and mediated communication channels to obtain sufficient social support.

Rawlins (1994) claims that there are three types of friendships among mature adults: active, dormant, and commemorative. *Active* friends still give one another support, while *dormant* friends do not maintain regular contact, but the expectation exists that the friend would be there if needed. Individuals do not maintain contact with *commemorative* friends, but these individuals are considered important in their life history. Greater access to computer-mediated communication should be affecting how individuals perceive and interact with friends from each of these categories. For instance, Pew Internet (2002b) found that individuals were using the Internet to communicate with a wider range of family members than they had had contact with previously, what they called the "clicking cousin" effect. In the past, long-distance friendships may have been more likely to be commemorative than currently, when individuals use computer-mediated communication channels to reconnect with these old friends. Adams (1998) warns that not all uses of computer-mediated channels may lead to closer friendships. Individuals' expectations for these old friendships may be violated if they try to restart contact, perhaps through email, but the old friend does not respond. A friendship may go from "fading away" (Hays, 1988) to being considered over, if one individual does not respond to the other's overtures. Future research should examine current long-distance friendships. Are individuals today more likely to perceive these friendships as active rather than dormant? Do individuals reassess their degrees of closeness when they interact with a dormant or commemorative friend through computer-mediated means? What happens when individuals seek to revive a commemorative friendship? What ideals do individuals have for long-distance and geographically close friends since the advent of so many possible channels for communication? How do these ideals coincide with the realities of using computer-mediated channels to maintain relationships?

Another potential conclusion from this review is that differences between male and female friendships may have lessened in some ways. For instance, the current authors argue that male and female long-distance friendships should be more similar than male and female geographically close friendships. Research based on geographically close friends finds that female friendships tend to focus on self-disclosure, while male friendships focus on shared activities (Swain, 1989). As women can continue to engage in self-disclosure, but men may have a harder time engaging in shared activities (although they may be able to engage in activities such as games over the Internet), some researchers have suggested that males may not be able to sustain long-distance friendships (Rohlfing, 1995). However, more current friendship research by the authors has shown little gender differences among long-distance friends (Becker et al., 2009; Johnson, Becker, et al., 2009; Johnson, Haigh, et al., 2009). Johnson, Haigh, et al. (2009) claim that when face-to-face contact is less and individuals have to rely on other channels of communication, differences between male and female friendships lessen. There is need for formative research about long-distance male friendships. Such research should explore what male long-distance friends expect from each other, what channels they use to communicate, and how often. As Johnson, Haigh, et al. (2004) found that males were more likely than females to report that they had met their long-distance friends online and Adams (1998) claims that new technologies allow more opportunities to make friends with individuals who never lived geographically close, research should examine how males develop friendships with individuals they meet over CMC as well as whether/how males continue friendships that were formed geographically close once they become long-distance. In addition, cross-sex long-distance friendships should be examined, as most of the current research has focused on same-sex friendship (e.g., Johnson, Becker, et al., 2009; Johnson, Haigh, et al., 2009).

Conclusions

Individuals in Western cultures have a very idealized view of friendship, as a voluntary relationship composed of unrelated similar individuals that is not focused on practical help (Bell & Coleman, 1999). However, the realities of friendships can be much different. In the same way, there can be idealized views of ways in which computer-mediated communication will benefit interpersonal relationships such as friendships. How these ideals coincide with the realities of how individuals actually communicate using these channels will determine their ultimate effects on how people conduct relationships in the modern era.

This interplay between ideal and real is illustrated by the research on geographically close and long-distance friends. Traditionally, researchers expected long-distance friends to be rare and less close (Stafford, 2005) because they were so fixated on the idea of frequent, face-to-face communication among friends. Today, on the other hand, there actually may be the expectation that long-distance friends will remain close. Researchers can have a better sense of how individuals are actually conducting friendships by considering friendships as flexible rather than fragile and considering the myriad of ways that friends communicate rather than automatically privileging face-to-face contact.

In the same way, researchers need to consider the ideal use of computer-mediated communication technology versus the real ways that individuals are actually using these tools in their interpersonal relationships. For instance, Pew Internet (2002b) found that after an initial increase in contact with family members through email, there was a decrease in contact using this channel over time. Even though the accessibility of long-distance relationships has changed with these new technologies, with less cost, more ease in use, and ever new and expanding methods of communicating with our friends, the actual use of these channels may lead to either greater or less closeness in friendships. For instance, the expectation used to be that friendships would "fade away" (Hays, 1988) when individuals moved away. People can no longer use this excuse, so what happens when individuals have the ability to communicate but do not use it (Adams, 1998)? How do individuals negotiate their social lives when they keep meeting people and adding contacts but no longer have as ready an excuse to drop previous relationships or friendships? What happens when expectations for computer-mediated communication and long-distance friendship do not match with reality? These questions and others guarantee that interpersonal communication and CMC scholars will have many fruitful areas for research as we seek to understand friendships in the current age.

References

Adams, R. G. (1985-1986). Emotional closeness and physical distance between friends: Implications for elderly women living in age-segregated and age-integrated settings. *International Journal of Aging and Human Development, 22*, 55–76.

Adams, R. G. (1998). The demise of territorial determinism: Online friendships. In R. G. Adams & G. Allan (Eds.), *Placing friendship in context* (pp. 153–182). Cambridge, UK: Cambridge University Press.

Altman, I., & Taylor, D. A. (1973). *Social penetration: The development of interpersonal relationships.* New York: Holt, Rinehart and Winston.

Altman, I., Vinsel, A., & Brown, B. (1981). Dialectic conceptions in social psychology: An application to social penetration and privacy regulation. *Advances in Experimental Social Psychology, 14*, 107–160.

Baxter, L., & Montgomery, B. M. (1996). *Relating: Dialogues and dialectics.* New York: Guilford.

Baym, N. K., Zhang, Y. B., Kunkel, A., Ledbetter, A., & Lin, M.-C. (2007). Relational quality and media use in interpersonal relationships. *New Media & Society, 9*, 735–752.

Becker, J. A. H., Johnson, A. J., Craig, E. A., Gilchrist, E. S., Haigh, M. M., & Lane, L. T. (2009). Friendships are flexible, not fragile: Turning points in geographically close and long distance friendships. *Journal of Social and Personal Relationships, 26*, 347–369.

Bell, S., & Coleman, S. (1999). The anthropology of friendship: Enduring themes and future possibilities. In S. Bell & S. Coleman (Eds.), *The anthropology of friendship* (pp. 1–19). New York: Berg.

Blieszner, R., & Adams, R. G. (1992). *Adult friendship.* Newbury Park, CA: Sage.

Boneva, B., Kraut, R., & Frohlich, D. (2001). Using email for personal relationships. *American Behavioral Scientist, 45*, 530–549.

Boneva, B. S., Quinn, A., Kraut, R. E., Kiesler, S., & Shklovski, I. (2006). IMing, text messaging, and adolescent social networks. In R. Kraut, M. Brynin, & S. Kiesler (Eds.), *Computers, phones, and the Internet: Domesticating information technology* (pp. 201–218). New York: Oxford University Press.

boyd, d. m., & Ellison, N. B. (2008). Social network sites: Definition, history, and scholarship. *Journal of Computer-Mediated Communication, 13*, 210–230.

Canary, D. J., & Stafford, L. (1992). Relational maintenance strategies and equity in marriage. *Communication Monographs, 59*, 239–267.

Canary, D. J., & Stafford, L. (1994). Maintaining relationships through strategic and routine interaction. In D. J. Canary & L. Stafford (Eds.), *Communication and relational maintenance* (pp. 3–22). San Diego, CA: Academic.

Cramer, D. (1998). *Close relationships: The study of love and friendship.* New York: Arnold.

Davis, M. (1973). *Intimate relations.* New York: Free Press.

Dindia, K. (1994). A multiphasic view of relationship maintenance strategies. In D. Canary & L. Stafford (Eds.), *Communication and relational maintenance* (pp. 91–112). New York: Academic.

Donath, J., & boyd, d. (2004, October). Public displays of connection. *BT Technology Journal, 22,* 71–82.

Ellison, N. B., Steinfield, C., & Lampe, C. (2007). The benefits of Facebook "friends": Social capital and college students' use of online social network sites. *Journal of Computer-Mediated Communication, 12,* Retrieved July 15, 2009, http://jcmc.indiana.edu/vol12/issue4/ellison.html

Facebook.com. (2010). *Press Room Statistics.* Retrieved from http://www.facebook.com/press/info.php?statistics

Fehr, B. (1999). Stability and commitment in friendships. In J. M. Adams & W. H. Jones (Eds.), *Handbook of interpersonal commitment and relationship stability* (pp. 239–256). New York: Kluwer/Plenum.

Fehr, B. (2000). The life cycle of friendship. In C. Hendrick & S. S. Hendrick (Eds.), *Close relationships: A sourcebook* (pp. 71–82). Thousand Oaks, CA: Sage.

Fehr, B. (2004). Intimacy expectations in same-sex friendships: A prototype interaction-pattern model. *Journal of Personality and Social Psychology, 86,* 265–284.

Guldner, G. T., & Swensen, C. H. (1995). Time spent together and relationship quality: Long-distance relationships as a test case. *Journal of Social and Personal Relationships, 12,* 313–320.

Hays, R. B. (1988). Friendship. In S. W. Duck (Ed.), *Handbook of personal relationships* (pp. 391–408). Chichester, UK: John Wiley.

Johnson, A. J. (1999, February) *If I do not see you every day, are you still my friend? Characteristics of long-distance friendships.* Paper presented to the Interpersonal Communication Division of the Western States Communication Association for their annual meeting at Vancouver, Canada.

Johnson, A. J. (2000, February) *A role theory approach to examining the maintenance of geographically close and long-distance friendships.* Paper presented to the Interpersonal Interest Group of the Western States Communication Association for their annual meeting in Sacramento, CA.

Johnson, A. J. (2001). Examining the maintenance of friendships: Are there differences between geographically close and long-distance friends? *Communication Quarterly, 49,* 424–435.

Johnson, A. J., Becker, J. A. H., Craig, E. A., Gilchrist, E. S., & Haigh, M. M. (2009). Changes in friendship commitment: Comparing geographically close and long-distance young-adult friendships. *Communication Quarterly, 57,* 395–415.

Johnson, A. J., Haigh, M. M., Becker, J. A. H., Craig, E. A., & Wigley, S. (2004, November). *College students' use of email to maintain relationships.* Paper presented at the annual meeting of the National Communication Association, Chicago, IL.

Johnson, A. J., Haigh, M. M., Becker, J. A. H., Craig, E. A., & Wigley, S. (2008). College students' use of relational management strategies in email in long-distance and geographically close relationships. *Journal of Computer-Mediated Communication, 13,* 381–404, http://www3.interscience.wiley.com/cgi-bin/fulltext/119414146/HTMLSTART

Johnson, A. J., Haigh, M. M., Craig, E. A., & Becker, J. A. H. (2009). Defining and measuring relational closeness: The test case of close long distance friendships. *Personal Relationships, 16,* 631–646.

Johnson, A. J., Wittenberg, E., Haigh, M., Wigley, S., Becker, J., Brown, K., & Craig, E. (2004). The process of relationship development and deterioration: Turning points in friendships that have terminated. *Communication Quarterly, 52,* 54–67.

Johnson, A. J., Wittenberg, E., Villagran, M. M., Mazur, M., & Villagran, P. (2003). Relational progression as a dialectic: Examining turning points in communication among friends. *Communication Monographs, 70,* 230–249.

Johnson, M. P. (1991). Commitment to personal relationships. In W. H. Jones & D. Perlman (Eds.), *Advances in personal relationships* (Vol. 3, pp. 117–143). London: Jessica Kingsley.

Kelley, H., Berscheid, E., Christensen, A., Harvey, J., Huston, T., Levinger, G., ... Peterson, D. (1983). *Close relationships*. New York: Freeman.

Kim, H., Kim, G. J., Park, H. W., & Rice, R. E. (2007). Configurations of relationships in different media: Ftf, email, instant messenger, mobile phone, and SMS. *Journal of Computer-Mediated Communication, 12*, article 3. Retrieved July 15, 2009, from http://jcmc.indiana.edu/vol12/ issue4/kim.html

Knapp, M. L. (1984). *Interpersonal communication and human relationships*. Boston: Allyn & Bacon.

Lange, R., & Lampe, C. (2008, May). *Feeding the privacy debate: An examination of Facebook.* Paper presented at the annual meeting of the International Communication Association, Montreal, Canada.

Ledbetter, A. M. (2008). Media use and relational closeness in long-term friendships: Interpreting patterns of multimodality. *New Media & Society, 10*, 547–564.

Ledbetter, A. M. (2009). Family communication patterns and relational maintenance behavior: Direct and mediated associations with friendship closeness. *Human Communication Research, 35*, 130–147.

O'Sullivan, P. (2000). What you don't know won't hurt me: Impression management functions of communication channels in relationships. *Human Communication Research, 26*, 403–431.

Pew Internet & American Life Project. (2002a). *The Internet goes to college: How students are living in the future with today's technology.* Retrieved July 15, 2009, from http://www.pewInternet.org/ pdfs/PIP_College_Report.pdf

Pew Internet & American Life Project. (2002b). *Getting serious online: As Americans gain experience, they use the Web more at work, write emails with more significant content, perform more online transactions, and pursue more activities online.* Retrieved July 15, 2009, from http://www.pewInternet. org/pdfs/PIP_Getting_Serious_Online3ng.pdf

Rabby, M. K. (1997, November). *Maintaining relationships via electronic mail.* Paper presented at the annual meeting of the National Communication Association, Chicago, IL.

Rabby, M. K., & Walther, J. B. (2003). Computer-mediated communication effects on relationship formation and maintenance. In D. J. Canary & M. Dainton (Eds.), *Maintaining relationships through communication* (pp. 141–162). Mahwah, NJ: Lawrence Erlbaum.

Rawlins, W. K. (1994). Being there and growing apart: Sustaining friendships during adulthood. In D. J. Canary & L. Stafford (Eds.), *Communication and relational maintenance* (pp. 275–294). New York: Academic.

Rohlfing, M. E. (1995). "Doesn't anybody stay in one place anymore?" An exploration of the under-studied phenomenon of long-distance relationships. In J. T. Wood & S. Duck (Eds.), *Under-studied relationships: Off the beaten track* (pp. 173–196). Thousand Oaks, CA: Sage.

Rose, S. M. (1984). How friendships end: Patterns among young adults. *Journal of Social and Personal Relationships, 1*, 267–277.

Rusbult, C. E. (1980). Satisfaction and commitment in friendships. *Representative Research in Social Psychology, 11*, 96–105.

Stafford, L. (2005). *Maintaining long-distance and cross-residential relationships*. Mahwah, NJ: Lawrence Erlbaum.

Stafford, L., Kline, S., & Dimmick, J. (1999). Home email: Relational maintenance and gratification opportunities. *Journal of Broadcasting and Electronic Media, 43*, 659–669.

Stanley, S. M., & Markman, H. J. (1992). Assessing commitment in personal relationships. *Journal of Marriage and the Family, 54*, 595–608.

Steinfield, C., Ellison, N., & Lampe, C. (2008, May). *Net worth: Facebook use and changes in social capital over time.* Paper presented at the annual meeting of the International Communication Association, Montreal, Canada.

Swain, S. (1989). Covert intimacy: Closeness in men's friendships. In B. Risman & P. Schwartz (Eds.), *Gender in intimate relationships: A microstructural approach* (pp. 71–86). Belmont, CA: Wadsworth.

Tognoli, J. (2003). Leaving home: Homesickness, place attachment, and transition among residential college students. *Journal of College Student Psychotherapy, 18*, 35–48.

Tong, S. T., Van Der Heide, B., Langwell, L., & Walther, J. B. (2008). Too much of a good thing? The relationship between number of friends and interpersonal impressions on Facebook. *Journal of Computer-Mediated Communication, 13*, 531-459, http://www3.interscience.wiley.com/cgi-bin/fulltext/119414155/HTMLSTART

Utz, S. (2007). Media use in long-distance friendships. *Information Communication and Society, 10*, 694–713.

Valkenburg, P. M., & Peter, J. (2009). The effects of instant messaging on the quality of adolescents' existing friendships: A longitudinal study. *Journal of Communication, 59*, 79–97.

Walster, (Hatfield), E., Berscheid, E., & Walster, G. W. (1973). New directions in equityresearch. *Journal of Personality and Social Psychology, 25*, 151–176.

Walther, J. B., & Parks, M. R. (2002). Cues filtered out, cues filtered in: Computer mediated communication and relationships. In M. L. Knapp & J. A. Daly (Eds.), *Handbook of interpersonal communication* (3rd ed., pp. 529–563). Thousand Oaks, CA: Sage.

Walther, J. B., Van Der Heide, B., Kim, S.-Y., Westerman, D., & Tong, S. T. (2008). The role of friends' appearance and behavior on evaluations of individuals on Facebook: Are we known by the company we keep? *Human Communication Research, 34*, 28–49.

Wang, H., & Andersen, P. A. (2007, May). *Computer-mediated communication in relationship maintenance: An examination of self-disclosure in long-distance friendships.* Paper presented at the annual meeting of the International Communication Association, San Francisco, CA.

Wiseman, J. P. (1986). Friendship: Bonds and binds in a voluntary relationship. *Journal of Social and Personal Relationships, 3*, 191–211.

Wood, J. T. (1995). *Relational communication: Continuity and change in personal relationships.* Belmont, CA: Wadsworth.

Wright, K. B. (2004). On-line relational maintenance strategies and perceptions of partners within exclusively internet-based and primarily internet-based relationships. *Communication Studies, 55*, 239–253.

CHAPTER THIRTEEN

A Cross-Contextual Examination of Technologically Mediated Communication and Social Presence in Long-Distance Relationships

Katheryn C. Maguire

Stacey L. Connaughton

Distanced relationships (DRs), across several contexts, are an ever-increasing occurrence. Up to one million marriages and as many as one-third of college premarital relationships experience long-term separations annually (Aylor, 2003). Distanced relationships are also common in organizational realms that utilize distanced work arrangements (i.e., remote teams, telecommuters) (e.g., Connaughton & Shuffler, 2007; Gibbs, 2002; Rosenfeld, Richman, & May, 2004). Although researchers in both relational and organizational contexts argue that DRs are fundamentally different from proximal relationships (PRs) (e.g., Aylor, 2003; Bell & Kozlowski, 2002; Beranek & Martz, 2005), we argue that DRs and PRs are actually quite similar in that unit members in both situations must depend on one another to achieve relational and task goals. The difference, we suggest, lies in the preferred and/or available means of interaction. Whereas individuals in PRs engage in frequent face-to-face (FtF) contact, those in DRs rely on technologically mediated communication (TMC) to accomplish their goals.

The question of how to successfully bridge temporal and/or geographic separation is at the heart of another body of research related to TMC, called *presence*, or the perceptual illusion of nonmediation (Lombard & Ditton, 1997). According to Lee (2004), feelings of presence lie at the center of all mediated experiences. To be sure, the "desire to overcome the limit of human sensory channels through the use of technological devices is a major impetus for the development of media and reality-simulation technologies" (p. 27). *Social presence*, a specific type of presence, is particularly relevant for unit members in DRs as perceptions of social presence can trigger psychological experiences of closeness, connectedness, and being with others in a mediated context (Bente, Rüggenberg, Krämer, & Eschenburg, 2008). Communication technologies that facilitate mediated interaction between distanced users are often designed with social presence in mind, thereby allowing users "to modulate social presence for a wide range of activities including getting to know someone, exchanging information or goods, problem solving and making decisions, exchanging opinions, generating ideas, resolving conflicts, or maintaining friendly relations" (Biocca, Harms, & Burgoon, 2004, p. 458).

Although some researchers who examine DRs in organizational contexts have been employing the concept of social presence in their research (e.g., Lowry, Roberts, Romano, Cheney, & Hightower, 2006), they often do so in a limited way, tending to connect social presence with media richness (Daft & Lengel, 1986; Short, Williams, & Christie, 1976), or the extent to which a *medium* is perceived as sociable or personal when used to interact with others (Lombard & Ditton, 1997). Instead, we follow more recent theorizing about social presence by viewing it as a psychological mechanism experienced by the users that is related to, but not dependent on TMC (Biocca et al., 2004; Lee, 2004). Furthermore, despite its applicability to romantic DRs, very few scholars have incorporated social presence into their theorizing about DRs. Instead, they tend to use the term *copresence* to differentiate between times when relational partners are co-located and when they are apart (e.g., Merolla, 2010a; Sahlstein, 2004; Stafford & Merolla, 2007). Whereas periods of physical copresence may facilitate feelings of social presence (Zhao, 2003), we examine ways in which TMC facilitates feelings of social presence so members can experience connectedness to one another across the distance.

In this chapter, we bring together theory and research from the DR and social presence literatures in an effort to understand how unit members maintain their distanced relationships through the use of communication technologies. Based on Rogers' (1986) definition of "communication technology" (i.e., "the equipment, organizational structures, and social values by which individuals collect, process, and exchange information with other individuals," p. 2), we conceptualize "technology" to include both computer-mediated communication (CMC) as well as more "traditional" channels of mediated interaction (e.g., snail mail, telephone). We argue that unit members utilize TMC that enhances (or diminishes, depending on the circumstances) feelings of social presence to maintain their relationship, that is, to sustain the existence of the relationship as well as qualities thought to be important to the relationship (Dindia & Canary, 1993). The employment of TMC to maintain relationships may be particularly important in romantic DRs as maintenance strategies can be used to "fight the relational entropy facilitated by distance" (Ficara & Mongeau, 2000, p. 4) and have been associated with lower levels of relationship-uncertainty and higher levels of trust (Dainton & Aylor, 2001). Relationship maintenance is important to team contexts as well (McGrath, 1991), in that communication is central to distributed team processes (Gibson & Gibbs, 2006). As Gibbs and colleagues (2008) argue, "Although teaming, in general, may be conceptualized as constituted through communication, the constitutive role of communication may be even more pronounced in virtual teams because communication facilitates the team's existence" (p. 192). Given the constitutive force of communication, interpersonal interactions must be effectively managed and maintained, for if not, other team processes may be adversely affected (Maruping & Agarwal, 2004).

To advance our argument, we review research on DRs in romantic and team contexts. We chose to focus on the distributed teams and romantic contexts for two reasons. First, placing these literatures in conversation with each other responds to Stafford's (2004) call for more collaboration across disciplines and subdisciplines to advance our knowledge of DRs. Second, both romantic DRs and distributed teams are characterized by interdependence and a shared commitment towards some end. Indeed, interpersonal communication and the existence of more "personal" relationships in the teams context is critical for their success as they can "lay the foundation for the effectiveness of other processes" (Marks, Mathieu, & Zaccaro, 2001, p. 368) such as trust and leadership. Given these commonalities, both literatures have yielded complementary empirical findings with regard to distance, and researchers in both areas are interested in understanding how distance affects communication, unit outcomes, and relationship maintenance through the employment of TMC.

The Context of Long-Distance Relating

Two core elements unite the various conceptual and operational definitions of distanced relationships advanced in the romantic and teams literatures: (a) some degree of geographic and/or temporal separation between relational members (e.g., Gibson & Gibbs, 2006; Merolla, 2010a), and (b) a heavy, sometimes exclusive reliance on TMC (e.g., Connaughton & Shuffler, 2007; Stafford, 2005). Although separation and use of TMC occur in PRs as well, we argue that it is the *chronic continuance* of separation and the *heavy reliance* on TMC that differentiate DRs from PRs. To capture these key elements, we define DRs as *committed relationships in which the unit members perceive they are unable to interact FtF for prolonged periods of time on a regular and/or frequent basis to accomplish their goals*. In this definition, "unit" refers to romantic relationship partners in the romantic context and team members in the workplace context. In the following sections, we discuss these core elements, explicate variation in how the term *relationship* itself can be conceived, and interrogate the TMC and presence literatures to determine their role in the maintenance of romantic and workplace DRs.

Spatial Separation

One core element that characterizes DRs is geographic separation resulting from situational constraints (e.g., living and/or working in different cities). Many studies view distance as a binary differentiation between long-distance and geographically close (proximate; co-located) relationships (Cramton, Orvis, & Wilson, 2007; Merolla, 2010a). Whereas some researchers who study romantic DRs set a minimum number of miles for a romantic relationship to be considered long distance (e.g., 100 miles, Carpenter & Knox, 1986) or require that the relational partner be outside a particular area or in another part of the state (Helgeson, 1994; Stephen, 1986), other scholars (e.g., Dainton & Aylor, 2001; Sahlstein, 2004) recognize that imposing set geographic parameters unnecessarily limits the types of relationships that could be classified as distanced. Similarly, in teams contexts, scholars have argued that distance is socially constructed and is defined differently by various team members (see Leonardi, Jackson, & Marsh, 2004). Researchers also recognize that there are various degrees of geographic distance (Connaughton & Shuffler, 2007; Scott & Timmerman, 1999) and that virtuality is a multidimensional phenomenon with the degree of geographic distance among team members as one component (Gibson & Gibbs, 2006; Kirkman & Mathieu, 2005).

Temporal Separation

Time, a second defining element in DRs, constitutes a multifaceted phenomenon that can be conceptualized in a number of ways (Merolla, 2010a; Montoya-Weiss, Massey, & Song, 2001). For one, time can be viewed in relation to geographic factors. Dainton and Aylor (2001) define long-distance dating relationships as ones in which relational partners are unable to spend time together FtF on most days. They found that those in long-distance romantic relationships (LDRRs) with some FtF contact felt more certain and trusting of their partners than those in LDRRs with little or no FtF contact. Similarly, distributed teams researchers point to time zone differences among members of global virtual teams as a factor that may shape the way in which work is accomplished (e.g., one team member located in Hong Kong and another in New York City; see Connaughton & Daly, 2003, 2004a; Gibbs, 2002). Given such constraints, issues of temporality may affect the ways in which teams operate (Montoya-Weiss et al., 2001).

Another useful way to conceptualize time in DRs involves expectations for the continuance of the relationship. For instance, instead of looking at frequency of visits as a measure of time apart, Maguire (2007a) conceptualized time apart in terms of expectations for continuing in a DR, finding that participants who felt certain about reuniting in the same city with their romantic partner were more satisfied and less distressed than those who felt uncertain about reuniting in the same city. Likewise, in team contexts, although some individuals may work remotely or interact virtually sometimes and other times are co-located, other teams function solely as distributed teams, with no expectation that their units will become co-located in the future (Connaughton & Shuffler, 2007). Scholars have also highlighted the degree of permanence as being a critical feature of distributed teams (i.e., temporary, ad hoc teams) and that issues of temporality may influence relationships in these teams in various ways (Connaughton & Shuffler, 2007; Saunders & Ahuja, 2006) as well as the nature of communication among individuals (DeSanctis & Monge, 1999).

Separation and Varying Views on "Relationship"

The two core elements of distance (i.e., geographic and temporal separation), and the subsequent lack of FtF contact that is often associated with them, may lead some unit members to question whether they are in "real" relationships. Stafford (2005) offers an important distinction that addresses this issue. She argues that there are two ways to define what it means to be in a relationship: (a) one based on interaction and (b) another based on perception. With regard to the former, Stafford writes: "in US culture, close relational bonds of any type are conceived as contingent on interaction, companionship, and intimacy, all of which presume a rather high degree of FtF contact" (p. 16). In support of this view, Burgoon and colleagues (2002) report a correspondence between physical and psychological closeness in the nonverbal literature, where "Physical proximity promotes psychological closeness, and physical distance conveys psychological distance" (p. 662). As such, DRs are "an inherent oxymoron; relationships are conceived as existing only when participants are interacting in the same physical space" (Stafford, 2005, p. 5), thus making it seem difficult, if not impossible, to maintain a satisfactory relationship.

The tendency to equate interaction with closeness—both physical and psychological—is problematic for a couple of reasons. First, equating the two (interaction and closeness) ignores the fact that interactivity—the interdependent exchange of messages—is a neutral concept (Burgoon et al., 2002) and not always beneficial. For example, whereas high levels of interactivity may lead to negative outcomes such as the inability to detect deception, low levels may lead to positive outcomes such as thoughtful, independent reflection (Burgoon et al., 2002). Indeed, temporary avoidance of one's partner is often considered a romantic relationship maintenance behavior (Canary, Stafford, Hause, & Wallace, 1993; Dainton & Stafford, 1993). Second, interactional accessibility or interactivity is not dependent on FtF settings. To the contrary, the use of TMC can facilitate a sense of shared meaning and make romantic and team relationships feel more "real." In order for this to happen, unit members need to be aware of the cultural frame surrounding the term *relationship* and recognize that DRs can be successful despite the separation. In doing so, they create a new frame for this different, but no less valid, type of relationship where maintenance activities can be successfully enacted, and a heightened sense of social presence can be achieved, through TMC.

A second way to define *relationship* is from a cognitive perspective: relationships are continued in our minds (Stafford, 2005). Sigman (1991) states that relationships do not end once FtF events or conversational engagements are terminated. "In a sense, relationships are 'larger' than the physical presence or interactional accessibility of the participants. Social relationships can therefore be said to be continuous, or to be oriented to and produced as such by relationship partners" (p. 108). Following this line of thinking, relationships continue through the use of maintenance behaviors that serve to define the relationship as continuous during periods of separation and/or interactional inaccessibility (Merolla,

2010a; Sigman, 1991). Such behaviors increase perceptions of relational connectedness, and have been associated with important relationship outcomes (e.g., increased satisfaction; Gilbertson, Dindia, & Allen, 1998) without the need to have one's partner physically or virtually copresent. In recognition that relationships have both a behavioral and a cognitive component, we suggest that successful DRs utilize TMC, opportunities for FtF meetings, and ways to stay cognitively oriented to the relationship during times of interactional separation (Sigman, 1991; Stafford, 2004) to maintain a sense of social presence with other unit members.

Social Presence, Relationship Maintenance, and TMC

Defining Social Presence

In a review and explication of the presence literature, Lee (2004) identified several ways to define and refer to *presence*. For example, *telepresence* refers to "the sense of being physically transported to a remote work space via teleoperating systems" (p. 29). Similarly, the terms *virtual presence* and *mediated presence* refer to the technologies needed to create this sense of being somewhere else. To avoid a technologically deterministic definition of presence, Lee defines presence as a "perceptual process" (p. 30) and "a psychological state in which the virtuality of experience is unnoticed" (p. 32). Goffman (1963) used the term *copresence* to describe the state when social actors are accessible, available, and influenced by one another—a state relevant and applicable to DRs. Yet, the term has been often been used as a synonym for FtF interaction (e.g., Sigman, 1991). Then again, Zhao (2003) states, "An individual's sense of being with others is basically a psychological phenomenon, which may or may not correspond to the actual state of copresence" (p. 450).

Another commonly used term to refer to interaction between social actors is *social presence*. Social presence theory underscores the importance of physical and psychological connection to another (Biocca, 1997) and argues that intimacy, immediacy, and involvement (Palmer, 1995; Rice, 1993) and social connectedness are critical aspects of presence (Rice, Chang, & Torobin, 1992; Short et al., 1976). Early work by Short and colleagues (1976) and Daft and Lengel (1986) measure social presence in terms of a medium's ability to convey the actual presence of communicators: rich communication channels (e.g., FtF communication) allow for more social context cues to communicate a sense of presence, while leaner communication channels (e.g., text-based communication) filter out these cues and limit a feeling of presence. Instead of focusing on characteristics of the medium, Lee (2004) defines social presence as "a psychological state in which virtual social actors are experienced as actual social actors in either sensory or nonsensory ways" (p. 45). While this definition has great pertinence to the study of virtual actors such as anthropomorphic robots, it is less heuristic to understanding mediated interactions between human beings. Instead, we will use the term *social pres-*

ence to describe the sense of connectedness with others that can happen with the assistance of communication technologies (Nowak, 2001).

Before we discuss the role of social presence, enabled by TMC, in the maintenance and management of DRs, it is important to realize that social presence may not always be the best course of action. Just as high levels of interactivity are not always beneficial, too much social presence can lead to negative outcomes (Biocca et al., 2004). There may be times when unit members need to be "relationally absent" from one another so they can be autonomous individuals and focus on their own personal or professional goals (Baxter & Montgomery, 1996). In romantic relationships, if unit members are too socially present (i.e., too cognitively oriented to and dependent on the relationship), they may have difficulty detaching from their partners so they can adjust and cope with the separation (Guldner, 2004). In Sigman's (1991) words, "couples who are 'always in each other's presence' may be considered pathological, at least in Western society, especially beyond a certain level or stage of romantic involvement" (p. 111). Furthermore, Biocca et al. (2004) report that social presence can make people vulnerable to manipulation, deception, and mindless processing. Without the ability to regulate a sense of presence, then, unit members may become prone to groupthink or even organizational control, thereby eliminating the potential benefits of being in a DR. Thus, it is important to see social presence as a variable existing along a continuum that can be dialed up or down as needed (Biocca et al., 2004). This notion is similar to how scholars are beginning to view distance more generally, such as acknowledging various degrees of virtuality (space, time, modality) or degrees of dispersion (partially vs. fully distributed) (see Connaughton & Shuffler, 2007). As such, unit members need to be mindful of the benefits and drawbacks of social presence so they can adjust their communication according to either increase or decrease perceptions of social presence.

Social Presence and the Maintenance of DRs

Previous research has established that social presence plays an important role in the maintenance and management of DRs. Avolio and Kahai (2003), for instance, argue that communicating social presence is a key aspect of leadership in distributed work contexts. They note that one important difference between these and co-located work settings may be whether team members are able to feel their distanced leader's presence. Similarly, participants in a study by White and White (2005) were able to use TMC to "maintain an ongoing symbolic proximity or copresence with people who shared a common history and mutual experiences" (Conclusion section, para. 4).

One way to increase a sense of presence, of course, is to be physically copresent and spend some time FtF. Lee (2004) claims that for a mediated experience to be considered "authentic" the social actors need to feel that their virtual interactions are connected to actual interactions. Thus, physical copresence, in terms of FtF meetings, can help sustain perceptions of social presence during separation. Evidence exists in the romantic DR research to support this claim. Helgeson (1994) reported that respondents who were still dating their partner at the end of the semester had more contact with their partner during the previous summer than those who broke up. Holt and Stone (1988) found that the farther apart (over 250 miles) and the longer the time between visits (more than six months), the lower the satisfaction. This was not as detrimental to couples who, at the same distance, saw each other within a six month time period. Similarly, Dainton and Aylor (2001, 2002) believe that DRs with some FtF contact more closely resembled geographically proximate relationships than DRs with no FtF contact.

In the organizational context, however, there is debate about whether FtF is a necessary ingredient for team effectiveness. Some scholars, for instance, have critiqued the assumption that FtF interaction is individuals' preferred medium (Rice, 1984; Rice & Gattiker, 2001) and others have argued that FtF might enable relationships initially, but then over time, mediated channels allow members to accomplish tasks (Zack, 1994) and communicate effectively (Alge, Wiethoff, & Klein, 2003). In a study of online collaboration, Ramirez and Zhang (2007) found that remaining online, and not switching to FtF interaction, "heightened evaluation of relational communication and other qualities of the interaction" (p. 302). Gibson and Gibbs (2006) found that creating a psychologically safe communication climate can help lessen the impact of geographic distribution on team processes and outcomes. Given that FtF interaction is only one way to maintain DRs, our discussion now turns to TMC as another way to maintain relationships, in part, by facilitating the experience of social presence. In the following sections, we examine TMC in both the distributed teams and romantic relationship contexts, and identify other issues, such as the mode of TMC and individual difference variables, that may influence the use of social presence as a way to maintain DRs.

TMC and the maintenance of distributed teams. When considering relationships between TMC and the maintenance of distributed teams, studies of distributed teams can be categorized into two general groupings: (a) research which ties social presence to the medium (technology) with the (implicit) assumption that TMC restricts communicative possibilities and relationships; and (b) research which allows for the possibility that distribution and/or mediation do not necessarily restrict relational development. This latter body of work often examines other relational mechanisms (i.e., trust, interaction) that can serve to maintain relationships among team members.

Some researchers tie perceptions of social presence (and interpersonal relationships) to the medium. As Cramton et al. (2007) aptly note, research on interpersonal relations in distributed groups often starts with the assumption that "communication mediated by technology carries less social information than face-to-face communication, affecting the development of interpersonal relationships in various ways" (p. 526). This work draws on that of Sproull and Kiesler (1986) which posits that TMC filters out some cues and thus makes it more difficult to achieve some relational and task goals. These assumptions are echoed in the body of distributed teams literature which argues vehemently that FtF communication is the preferable and best mode of communication. Drawing on media richness theory (Daft & Lengel, 1986), this research (implicitly) argues that FtF interaction is superior to TMC and that, because of this, teams should do what they can to come together physically from time-to-time if not throughout their existence (i.e., Maznevski & Chudoba, 2000) so as not to suffer the process losses that are argued to be inextricably tied to distributed teams. Indeed, distributed teams researchers sometimes use language which depicts distribution and TMC as challenges to be overcome (see Connaughton & Shuffler, 2007, for this discussion).

Although not as blatantly skeptical of TMC, other researchers contend that one must consider the technology in relation to the task being performed. Maruping and Agarwal (2004), for example, present a conceptual model in which "The underlying causal mechanism...is one of task-technology fit. The greater the congruence between the communication requirements of the task (its nature and timing) and the specific functionalities of the selected medium, the more effective the task can be managed" (p. 979). Similarly, Connaughton and Daly (2003, 2004a) have argued that distanced leaders must carefully consider the medium they choose when executing certain task and relational functions (i.e., visioning, building trust) for, based on interviews with several industry positional leaders of remote teams, certain media are perceived to be more effective than others at achieving various goals.

A twist on this general theme is the notion that the medium is the enabler of social presence. Although they do not explicitly refer to social presence in this passage, note the language of social presence articulated by Biocca et al. (2004) and Burgoon and colleagues (2002) as Leonardi, Jackson, and Marsh (2004) explain: "Through email, telephone, chat rooms, and so forth, the distance worker *connects to* and *interacts with* other members, who provide socialization on how to be not only a member of the organization but also a distance worker for the organization" (emphasis added, p. 168). Although specific technologies are mentioned, they are cast as enablers of connection; what is deemed important for relationships to exist (and be maintained) is that distanced workers can *connect to* and *interact with* other members.

This line of thinking provides an appropriate transition to the second general strand of research in the distributed teams literature: that which allows for the possibility that distribution and/or mediation do not necessarily restrict relationships. At one extreme is an argument that TMC can possess the same characteristics as FtF interaction. Note how Avolio and Kahai (2003) state this claim in the context of leadership over distance:

> We are fairly confident that leadership mediated by information technology can exhibit exactly the same content and style as traditional face-to-face leadership, especially

> as virtual interactions become more visual. The critical differences may be in which is meant by "feeling the leader's presence," as well as the reach, speed, permanence, and perception of a leader's communication. (p. 327)

Here, Avolio and Kahai underscore two important points: (a) how team members think about (the leader's) presence matters (in other words it may be different than how we think of physical presence), and (b) other mechanisms and processes (i.e., leader's communication) may be as important to leader-member relationships, if not more critical, than the technology itself.

The argument that other mechanisms and processes can help to foster relationships over distance is advanced by other distributed teams researchers as well. Researchers have identified trust (Jarvenpaa & Leidner, 1999; Jarvenpaa, Shaw, & Staples, 2004), communication (frequency, spontaneity, content, and mode; Connaughton & Shuffler, 2007; Cramton et al., 2007; DeSanctis & Monge, 1999; Hinds & Mortensen, 2005), and interpersonal and task conflict management (Hinds & Mortensen, 2005; Montoya-Weiss et al., 2001) as key relational processes in distributed teams. Given time, which is not always available (i.e., temporary, ad hoc teams), researchers have found that interpersonal relationships among distributed team members can develop and that individuals will exchange social messages as well as task-oriented messages over TMC (Chiadambaram, 1996; Jarvenpaa & Leidner, 1999). And as noted previously, some researchers question whether distance (distribution) is the central variable to understanding effective teaming over distance at all (i.e., Gibson & Gibbs, 2006).

TMC and the maintenance of romantic DRs. Studies of TMC in the romantic relationship research also fall along two lines: (a) the use of TMC as a maintenance strategy in and of itself, and (b) the use of TMC to convey relationship maintenance messages. First, many researchers recognize the importance of TMC in the maintenance of romantic DRs (Aylor, 2003; Carpenter & Knox, 1986; Dainton & Aylor, 2002; Gerstel & Gross, 1983; Holt & Stone, 1988; Kirschner & Walum, 1978; Maguire, 2007b; Merolla, 2010b; Stephen, 1986). Indeed, Canary et al. (1993) as well as Dainton and Stafford (1993) include mediated communication as a strategy used to maintain close relationships, many of which are DRs, and Maguire and Sahlstein (in press) cited relationship maintenance, often employed via TMC, as a way that military wives cope with stress during a wartime deployment. Commonly reported modes of TMC include letters, phone calls, diaries, and tapes (Gerstel & Gross, 1983; Carpenter & Knox, 1986; Holt & Stone, 1988; Kirschner & Walum, 1978). The use of CMC as a maintenance strategy has also gained prominence in the literature as an efficient, affordable means of communicating while apart (Rabby & Walther, 2003; Walther & Parks, 2002). Using TMC helps relational partners to keep track of each other's daily experiences during separation (Maguire, 2007b; Westefeld & Liddell, 1982), and keeps the relationship at a desired level (e.g., satisfactory) (Ficara & Mongeau, 2000; Maguire, 2007a).

Second, a small but growing body of research has examined the extent to which CMC allows relational partners to employ both strategic and routine maintenance strategies in their romantic DRs. Researchers in this area often employ Stafford and Canary's (1991) typology of relationship maintenance in their investigation: (a) openness (i.e., direct discussion of the relationship), (b) assurances (i.e., implications of a future together and commitment to one another), (c) positivity (i.e., behaviors that make the interactions cheerful and pleasant), (d) sharing tasks (i.e., behaviors indicating a willingness to take equal responsibility in the relationship), and (e) use of social networks. Through the use of these five strategies, relational partners are reminded of the importance of the bond that they share as well as the strength of the connection between them. Merolla's (2010b) study of wives' communication during a military deployment identifies several additional maintenance activities accomplished through TMC as well, including topic avoidance, intimacy and affection, future planning, and faith. In essence, the use of such strategies can be seen to increase perceptions of social presence when the concept is viewed as psychological involvement with others (Biocca et al., 2004).

For instance, Janan Johnson, Haigh, Becker, Craig, and Wigley (2008) sought to determine the extent to which individuals in PRs and DRs use email to maintain their relationships, finding that assurances, openness, positivity, and discussing social networks were commonly employed by those in romantic DRs. Similarly, Sidelinger, Ayash, Godorhazy, and Tibbles (2008) found that the more relational partners used CMC to engage in maintenance activities, the higher the reports of relationship satisfaction. Finally, Dainton and Aylor (2002) examined the relationship among communication channel type (FtF, telephone, Internet, and written letters) and relationship maintenance. Results indicated that the use of telephone was positively correlated with three maintenance behaviors (i.e., assurances, openness, and shared tasks), the Internet was positively correlated with two strategies (i.e., positivity and shared networks), and written letters with one (i.e., assurances). They concluded that mediated communication plays a prominent role in the maintenance of distance relationships.

Whereas many of the previous scholars did not specify the type of DR, it is likely that most of them were examining DRs with at least some FtF contact or that had started as a PR that eventually "migrated" to the online world (Rabby, 2007). One subset of research focuses on the maintenance of a specific type of romantic DR: the virtual, or online romance. According to Rabby (2007), individuals in virtual relationships generally have never met each other in person and as such, their relationship exists only in the online world. They initiated their relationship through CMC and maintain enough regular contact that they consider it to be a "real" relationship. Although studying strangers engaged in cybersex, Jones' (2008) findings that text, combined with images transmitted through a web camera, helped users increase their sense of "presence" in the interaction, indicates that relational partners could participate in some form of "mediated physical intimacy" with each other while apart.

Merkle and Richardson (2000) suggest that virtual relationships are fundamentally different from offline relationships in the way they develop (i.e., they start with exchanges, then self-disclosure, then FtF meeting if at all); yet, they often are seen as satisfactory relationships. In one of the only studies to compare relationship maintenance in virtual relationships with other types of DRs (e.g., those with periodic FtF meetings), Rabby (2007) reported that although individuals in virtual relationships reported lower use of relationship maintenance strategies in comparison to other relationship types, the differences were sometimes not significant, and often times the mean value of the reported use of relationship maintenance was still relatively high. If the partners were committed to each other, these differences practically disappeared. Furthermore, Nice and Katzev (1998) found that individuals in virtual romances rated their online relationship as equal to or superior to those they had established offline on measures of strength, satisfaction, and ease of communication. Then again, Scott et al. (2006) found that the potential benefits of CMC, such as increased access to each other, did not translate into increased reported intimacy in the participants' virtual relationships, and Scott, Mottarella, and Lavooy (2006) reported stronger levels of intimacy in their FtF relationships than their online relationships. To explain the disparity, Whitty (2008) points out that the existence of, and satisfaction with, virtual DRs may depend on the user, as shy and lonely individuals may find cyberspace a unique place for them to experience and learn about relationships and sexuality.

Comparison of TMC channels. From a media richness perspective, social presence is facilitated through different forms of TMC, some of which allow for rapid, synchronous communication (e.g., chat) whereas others allow visual cues to be transmitted through the use of webcams, or even the creation of avatars (i.e., computer images that represent the user; Rabby & Walther, 2003). Although all of these technologies have the potential to increase perceptions of social presence (Daft & Lengel, 1986), some research suggests that TMC with greater media richness (e.g., video conferencing vs. text-only email) is associated with greater perceptions of social presence, due to a higher level of embodiment (i.e., involvement of human bodies in the process of communication and the inclusion of nonverbal cues in the interaction, Biocca et al., 2004; Zhao, 2003). Given that forms of TMC differ in terms of synchronicity as well as number and type of cues available to users, a number of studies have compared various forms of TMC—regardless of type of relationship—to determine which offer the most functionality and/or sense of presence for users.

Some studies examine patterns of channel use within and among different types of relationships. For instance, Dainton and Aylor (2002) found that FtF interaction in romantic relationships was positively related to telephone use, and Internet use was positively related to writing letters, suggesting that these channels grouped together based on the level of synchrony and number of available cues. In a study by Baym, Zhang, and Lin (2004), whereas phone use and FtF contact were associated with more intimate relationships, individuals in DRs (romantic or otherwise) used the telephone as much as they used the Internet, and were more likely to use written letters and the Internet than those in PRs. Similarly, Boase (2008) found that FtF and land-line phone contact were the most common ways of connection with personal networks, despite the popularity of email and mobile phones. Whereas some individuals may have preferences for written forms of communication such as email or letters, others may prefer the immediacy of the telephone to maintain or manage their personal relationships.

Other studies focus on the personal and task needs fulfilled through various modes of TMC. In one of the earlier media comparison studies, Westmyer, DiCioccio, and Rubin (1998) found that users favored FtF interaction in general, and oral means of communicating (e.g., telephone) to fulfill a wider spectrum of their and their interaction partner's needs than written channels (e.g., email). In a study of the use of telephone, text messaging, and email during travel (i.e., transitional DRs), White and White (2005) found that participants used text messaging to "play" with their distant friends, email to get connected with their own thoughts as they send updates about their journey to their social network, and the telephone to communicate emotions to those they left behind. The primacy of the phone has been cited in other research as well, as it can provide the optimal balance of richness and presence for natural, satisfying relationships over both CMC and FtF conversations (Connell, Mendelsohn, Robins, & Canny, 2001).

Then again, Baym, Zhang, Lin, Kunkel, Lin, and Ledbetter (2007) reported that the estimated proportion of telephone and Internet use was not significantly associated with relational quality, concluding that TMC does not appear to improve, or detract from, relational quality. Similarly, in their comparison of gratifications obtained and gratification opportunities across different media, Dimmick, Kline, and Stafford (2000) found that whereas the telephone fulfilled consumers' needs, email provided greater gratification opportunities. Finally, Ramirez, Dimmick, Feaster, and Lin (2008) found that participants used their land-line phone and email less when they started using instant messaging (IM), suggesting that IM fulfilled a broader spectrum of needs than the other two forms of TMC; however, the mobile phone still emerged as superior in fulfilling users' needs over IM, due to its mobility and ability to convey emotion.

Other relevant factors. There are other factors that can come into play when determining the extent to which unit members are able to experience social presence when using TMC. According to Biocca et al. (2004), social presence is a transient phenomenological state that not only varies with the medium (as previously discussed), but also knowledge of relational partners and social context. Indeed, Bregman and Haythornthwaite (2003) claimed that relation, or the nature of the tie between the speaker as determined by the frequency and duration of contact and level of intimacy in the relationship, is associated with presence: the stronger the tie, the greater the motivation to share and reciprocate exchanges and thus, experience presence. Individual difference variables may also affect feelings of social presence. For instance, Nowak (2003) reported that female participants believed CMC had a greater potential to elicit perceptions of social presence than male participants; however, the differences were very small, indicating that sex might not have much of an effect on experiences of social presence. Lombard and Ditton (1997) and Lee (2004) list a number of other factors that could influence a user's experience of presence, including a willingness to suspend disbelief (i.e., overlook the signs that they are using an electronic means to communicate with others) as well as familiarity with a medium.

To start, adjustments of expectations may help users increase their sense of presence and their ability to maintain satisfactory relationships, given some scholars' doubts about the ability of TMC to adequately maintain close relationships (e.g., Fortunati, 2005). According to Clark (2007), "if we expect these technologies to deliver, at a distance, the very same kinds of sensory input and interactive potential that we encounter in 'normal' daily life, they will almost certainly continue to disappoint" (p. 427). Adjustment of expectations regarding the relationship itself may also help users achieve social presence in their DRs, given the cultural beliefs that honor FtF interaction as the only way to achieve intimacy in North American close relationships (Stafford, 2005). An alteration in the definition of what constitutes intimacy and closeness in a relationship may be necessary for relational partners in DRs to have fulfilling relationship experiences and experience social presence in their mediated interactions.

Second, communication competence, in terms of a user's ability to make use of the available cues in a given channel, has also been shown to increase reports of social presence (Wrench & Punyanunt-Carter, 2007). Similarly, Walther and Bazarova (2008) found that social competence (i.e., social sensitivity, emotional sensitivity, social control, emotional control, social expressivity, and social manipulation) can also help users overcome the limitations of low-bandwidth TMC channels. They determined that members of a collaborative team experience greater propinquity when faced with few channel choices–in a sense, they either made the best of the available channel, or adjusted their expectations of the channel so it could still facilitate feelings of electronic closeness. Yet, when faced with a difficult task, communicator skill was not able to overcome the propinquity-lessening effect of task complexity.

Conclusion

As is evident throughout this review, individuals involved in workplace or romantic DRs are able to compensate for the temporal and/or geographic separation resulting from their situation and thus maintain satisfactory, functional relationships. In order to accomplish this goal, unit members employ various modes of

TMC to re-establish a sense of connection with each other through increased perceptions of social presence. At the same time, unit members may choose channels of communication with fewer interaction cues to purposefully decrease this sense of social presence to accomplish their individual goals. In doing so, they are able to balance competing needs for both autonomy and connection in their relational lives (Baxter & Montgomery, 1996) and achieve work outcomes (i.e., satisfaction, productivity) in their teams.

The bridging of the romantic and teams literatures offers some interesting theoretical insight into the role of social presence in the maintenance of DRs, as well as the relational nature of the social presence concept. First, it appears that social presence is the mechanism by which the use of TMC, in and of itself, serves as a relationship maintenance strategy. Although important relational communication activities take place through TMC, such as conflict management in teams (Hinds & Mortensen, 2005) and assurances of the future in romantic relationships (Janan Johnson et al., 2008), the sheer act of engaging in TMC creates a "virtual proximity" in which unit members can experience a sense of copresence or shared space. Indeed, Ylinen and Valo (2004) found that DR partners experience a sense of copresence by just having a computer connection open (e.g., a chat window), which allowed them to feel closer to their partner because they could access them when and if needed. As stated by White and White (2005),

> When travelers have ready access to, and use mobile and other electronic communication services, the liminal experience of being both present and absent in each other's lives is transformed into a continuing engagement with established relationships and an ongoing connection to people back home." (Conclusion section, para. 5).

This rationale is the impetus for the creation of product prototypes such as *LumiTouch*, a picture frame that lights up when one's partner touches his/her own frame from a remote location thereby making emotional communication "tangible" (Chang, Resner, Koerner, Wang, & Ishii, 2001) and the *Portal Frame* (Bergstrom & Karahalios, 2006), a type of picture frame that allows someone to "keep track" of his/her partner's activities without direct interaction by displaying a different picture, depending on his/her activities (e.g., a picture of him/her working during business hours; a picture of him/her eating if out to dinner). Both technologies were designed with social presence in mind.

That is not to discount the importance of the communicative activities that take place during TMC. In FtF relationships, unit members have the ability to interact FtF, where behavioral engagement, mutual involvement, and cognitive orientation are easier to gauge because of the number of nonverbal cues available in the interaction (Burgoon et al., 2002). In a DR, however, it may be more difficult for unit members to feel they are in a "real" relationship because they cannot physically see or touch their partner, which is an important way to increase perceptions of presence (Zhao, 2003). Instead, it is through TMC that the relationship becomes more "real" as unit members become behaviorally engaged with one another and exchange messages that reify their connections with, and commitment to, one another.

According to Duck and Pittman (1994), "it is the communication between two partners that creates the mental representation of the relationship and indeed in some sense creates the relationship" (p. 679). Stephen (1986) also believes that communication creates the relationship, in that "coordinated or interdependent interaction is not possible unless at least some of the [negotiated shared meanings] are held in common by interactants" (p. 192). These shared meanings are either confirmed or disconfirmed through future interactions and are not constrained by physical proximity. On the contrary, Stephen found that the relationship between the frequency of talk and the degree of symbolic interdependence was *stronger* for DRs than for nonseparated couples. Although participants in DRs conversed less than those in PRs, they were still able to maintain a shared sense of reality in order to keep their mental image of the relationship alive. In short, whereas TMC is the medium through which messages that reinforce the mental image of the relationship are communicated, social presence is the mechanism by which the connection between unit members in DRs is actually felt and maintained.

Second, by examining the connection between social presence and the maintenance of DRs, it is possible to further explore the relational nature of the social presence concept. In doing so, "we may arrive at a better understanding of how humans arrive at that sense of mutuality that underpins all communication between people and that is a prerequisite to establishing common ground" (Biocca et al., 2004, p. 459). Social presence scholars often use relationship concepts like immediacy, intimacy, mutual understanding, reciprocal orientation, and psychological involvement in reference to the experience of social presence (Biocca et al., 2004) or copresence (Zhao, 2003; Zhao & Elesh, 2008). Given that one of the original formulations of social presence dealt with human-to-human interaction across various types of media in an organizational context (i.e., Social Presence Theory, Short, et al., 1976), it makes sense that such relational constructs would be used to describe social presence. More recent scholars who use the term social presence, however, have expanded the concept to include human-machine interaction, such as the human-to-computer or human-to-robot interaction (Biocca & Harms, 2002), both of which are outside the scope of this chapter.

Instead of arguing whether or not human-computer or human-robot interaction can lead to a "real" relationship, we instead advocate for a new term—*relational presence*—to differentiate experiences of interconnectedness (i.e., social presence) in the context of technologically mediated, human-to-human relationships, such as those examined in this chapter, from experiences of social presence in other forms of quasi-social relationships (Biocca & Harms, 2002). Zhao (2003) made a similar distinction in her explication of the term *copresence* with the term *corporeal telecopresence* to differentiate human-to-human interaction via TMC from other forms of interaction (e.g., human-to-robot). Our view of relational presence is situated in Sigman's (1991) claim that social relationships are "'larger' than the physical presence or interactional accessibility of the participants" (p. 108). We suggest that unit members experience varying degrees of relational presence, aided by TMC, but *independent* of interactivity. We offer this construct in the spirit of conceptual clarity and hope that it assists fellow researchers understand the relational side of the presence construct. Future research should explore of the role of relational presence in the context of DRs by identifying the underlying dimensions of the construct, creating a way to measure relational presence, and determining its association with important task and relationship outcomes.

References

Alge, B. J., Wiethoff, C., & Klein, H.J. (2003). When does the medium matter? Knowledge-building experiences and opportunities in decision-making teams. *Organizational Behavior and Human Decision Processes, 91*, 26–37.

Avolio, B. J., & Kahai, S. S. (2003). Adding the "E" to e-leadership: How it may impact your leadership. *Organizational Dynamics, 31*, 325–338.

Aylor, B. (2003). Maintaining long-distance relationships. In D. J. Canary & M. Dainton (Eds.), *Maintaining relationships through communication: Relational, contextual, and cultural variations* (pp. 127–140). Mahwah, NJ: Lawrence Erlbaum.

Baxter, L. A., & Montgomery, B. M. (1996). *Relating: Dialogues & dialectics.* New York: Guilford.

Baym, N. K., Zhang, Y. B., & Lin, M. C. (2004) Social interactions across media: Interpersonal communication on the Internet, face-to-face, and the telephone. *New Media & Society, 6*, 299–318.

Baym, N. K., Zhang, Y. B., Lin, M. C., Kunkel, A., Lin, M., & Ledbetter, A. (2007). Relational quality and media use. *New Media & Society, 9*, 735–752.

Bell, B. S., & Kozlowski, S. W. J. (2002). A typology of virtual teams: Implications for effective leadership. *Group & Organization Management, 27*, 14–49.

Bente, G., Rüggenberg, S., Krämer, N., & Eschenburg, F. (2008). Avatar-mediated networking: Increasing social presence and interpersonal trust in net-based collaborations. *Human Communication Research, 34*, 287–318.

Beranek, P. M., & Martz, B. (2005). Making virtual teams more effective: Improving relational links. *Team Performance Management, 11*, 200–213.

Bergstrom, T., & Karahalios, K. (2006). Communicating more than nothing. *Extended Abstracts of CHI 2006*, 532–537.

Biocca, F. (1997). The cyborg's dilemma: Progressive embodiment in virtual environments. *Journal of Computer Mediated Communication*, 3: Retrieved July 19, 2009, from http://www.ascusc.org/jcmc/vol3/issue2/ biocca2.html

Biocca, F., & Harms, C. (2002, October). Defining and differentiating copresence, social presence, and presence as transportation. Paper presented at the 5th Annual International Workshop on Presence, Porto, Portugal. Retrieved from http://www.temple.edu/ispr/prev_conferences/proceedings/2002/Final papers/Presence2002.

Biocca, F., Harms, C., & Burgoon, J. K. (2004). Toward a more robust theory and measure of social presence: Review and suggested criteria. *Presence, 12*, 456–480.

Boase, J. (2008). Personal networks and the personal communication system. *Information, Communication & Society, 11*, 490–508.

Bregman, A., & Haythornthwaite, C. (2003). Radicals of presentation: Visibility, relation, and co-presence in persistent conversation. *New Media & Society, 5*, 117–140.

Burgoon, J. K., Bonito, J. A., Ramirez, Jr., A., Dunbar, N. E., Kam, K., & Fischer, J. (2002). Testing the interactivity principle: Effects of mediation, propinquity, and verbal and non-verbal modalities in interpersonal interaction. *Journal of Communication, 52*, 657–677.

Canary, D. J., Stafford, L., Hause, K. S., & Wallace, L. A. (1993). An inductive analysis or relational maintenance strategies: Comparisons among lovers, relatives, friends, and others. *Communication Research Reports, 10*, 5–14.

Carpenter, D., & Knox, D. (1986). Relationship maintenance of college students separated during courtship. *College Student Journal, 28*, 86–88.

Chang, A., Resner, B., Koerner, B., Wang, X., & Ishii, H. (2001). LumiTouch: An emotional communication device. *CHI '04 Extended Abstracts on Human Factors in Computing Systems*, 313–314.

Chiadambaram, L. (1996). Relational development in computer-supported groups. *MIS Quarterly, 18*, 143–165.

Clark, A. (2007). A sense of presence. *Pragmatics & Cognition, 15*, 413–433.

Connaughton, S. L., & Daly, J. A. (2003). Long distance leadership: Communicative strategies for leading virtual teams. In D. J. Pauleen (Ed.), *Virtual teams: Projects, protocols, and processes* (pp. 116–144). Hershey, PA: Idea.

Connaughton, S. L., & Daly, J. A. (2004a). Leading in geographically dispersed organizations: An empirical study of long distance leadership from the perspective of individuals being led from afar. *Corporate Communication: An International Journal, 9*, 89–103.

Connaughton, S. L., & Shuffler, M. (2007). Multinational multicultural distributed teams: A review and future agenda. *Small Group Research, 38*, 387–412.

Connell, J. B., Mendelsohn, G. A., Robins, R. W., & Canny, J. (2001, September). *Effects of communication medium on interpersonal perceptions: Don't hang up the telephone yet!* In Proceedings of the 2001 International ACM SIGGROUP Conference on Supporting Group Work,Boulder, CO. Retrieved July 20, 2009, from http://psyweb2.ucdavis.edu/labs/robins/lab/group.pdf

Cramton, C. D., Orvis, K. L., & Wilson, J. M. (2007). Situation invisibility and attribution in distributed collaborations. *Journal of Management, 33*, 525–546.

Daft, R. L., & Lengel, R. H. (1986). Organizational information requirements, media richness and structural design. *Management Science, 32*, 554–571.

Dainton, M., & Aylor, B. (2001). A relational uncertainty analysis of jealousy, trust, and maintenance in long-distance versus geographically close relationships. *Communication Quarterly, 49*, 172–199.

Dainton, M., & Aylor, B. (2002). Patterns of communication channel use in the maintenance of long-distance relationships. *Communication Research Reports, 19*, 118–129.

Dainton, M., & Stafford, L. (1993). Routine maintenance behaviors: A comparison of relationship type, partner similarity and sex differences. *Journal of Social and Personal Relationships, 10*, 255–271.

DeSanctis, G., & Monge, P. (1999). Communication processes for virtual organizations. *Organization Science, 10*, 693–703.

Dimmick, J., Kline, S., & Stafford, L. (2000). The gratification niches of personal e-mail and the telephone. *Communication Research, 27*, 227–248.

Dindia, K., & Canary, D. J. (1993). Definitions and theoretical perspectives on maintaining relationships. *Journal of Social and Personal Relationships, 10*, 163–173.

Duck, S., & Pittman, G. (1994). Social and personal relationships. In M. Knapp & H. Miller (Eds.), *Handbook of interpersonal communication* (2nd ed., pp. 676–725). Thousand Oaks, CA: Sage.

Ficara, L. C., & Mongeau, P. A. (2000, November). *Relational uncertainty in long-distance college student dating relationships.* Paper presented at the annual meeting of the National Communication Association, Seattle, WA.

Fortunati, L. (2005). Is body-to-body communication still the prototype? *Information Society, 21,* 53–61.

Gerstel, N., & Gross, H. E. (1983). Commuter marriages: Couples who live apart. In E. Macklin & R. Rubin (Eds.), *Contemporary families and alternative lifestyles: Handbook on research and theory* (pp. 180–193). Beverly Hills, CA: Sage.

Gibbs, J. L. (2002). *Loose coupling in global teams: Tracing the contours of cultural complexity.* Unpublished doctoral dissertation, University of Southern California.

Gibbs, J. L., Nekrassova, D., Grushina, S. V., & Wahab, S. A. (2008). Reconceptualizing virtual teaming from a constitutive perspective: Review, redirection, and research agenda. In C. S. Beck (Ed.), *Communication Yearbook, 32* (pp. 187–229). New York: Routledge.

Gibson, C. B., & Gibbs, J. L. (2006). Unpacking the concept of virtuality: The effects of geographic dispersion, electronic dependence, dynamic structure, and national diversity on team innovation. *Administrative Science Quarterly, 51,* 451–495.

Gilbertson, J., Dindia, K., & Allen, M. (1998). Relational continuity constructional units and the maintenance of relationships. *Journal of Social and Personal Relationships, 15,* 774–790.

Goffman, E. (1963). *Behavior in public places.* New York: Free Press.

Guldner, G. T. (2004). *Long distance relationships: The complete guide.* Corona, CA: JF Milne.

Hart, R. K., & McLeod, P. L. (2003). Rethinking team building in geographically dispersed teams: One message at a time. *Organizational Dynamics, 31,* 352–361.

Helgeson, V. (1994). Long-distance romantic relationships: Sex differences in adjustment and breakup. *Personality and Social Psychology Bulletin, 20,* 254–265.

Hinds, P. J., & Mortensen, M. (2005). Understanding conflict in geographically distributed teams: The moderating effects of shared identity, shared context, and spontaneous communication. *Organization Science, 16,* 290–307.

Holt, P. A., & Stone, G. L. (1988). Needs, coping strategies, and coping outcomes associated with long-distance relationships. *Journal of College Student Development, 29,* 136–141.

Huff, C., Sproull, L., & Kiesler, S. (1989). Computer communication and organizational commitment: Tracing the relationship in a city government. *Journal of Applied Social Psychology, 19,* 1371–1391.

Janan Johnson, A., Haigh, M. M., Becker, J. A. H., Craig, E. A., & Wigley, S. (2008). College students' use of relational management strategies in email in long distance and geographically close relationships. *Journal of Computer–Mediated Communication, 13,* 381–404.

Jarvenpaa, S. L., & Leidner, D. F. (1999). Communication and trust in global virtual teams. *Organization Science, 10,* 791–815.

Jarvenpaa, S. L., Shaw, T. R., & Staples, D. S. (2004). Toward contextualized theories of trust: The role of trust in global virtual teams. *Information Systems Research, 15,* 250–267.

Jones, R. H. (2008). The role of text in televideo cybersex. *Text & Talk, 28,* 453–473.

Kirkman, B., & Mathieu, J. (2005). The dimensions and antecedents of team virtuality. *Journal of Management, 31,* 700–718.

Kirschner, B. F., & Walum. L. R. (1978). Two-location families: Married singles. *Alternative Lifestyles, 1,* 513-525.

Lee, K. M. (2004). Presence, explicated. *Communication Theory, 14,* 27–50.

Leonardi, P. M., Jackson, M. K., & Marsh, N. (2004). The strategic use of "distance" among virtual team members: A multi-dimensional communication model. In S. H. Godar & S. P. Ferris (Eds.), *Virtual & Collaborative Teams: Process, Technologies, & Practice* (pp. 156-172). Hershey, PA: Idea Group.

Lombard, M., & Ditton, T. (1997). At the heart of it all: The concept of presence. *Journal of Computer Mediated-Communication* [On-line], 3 (2). Retrieved July 19, 2009, from http://www.ascusc.org/jcmc/ vol3/issue2/lombard.html

Lowry, P., Roberts, T., Romano Jr., N., Cheney, P., & Hightower, R. (2006). The impact of group size and social presence on small-group communication. *Small Group Research, 37*, 631-661.

Maguire, K. (2007a). Bridging the great divide: An examination of the relationship maintenance of couples separated during war. *Ohio Communication Journal, 45*, 131–158.

Maguire, K. C. (2007b). Will it ever end? A (re)examination of uncertainty in college student premarital long-distance romantic relationships. *Communication Quarterly, 55*, 415–432.

Maguire, K. C., & Sahlstein, E. (in press). In the line of fire: Family management of acute stress during wartime deployment. In F. Dickson & L. Webb (Eds.), *Families in crisis: Effective communication for managing unexpected, negative events.* Cresskill, NJ: Hampton Press.

Marks, M. A., Mathieu J., & Zaccaro S. J. (2001). A temporally based framework and taxonomy of team processes. *Academy of Management Review, 26*, 356–376.

Maruping, L. M., & Agarwal, R. (2004). Managing team interpersonal processes through technology: A task-technology fit perspective. *Journal of Applied Psychology, 89*, 975–990.

Maznevski, M.L., & Chudoba, K.M. (2000). Bridging space over time: Global virtual team dynamics. *Organization Science, 11*(5), 473.

McGrath, J. E. (1991). Time, interaction, and performance (TIP): A theory of groups. *Small Group Research, 22*, 147–174.

Merkle, E. R., & Richardson, R. A., (2000). Digital dating and virtual relating: Conceptualizing computer mediated romantic relationships. *Family Relations, 49*, 187–192

Merolla, A. J. (2010a). Relational maintenance and non-co-presence reconsidered: Conceptualizing geographic separation in close relationships. *Communication Theory, 20*, 169–193.

Merolla, A. J. (2010b). Relational maintenance during military deployment: Perspectives of wives of deployed US soldiers. *Journal of Applied Communication Research, 38*, 4-26.

Montoya-Weiss, M. M., Massey, A. P., & Song, M. (2001). Getting it together: Temporal coordination and conflict management in global virtual teams. *Academy of Management Journal, 44*, 1251–1262.

Nice, M. L., & Katzev, R. (1998). Internet romances: The frequency and nature of romantic online relationships. *CyberPsychology & Behavior, 1*, 217–223.

Nowak, K. (2001, May). *Defining and differentiating copresence, social presence, and presence as transportation.* Paper presented at the Presence 2001 Conference, Philadelphia, PA.

Nowak, K. (2003). Sex categorization in computer mediated communication (CMC): Exploring the utopian promise. *Media Psychology, 5*, 83–103.

Palmer, M. (1995). Interpersonal communication and virtual reality: Mediating interpersonal relationships. In F. Biocca & M. Levy (Eds.), *Communication in the age of virtual reality* (pp. 277–299). Hillsdale, NJ: Lawrence Erlbaum.

Rabby, M. K. (2007). Relational maintenance and the influence of commitment in online and offline relationships. *Communication Studies, 58*, 315–337.

Rabby, M. K., & Walther, J. B. (2003). Computer-mediated communication effects on relationship formation and maintenance. In D. Canary & M. Dainton (Eds.), *Maintaining relation-*

ships through communication: Relational, contextual, and cultural variations (pp. 141–162). Mahwah, NJ: Lawrence Erlbaum.

Ramirez, A., Dimmick, J., Feaster, J., & Lin, S. (2008). Revisiting interpersonal media competition: The gratification niches of instant messaging, e-mail, and the telephone. *Communication Research, 35*, 529–547.

Ramirez, A., & Zhang, S. (2007). When online meets offline: The effect of modality switching on relational communication. *Communication Monographs, 74*, 287–310.

Rice, R. E. (1984). Mediated group communication. In R. E. Rice (Ed.), *The new media: Communication, research, and technology* (pp. 129–154). Beverly Hills, CA: Sage.

Rice, R. E. (1993). Media appropriateness: Using social presence theory to compare traditional and new organizational media. *Human Communication Research, 19*, 451–484.

Rice, R., Chang, S., & Torobin, J. (1992). Communicator style, media use, organizational level, and use and evaluation of electronic messaging. *Management Communication Quarterly, 6*, 3–33.

Rice, R. E., & Gattiker, U. E. (2001). New media and organizational structuring. In F. M. Jablin & L. L. Putnam (Eds.), *The new handbook of organizational communication: Advances in theory, research, and methods* (pp. 544–581). Thousand Oaks, CA: Sage.

Rogers, E. M. (1986). *Communication technology: The new media in society*. New York: Free Press.

Rosenfeld, L. B., Richman, J. M., & May, S. K. (2004). Information adequacy, job satisfaction and organizational culture in a dispersed-network organization. *Journal of Applied Communication Research, 32*, 28–54.

Sahlstein, E. M. (2004). Relating at a distance: Negotiating being together and being apart in long distance relationships. *Journal of Social and Personal Relationships, 21*, 689–710.

Saunders, C. S., & Ahuja, M. K. (2006). Are all distributed teams the same? Differentiating between temporary and ongoing distributed teams. *Small Group Research, 37*, 662–700.

Scott, C. R., & Timmerman, C. E. (1999). Communication technology use and multiple workplace identifications among organizational teleworkers with varied degrees of virtuality. *IEEE Transactions on Professional Communication, 42*, 240–260.

Scott, V. M., Mottarella, K. E., & Lavooy, M. J. (2006). Does virtual intimacy exist? A brief exploration into reported levels. *CyberPsychology & Behavior, 9*, 759–761.

Short, J., Williams, E., & Christie, B. (1976). *The social psychology of telecommunications*. London: John Wiley.

Sidelinger, RJ, Ayash, G., Godorhazy, A., & Tibbles, D. (2008). Couples go online: Relational maintenance behaviors and relational characteristics use in dating relationships. *Human Communication, 11*, 341–355.

Sigman, S. J. (1991). Handling the discontinuous aspects of continuing social relationships: Toward research of the persistence of social forms. *Communication Theory, 1*, 106–127.

Sproull, L., & Kiesler, S. (1986). Reducing social context cues: Electronic mail in organizational communication. *Management Science, 32*, 1492-1512.

Stafford, L. (2004). Romantic and parent-child relationships at a distance. *Communication Yearbook, 28*, 37–85.

Stafford, L. (2005). *Maintaining long-distance and cross-residential relationships*. Mahwah, NJ: Lawrence Erlbaum.

Stafford, L. H., & Canary, D. J. (1991). Maintenance strategies and romantic relationship type, gender and relational characteristics. *Journal of Social and Personal Relationships, 8*, 217–242.

Stafford, L., & Merolla, A. (2007). Idealization, reunions, and stability in long distance dating relationships. *Journal of Social and Personal Relationships, 23*, 901–919.

Stafford, L., & Reske, J. R. (1990). Idealization and communication in long-distance premarital relationships. *Family Relations, 39*, 274–279.

Stephen, T. (1984). A symbolic exchange framework for the development of intimate relationships. *Human Relations, 37,* 393–408.

Stephen, T. (1986). Communication and interdependence in geographically separated relationships. *Human Communication Research, 13,* 191–210.

Townsend, A. M., DeMarie, S. M., & Hendrickson, A. R. (1998). Virtual teams: Technology and the workplace of the future. *Academy of Management Executive, 12,* 17–29.

Walther, J. B., & Bazarova, N. N. (2008). Validation and application of electronic propinquity theory to computer-mediated communication in groups. *Communication Research, 35,* 622–645.

Walther, J. B., & Parks, M. R. (2002). Cues filtered out, cues filtered in: Computer-mediated communication and relationships. In M. L. Knapp & J. A. Daly (Eds.), *Handbook of Interpersonal Communication* (3rd ed., pp. 529–563). Thousand Oaks, CA: Sage.

Westefeld, J. S., & Liddell, D. (1982). Coping with long-distance relationships. *Journal of College Student Personnel, 23,* 550–551.

Westmyer, S., DiCioccio, R., & Rubin, R. B. (1998). Appropriateness and effectiveness of communication channels in competent interpersonal interaction. *Journal of Communication, 48,* 27–48.

White, P., & White, N. (2005, August). Virtually there. First Monday, 10(8). Retrieved July 19, 2009, from http://www.uic.edu/htbin/cgiwrap/bin/ojs/index.php/fm/article/viewArticle/1267/1187

Whitty, M. T. (2008). Liberating or debilitating? An examination of romantic relationships, sexual relationships and friendships on the net. *Computers in Human Behavior, 24,* 1837–1850.

Wiesenfeld, B. M., Raghuram, S., & Garud, R. (1999). Communication patterns as determinants of organizational identification in a virtual organization. *Organization Science, 10,* 777–790.

Wrench, J., & Punyanunt-Carter, N. (2007). The relationship between computer-mediated-communication competence, apprehension, self-efficacy, perceived confidence, and social presence. *Southern Communication Journal, 72,* 355–378.

Ylinen, A., & Valo, M. (2004, November). *Experiencing and Expressing Co-Presence in Technologically Maintained Close Relationships.* Paper presented at the annual meeting of the National Communication Association Convention, Chicago, IL.

Zack, M. H. (1994). Electronic messaging and communication effectiveness in an ongoing work group. *Information and Management, 26,* 231–241.

Zhao, S. (2003). Toward a taxonomy of copresence. *Presence, 12,* 445–455.

Zhao, S., & Elesh, D. (2008). Co-presence as being with: Analyzing online connectivity. *Information, Communication, & Society, 11,* 565-583.

CHAPTER FOURTEEN

Healthcare Provider-Recipient Interactions: Is "Online" Interaction the Next Best Thing to Being There?

Theodore A. Avtgis

E. Phillips Polack

Sydney M. Staggers

Susan M. Wieczorek

Note: The authors would like to thank Dr. Matthew M. Martin for comments made on an earlier draft.

The patient-provider relationship is by far the most researched of all relationships within healthcare (see, e.g., Thompson, Dorsey, Miller, & Parrot, 2003). Whether under the guise of the search for enhanced healing, greater perceived satisfaction, reduced litigation, or better targeted services and interventions, the exchange of information and subsequent relational development has intrigued scholars and practitioners from virtually all healthcare disciplines, social sciences, and humanities. The influx of technology has affected all aspects of medicine including the interpersonal dynamics between the patient and the physician. The influence of computer-mediated interaction on patient care is something that is in the process of being assessed by all aspects of medical practitioners. That is, medical regulatory agencies, patient advocacy groups, insurance carriers, as well as the federal government (Medicaid, Medicare).

The goal of this chapter is to review the existing data on computer technology and interpersonal interaction within healthcare and discuss the implications concerning the use of computer technology in the delivery of healthcare and the relationships that develop between patients and providers. By provider, we refer to physicians, physician extenders (e.g., Physician Assistants [PAs], Nurse Practitioners [NPs]) either solo, in clinics, or in the hospital setting. All providers utilize various electronic forms of information, all intended to make the patient care more manageable and less expensive.

Overview

In 1971, computer engineer Ray Tomlinson's description of electronic mails between his own computers changed the ecology of media forever (Car & Sheikh, 2004). Predicting the influence of such media on society, McLuhan (1964) stated:

> The new media in technologies by which we amplify and extend ourselves constitutes huge collective surgery carried out on the social body with complete disregard for antiseptics. If the operations are needed, the inevitability of infecting the whole system during the operation has to be considered. For operating on society with a new technology, it is not the incised area that is most affected, the area of impact and incision is numb. It is the entire system that is changed. (p. 64)

This begs two questions: How have emails changed the landscape of the provider-patient relationship that was previously constructed and maintained through face-to-face communication? How have emails encroached on the previously time-constrained relationship between provider and patient? Through email, physicians can control patient access through preprogramming their email to only allow patients to respond with a certain number of characters or other types of controlling mechanisms that filter out "extraneous" medical or social information (Mandl, Kohane, & Brandt, 1998).

From an historic perspective, Starr (1984) in his Pulitzer Prize-winning novel chronicles the popularization of the telephone thus reducing the need for the telegraph or tracking down the "peripatetic practitioner on foot." In the history of machine-mediated communication, the telephone allowed messages to be transferred synchronously through time and over distance. Although it promised to improve physician accessibility and efficiency, concerns for patient abuse of physician time, safety of telephone diagnoses, and privacy issues arose (Mandl et al., 1998). Despite some concern about '*party line*' eavesdropping and other security issues, by the mid-1920s the telephone became as commonplace and necessary in physician practices as the stethoscope (Aronson, 1977). This form of mediated communication enabled the provider to enter the patient's home for private and immediate consultation, a benefit which far outweighed any security concerns.

By the early 1970s, email first entered the healthcare setting through hospital information systems (Bleich, Beckley, & Horowitz, 1985). Within a twenty-year span and through the dawn of the Internet revolution, published accounts of emails emerged between clinicians as early as the 1980s (Bergus, Sinift, Randall, & Rosenthal, 1998; Branger, van der Wouden, & Schudel., 1992; Sands, Safran, Slack, & Bleich, 1993, 1999) and between providers and patients in the 1990s (Moyer, Stern, Katz, & Fendrick, 1999; Neill, Mainous, Clark, & Hagen, 1994). Today, evidence-based studies (i.e., studies based on empirical data) reveal the positive effects emails have on patient satisfaction, safety, and quality of care (Brooks & Menachemi, 2006; Houston, Sands, Nash, & Ford, 2003; Lin, Wittevrongel, Moore, Beaty, & Ross, 2005; Roter, Larson, Sands, Ford, & Houston, 2008; Spielberg, 1999; Tjora, Tran, & Faxvaag, 2005). These benefits transcend issues as to whether or not email exchanges between patient and provider should be the primary means of interaction or utilized as complementary to face-to-face communication.

Demographics and CMC/Health Issues

Given that email use and perceptions of medicine are not static constructs among consumers, a brief discussion of demographic trends in healthcare and technology use is warranted. Healthcare's technological landscape in the 21st century has markedly advanced with wide-spread Internet use throughout the United States. Now more than ever, health information, telemedicine, and basic emailing have become commonplace to millions of patients from all walks of life. The technology available to patients for communicating with their physician has begun to transcend traditional demographic trends regarding the use of technology. According to the most recent census reports comparing populations from 1997–2007 (United States Census Bureau, 2007), 62% of households report having Internet access in the home while 64% of individuals over 18 report using the Internet from remote locations. This represents a tripling effect since 1997 when only18% had household access and 22% had access from remote locations. Those who use the Internet, do so through high-speed access (82%) as opposed to dial-up technology (17%).

More specifically, the census data indicates varied usage rates according to state, educational level, race, and age groups. States having the highest Internet usage rates are Alaska (78.5%) and New Hampshire (82.6%) while Mississippi (52.8%) and West Virginia (56.5%) have the lowest rates. Individuals aged 25 and older with bachelor's degrees report an 87% online usage rate from remote locations while those with only some college education (74%), only a high school diploma (9%), and no high school diploma (19%) report a significantly lower rate. Similar usage disparities are found among races with 74% of the Asian population using the Internet followed by 69% of whites, 51% of blacks, and 48% of Hispanics. Finally age comparisons show that 73% of those 18–34 years old, 56% of those aged 3–17, and 35% aged 65 and older used the Internet. It is im portant to note that according to the Pew Research Center's Internet & American Life Project spanning 2006–2008 (Jones & Fox, 2009), the largest increase of Internet use is seen in the 70–75 year-old age group with 26% of that group online in 2005 and 45% of that same group online in 2008. Pew anticipates this to be the fastest growing group in the United States in years to come.

Certainly the digital usage patterns which once divided consumers by location, educational level, age, and race have narrowed considerably in recent years. Further, evidence of this trend is represented by a marked increase in Internet usage amongst older consumers who seek healthcare information via the web (Campbell & Wabby, 2003; Dehn, 2002; Ketteridge, Delbridge, & Delbridge,

2005; Mo, Malik, & Coulson, 2009; Rannefeld, 2004-2005; Slack, 2004; Stalberg, Yeh, Ketteridge, Delbridge, & Delbridge, 2008).

In the 2008 Tracking Report by the Center for Studying Health System Changes, Tu and Cohen (2008) investigated how people acquired healthcare information. Results indicate that in 2007, 56% of American adults (122 million people) sought information about personal health concerns through health-related web sites. This is up significantly from 38% (72 million people) in 2001. Consumer use of the Internet for health information now rivals patient use of more traditional and long-standing sources such as books, magazines, and newspapers (33%) as well as friends and relatives (31%). Further, Internet access of medical information is reduced for elderly and less-educated populations. More specifically, seniors are half as likely as others (aged 18–49) to turn to the Internet for information about personal health concerns (18% vs. 36%) (Tu & Cohen, 2008). In addition to the age demographic, there are other notable exceptions, such as African Americans and Hispanic consumers, who were more likely to be influenced by Internet information than were Caucasian consumers. Only half of Caucasians reported that the Internet information altered their overall approach to health maintenance (Tu & Cohen, 2008). Some reasons for this trend are the increased availability of high-speed Internet residential access (Horrigan & Smith, 2007) and an increase in web-based health sites for consumers (Noonan, 2007). Based upon this evidence of wide-spread Internet use and health-information interest, there is no reason not to believe that as remote and underserved areas increase their access levels, improved Internet-based consumerism will continue to escalate in numbers and demographic representation.

Patient Perspectives

According to Suggs (2006), patients expect providers to meet very specific goals during the course of their care. These include maintaining and improving health, preventing disease and illness, improving lifestyle behaviors, reducing risk factors for disease, increasing compliance with a medication or treatment plan, improving self-management strategies, providing social support, helping with decisions about health, and locating and retrieving reliable health information. Each of these goals may be achieved through face-to-face communication between patient and provider. However, these goals are also being realized through online information systems and email contact.

According to the American Medical Informatics Association (AMIA), these health goals of the patient must be maintained in an effort to achieve the patient's trust, safety, and overall health. In the medical relationship therefore emails act as "computer-based communication between clinicians and patients, within a contractual relationship in which the health care provider has taken on an explicit measure of responsibility for the client's care" (Kane & Sands, 1998, para. 4).

Care includes medical advice, treatment, and information exchanged professionally between physicians and their patients through electronic interactions.

Although all patients may have the right to communicate through multiple media channels, not all members of the population have equal access, the financial means, or are technologically savvy to take advantage of this right. Even though high-speed access more than doubled in nearly all American homes, regardless of age (Jones & Fox, 2009), only 51% of the total population even has computer access at home, 47% have email, 35% rely on friends or family outside the home for access, 15% share email accounts with friends and family, and 12% neither use nor have access to the Internet or email (Pelletier, Sutton, & Walker, 2007). These technological usage factors (lack of access exclusivity) can compromise patient privacy which further exacerbates the utilization of electronic interactions between patient and provider.

Financially, some patients cannot afford to use the Internet (Rideout, Neuman, Kitchman, & Brodie, 2005) while others struggle with even the most basic health-literacy skills (Gazmararian, Baker, & Williams. 1999; Williams, Parker, Baker, 1995). Approximately one-third of the American population has difficulty with understanding general health explanations or instructions provided by their physician (Polack, Richmond, & McCroskey, 2008). This barrier affects some populations more than others. For example, the elderly, African Americans, and Hispanics are especially affected by health literacy (Tu & Cohen, 2008). Consideration of a patient's level of understanding during treatment is not routinely assessed by many healthcare providers (Silverman, Kurtz, & Draper, 2005). Disparities such as these marginalize populations, creating a "digital divide" that cannot be ignored in the pursuit of equal treatment and care that is desired for all patients in the 21st century (Lenhart et al., 2003). At this point in time, electronic interactions between patient and provider, given the necessary access and equipment, can actually contribute to marginalizing a segment of the American population that does not have access to such benefits.

Attempts at rectifying this disparity continue as assessment techniques for measuring health and e-health literacy levels have been created (Norman & Skinner, 2006; Shaw, Ibrahim, Reid, Ussher, & Rowlands, 2009). Training programs attempt to provide users with basic e-literacy skills in hopes of narrowing the digital divide (Macias & McMillan, 2008; Shaw, Gustafson, Hawkins, McTavish, McDowell, Pingree, & Ballard, 2006). Even in light of these efforts, the e-accessibility gap continues to grow as government mandates for increased technology in medicine place further demands on patients and physicians alike. Simply put, not all people choose to communicate with the physician or healthcare provider through email. Some are limited by access, financial burdens, and literacy levels. Physicians and patients must share their willingness and ability to engage in electronic communication in advance. Those who do choose to use emails and obtain information from online resources have the advantage of enhanced, patient-centered care. As Roter et al. (2008) found, email messages between patients and physicians mimic the same communication dynamics as traditional medical dialogue, whether it be face-to-face or over the telephone. Although physicians respond through emails with shorter dialogues than patients, the email facilitates informal tasks and enhances emotional support and partnership with physicians. In short Roter et al. state, "Email has the potential to support the doctor-patient relationship by providing a medium through which patients can express worries and concerns and physicians can be patient-centered in response." (p.80)

Provider Perspectives

There are unique challenges that a provider must contend with should computer-mediated interaction with patients be viable, safe, and effective. What if a patient has a stronger need to communicate with their physician via email rather than coming in for a face-to-face consult? This is in fact the case for 78% of surveyed patients (Pelletier et al., 2007). Consider the physician who has previously engaged exclusively in face-to-face encounter with patients, yet fully recognizes that the technological requirements of the business and practice of medicine require such technological acumen. Further, the desires of society in general mandate this transition and in relatively quick fashion (Mandl et al., 1998). Does this electronic patient-physician communication create *problems* or *promise* for the future of the practice of medicine? It is our contention that the transition to more email interaction will experience inherent growing pains as the logical result is a decrease in face-to-face interaction.

By moving away from a channel that provides the richness of nonverbal messages to one where garnering nonverbal cues is more difficult, can the provider effectively communicate with a patient and make decisions about the care of the individual patient without the nonverbal input that occurs in the face-to-face interaction?

Can the robust discussion that occurs in face-to-face communication occur with the use of mediated communication such as email? Do nonverbal messages add to the "current evidence" that is required in evidence-based medicine to make conscientious and explicit decisions about patient care? Should it be assumed that email interaction between patient and provider be used only as an accentuation to communication that is first predicated on previous face-to-face interaction? Or is email interaction enough, in and of itself, for proper presentation of illness and diagnosis? The answers to these types of questions will come through the millions of interactions, both face-to-face and mediated that occur and will occur between patient and provider.

Another conundrum for the clinician is dealing with the emergency patient. Whereas email is an asynchronous form of communication, concern exists if an email is sent about a life-altering or threatening condition and either not received, not read in a timely fashion, or misinterpreted. This asynchronous form of communication can be deleterious to the health and welfare of the patient. Perhaps a more comprehensive question of utilizing email technology between patient and provider should be not whether email is an effective means of health information exchange between patient and provider but when, where, and why is it an equally or more effective tool than traditional methods of interaction (i.e., face-to-face).

Cyber-Time: Financial and Legal

In the practice of Medicine, reimbursement for service is always an issue that eventually comes into the forefront for those who provide medical services. How is a provider to bill an email exchange? Is it billable by the time spent encoding the message, the actual production of the message (i.e., typing), or a combination of both? What billing standards are to be applied? Most physicians in this country practice in individual practice scenarios whereby their services rendered must be billable and although some health insurance companies such as Aetna/Cigna are looking at the issue of reimbursement as it remains a relatively unexplored and eventual massive problem in medical procedural accounting.

And the final question, what about the medical legal prudence of email transmittals? In no situation is the saying *you can't take back communication* more applicable. Once it is out there in cyberspace, it is indelible, thus all transmittals via electronic mail are "legal documents." Take the simple comforting message "things will work out, don't worry." If this message was sent from a physician to a patient, what legal significance does such a supportive message hold? Because email represents a physical piece of possible evidence, similar to all written documents, it is subject to interpretation. Does the utterance imply that the doctor will heal the patient? The physician understands the uncertainty being experienced by the patient and wants them to think positively? Or something in between?

Although there has been criteria offered as professional guidelines concerning how to use email as a mechanistic communication device with less emphasis on when email between physician and patient is appropriate, we propose the following *Email Appropriation Determination* (EAD) criteria for evaluating when the use of email between patients and providers is warranted and perhaps recommended:

- There should be a firm contract about email between provider and patient including such things as response time and expectations on the part of both parties as to the limitations of email versus face-to-face interaction. When the provider is not available, there should be an automatic response system on the computer notifying the patient of such nonavailability and directing them to an appropriate phone contact.
- Email should never be used for emergency purposes. All email transmittals on the part of the provider should have an included "footer" indicating that emergency transactions are never appropriate. In an effort to help the patient determine what is, and what is not an emergency, patients can select a message priority indicator at the onset of each transaction. An automatic loop for

those marked "urgent" may prevent patients from leaving a message and then direct them to call the physician's office immediately. This is much like a telephone answering service that directs patients to call an emergency number or go to the Emergency Room instead of leaving a message.

- Email should never be used for the delivery of bad news or for the notification of abnormal, confusing, or unexplained test results. Such standards and limitations must be established in advance with all who use the patient-provider email system.
- Anything said on email cannot be "taken back" on the part of either party and will become part of the patient's permanent, medical legal record. Therefore all messages should be encoded and sent in a mindful fashion with the understanding that all email messages may have legal precedence.
- Any reimbursement issues concerning billing for e-consultation must be negotiated between the physician, patient, and insurance companies through a written notification prior to any electronic contact with the patient.

The efficacy of such guidelines assumes that there is a mutual knowledge of each party's level of technological literacy, health literacy, and technological availability. As computer-based technology transforms the face of medicine, shared responsibility for effective communication between the patient and physician remains vital for good health. As the American Institute of Medicine (2001) states, patients deserve quality care whenever they need it and in many forms, not just via face-to-face interaction. Healthcare systems must be responsive at all times, and access to care should be provided over the Internet, by telephone, and by other means in addition to in-person visits.

Electronic media serves to define the relationship of its users and transforms the entire system of communication between patients and their healthcare providers. Asynchronous emails facilitate the interaction between the patient and provider. Multiple levels of communication which include emails are the key to trusting and effective medical relationships.

Healthcare and Technology

Over the past 20 years, there has been a profound change in the way that healthcare is consumed. The change is evident in recent headlines that read: "Doctors Will Make Web Calls in Hawaii" (Miller, 2009), "Electronic Prescribing Gathers Steam" (Gaudie, 2006), and "Massachusetts requires EHRs in hospitals by 2015" (Ferris, 2008). To date, the healthcare industry has been operating from a paradigm that places a greater focus on the patient, thus contributing to what is known as the consumer-driven market (Gupta, 2003). In this system, the consumer is in control and assumes a large degree of responsibility for his/her own healthcare (Frist, 2005).

To meet the consumers' demands in this marketplace, the healthcare industry has been rapidly advancing technological capabilities or health information technology (HIT) (Goldschmidt, 2005; Hillestad et al., 2005) to be more efficient through seamless and paperless medical information exchange via electronic medical records (EMRs). In addition to EMRs, there is a wide array of technology in place to coordinate care and cut costs. Examples of such technologies include, but are not limited to *personal health records* (PHRs), *telemedicine*, *practice management systems* (PMS), *e-prescribing*, and *decision support tools*, among others.

Technology–Benefits, Barriers, and Implications

Electronic Medical Records. The EMR is a computer-based system for storing an individual's health-related information (McGrath, Arar, & Pugh, 2007). More specifically, EMRs, are defined as:

> An application environment composed of the clinical data repository, clinical decision support, controlled medical vocabulary, order entry, computerized provider order entry, pharmacy, and clinical documentation applications. This environment supports the patient(s) electronic medical record across inpatient and outpatient environments, and is used by healthcare practitioners to document, monitor, and manage health care delivery within a care delivery organization (CDO). The data in the EMR is the legal record of what happened to the patient during their encounter at the CDO and is owned by the CDO. (Garets & Davis, 2006, p. 2)

There are a myriad of pros and cons that are still emerging concerning the use of EMRs given that the field of healthcare is in a state of rapid development with a myriad of state governments mandating that such systems be deployed as soon as possible (e.g., Ferris, 2008). The EMR has the "potential to revolutionize many aspects of documentation management for health care organizations of all types" (Jones, 2008, p. 41). As such, these electronic capabilities have allowed acceptance of practice management systems (PMSs) which are utilized in virtually all medical offices that concern billing at both federal and state levels (Jones).

However, Jones (2008) noted that there are barriers to the using of EMRs; most noteworthy is cost and complexity of implementing such systems. Presently, physicians bear the cost, but payers (i.e., health insurers) reap the benefits. A barrier is complexity whereby the process associated with migrating from paper to electronic modes of record keeping can be a challenge in and of itself (Mitchell, 2007). Further, there is the issue of privacy, which is a topic that leads to great anxiety for patients (Mandl, Szolovits, & Kohane, 2001). According to Mandl et al., this anxiety is warranted based on the fact that medical data is frequently shared with payers, government, and employers, to name but a few.

While organizations using these technologies boast about the advantages associated with the implementation and utilization of the technology, there are a variety of implications for health insurers, the consumer, as well as the healthcare system as a whole. For instance, EMRs provide accessible health records regardless of geographic location given the interoperability of such systems. The interoperability of EMR systems could facilitate the sharing of patient files between an individual's primary care provider and specialists, which can help to alleviate issues like adverse drug interactions. This type of medical event (i.e., adverse drug interaction) is often a result of a patient forgetting to tell one of his/her doctors about a medical allergy or a particular medication that they are presently taking (Jonietz, 2003). Thus, continuity of care is established to some degree through the use of EMRs (Chen, Garrido, Chock, Okawa, & Liang, 2009). Many EMRs also have the ability to identify possible adverse drug interactions, thus eliminating potential medical emergencies that a consumer may encounter had such a system not been in place (Jones, 2008). As such, it is believed that EMRs will "lead to major health care savings, reduce medical errors, and improve health" (Hillestad et al., 2005, p. 1103).

According to a study summarizing findings from the 2003 RAND Health Information Technology Project, it is estimated that at a 90% adoption rate of EMRs, savings could average over $77 billion per year (Hillestad et al., 2005). While savings would accrue for a variety of stakeholders, it is assumed that over time, these savings would eventually be realized by the payers. Therefore this is viewed as a strong incentive for payers to advocate the adoption of EMR systems (Hillestad). Moreover, another implication of EMR systems entails the inclusion of features like preventive care and disease management. Preventive care via EMRs would involve the integration of "evidence-based recommendations" for services like annual screening exams, which could keep consumers healthy. In terms of disease management, EMRs role would be to track the frequency of screenings of those with a chronic condition and alert physicians to offer any missing or necessary tests at the time of appointment, thus lessening the severity of chronic conditions and the complications associated with the maltreatment of such conditions. As with any administration of health records, there is always the possibility of compromising data. Whether by accident or malice, there are inherent ethical issues that will need to be addressed.

Personal Health Records. PHRs are similar in nature to EMRs; however, PHRs capture health information that is entered by the individual consumer. PHRs are reflective of the current consumer-driven market that has evolved. However, PHRs have not had the same momentum, nor the amount of attention that EMRs have received (Tang, Ash, Bates, Overhage, & Sands, 2006). It is believed that with increased integration and adoption, PHRs will be as seamless as EMRs in terms of system interfacing. The integration of PHRs and EMRs would provide great benefits to the patient, including more relevant data being made available more quickly. Likewise, the integration of PHRs with EMRs would permit more than sufficient backup in the event of a natural disaster (e.g., Hurricane Katrina). Yet, considered alone, PHRs offer a significant benefit to the patient as a vehicle through which the patient becomes empowered by taking control of his/her own health (Kahn, Aulakh, & Bosworth, 2009; University of California, San Francisco, 2009).

Barriers to using PHRs, as identified by Kahn et al. (2009) are: cost, issues related to privacy, information exchange problems, and overall design. Similarly, Denton (2001) found that privacy concerns were a barrier to PHA adoption in a study conducted with patients with spinal disorders. As to integration of PHRs with EMRs literacy is likely to become a critical factor because of potential reliability issues regarding patient-centered data. For instance, patients could reliably enter information such as height and weight, but more complex laboratory test reporting may be less reliable, thus problematic for establishing proper care. Health literacy in the case of PHRs then takes the form of a language and communication competency as well as a technological competency. Thus, there is a tripartite of literacy elements that must be addressed/assessed before any patient information can be seen as valid.

Implications associated with the PHR from the consumer perspective entail support for decision-making of his/her care and current conditions (Tang et al., 2006). Likewise, the ease and portability of such records is likely to help reduce medical errors and lead to better clinical outcomes when individuals move or change jobs and are forced to switch payers (Denton, 2001; Tang et al., 2006). While the integration of PHRs with EMRs is still in progress, the model proposed by Tang and colleagues suggests that the burden of the cost be assumed by payers, but could reap possible returns on investment. This is especially salient for the financial savings that implementing disease management component of PHR could bring.

Telemedicine. According to Wootton (2001), telemedicine is "an umbrella term that encompasses any medical activity involving an element of distance" (p. 557). Telemedicine is also referred to as e-health, online health, and telehealth and most commonly includes things such as patient-physician interaction via telecommunication. Examples that fall under the umbrella of telemedicine include *teleconsultation* (i.e., doctor and patient are in different locations, but connected via some type of communication link, like videoconferencing), *teleradiology* (i.e., radiological images transmitted via some type of communication link for appropriate interpretation/diagnosis), and *telepathology* (i.e., use of microscope or other equipment to make diagnosis) (Wootton, 1996).

Telemedicine has experienced recent advances given technological advancements. Many of these advancements have resulted in cost savings to the health care industry as a whole. Such technologies include Kaiser Permanente's clinical trial utilizing telenursing. Telenursing equipment, specific to this trial, included a low-resolution videophone and an electronic stethoscope. Similarly, Miller (2009) reported that the Hawaii Medical Association, a Blue Cross-Blue Shield licensee, will make house-calls via the Internet. Consumers simply login to the health plan's website where they can speak to a doctor for a fee. These technologies are advantageous for remote or rural areas where there is limited access to healthcare or healthcare professionals. Drawbacks to the use of telemedicine include depersonalization, overdependence on technology, clinical risk, and legal implications.

Similar to EMRs and PHRs the integration of telemedicine by insurers could not only provide access to remote and rural areas, but could also contribute to patient satisfaction and savings for payers. A study by Kaiser Permanente found that decreased office visits and increased telephone visits can act as a reasonable alternative for standard office visits when integrated with EMRs (Chen et al., 2009). Moreover, this study indicated that quality of care and patient satisfaction were upheld through the use of such technologies. These findings suggest that telemedicine is fast becoming an established practice and has a multitude of implications for patients that range from easily obtaining access to care and if required, access to in-person or emergency care.

Implications for Communication

Taken together, EMRs, PHRs, and telemedicine have implications for the field of communication as it relates to literacy, specifically health literacy. If eHealth is to be an effective tool in the delivery of medicine, there needs to be greater congruity between what is made available to be accessed and what is actually accessed by the patient. (Harvey, Skinner, & Simms, 2006, n.p.). For communication scholars, there is a need to educate insurers and consumers how to communicate and interpret health information, respectively. This is of particular importance since inadequate literacy levels have been associated with poor health outcomes (e.g., Feldman, Makuc, Kleinman, & Cornoni-Huntley, 1989). In this sense education is imperative for consumers seeking to engage in utilizing HIT.

Scholars in the field should also focus attention on the possible communicative issues that affect the delivery of HIT as a means of improving services through this medium. From this perspective, it would be a worthwhile endeavor to understand the inequitable access to technology for those of varying socio-economic statuses. Mandl et al. (1998) suggest that the distribution of HIT may broaden the scope of access and outcomes for social disparities.

To summarize, EMRs, PHRs, and telemedicine provide worthy technologies for consumers and payers alike. From the consumer point of view, these technologies provide empowerment, knowledge, and a say in medical decision-making; while payers experience return on investment of such technological advances. Moreover, it is likely that the integration of a variety of information systems will allow physicians to access the most recent and up-to-date information on a patient, thus reducing medical errors and providing appropriate care. The consumer-driven market is not likely to dissipate in the near future; thus, it is critical for researchers, industry professionals, and consumers alike to understand how the implications associated with the use and adoption of technology impact communication patterns and relationships.

Conclusion

This chapter presented various perspectives regarding the practice, consumption, and profession of medicine via computer-mediated communication. There are several themes that pervade the use of technology in medical care. Perhaps the most significant is compatibility. By compatibility, we are referring to interpersonal, person-technology, inter-technology, as well as message-receiver compatibility. All of these different levels of compatibility have been shown to be critical elements should interaction between patients and providers become meaningful and effective.

As evidenced in this chapter, there are many efforts under way to test various approaches to making computer-mediated communication as fundamental/mandatory technology in medical care. However, we should caution that by focusing on technology as a replacement to face-to-face interaction, such efforts will be doomed to failure if there is a latent assumption that there is some degree of equivalency or superiority in the technology as opposed to natural interaction between patient and provider.

Finally, one cannot address such issues without considering the ethical considerations associated with such technology. Technology in the administration of healthcare is something that should be used as an accentuation of human relationship not a substitute. Any technological advancement will serve to benefit, yet marginalize people simultaneously. Will the use of email, EMRs, PHRs, and other technology even the playing field for all consumers of healthcare? Although we cannot definitively answer this question, we can conclude that the playing field will be changed in profound ways. The degree to which the field is level is better left to ethicists.

References

American Institute of Medicine, Committee on Quality of Health Care in America. (2001). *Crossing the quality chasm: a new health system for the 21st century*. Washington, DC: National Academy Press.

Aronson, S. H. (1977). The lancet on the telephone, 1876–1975. *Medical History, 2*, 69–87.

Bergus, G. R., Sinift, S. D., Randall, C. S., & Rosenthal, D. M. (1998). Use of an email cubside consultation service by family physicians. *Journal of Family Practice*, 47, 357–360.

Bleich, H. L., Beckley, R. F., Horowitz, G. (1985). Clinical computing in a teaching hospital. *New England Journal of Medicine, 312*, 756–764.

Branger, P. J., van der Wouden, J. C., & Schudel, B. R (1992). Electronic communication between providers of primary and secondary care. *British Medical Journal, 305*, 1068–1070.

Brooks, R., & Menachemi, N. (2006, January). Physicians' use of email with patients: Factors influencing electronic communication and adherence to best practices. *Journal of Medical Internet Research, 8*(1), e2.

Campbell, R. J., & Wabby, J. (2003). The elderly and the internet: A case study. *The Internet Journal of Health.* 3,(1). Retrieved July 14, 2009, from http://www.ispub.com/journal/

the_internet_journal_of_health/ volume_3_number_1_22/article/ the_elderly_and_the_internet_a_case_study.html

Car, J., & Sheikh, A. (2004). Information in practice email consultations in health care: Scope and effectiveness. *British Medical Journal, 329,* 435–438.

Chen, C., Garrido, T., Chock, D., Okawa, G., & Liang, L. (2009). The Kaiser Permanente electronic health record: Transforming and streamlining modalities of care. *Health Affairs, 28,* 323–333.

Dehn, R. W. (2002). Using computers in clinical practice. *Physician Assistant* 26(8), 35–40.

Denton, I. C. (2001). Will patients use electronic personal health records? Responses from a real-life experience. *Journal of Healthcare Information Management, 15,* 251–259.

Feldman, J. J., Makuc, D. M., Kleinman, J. C., & Cornoni-Huntley, J. (1989). National trends in educational differentials in mortality. *American Journal of Epidemiology, 129,* 919–933.

Ferris, N. (2008, August 11). Massachusetts requires EHRs in hospitals by 2015. *Federal Computer Week.* Retrieved July 15, 2009, from http://www.fcw.com/Articles/2008/08/11/Massachusetts-requires-EHRs-in-hospitals-by-2015.aspx

Frist, W. H. (2005). Health care in the 21st century. *The New England Journal of Medicine, 352,* 267–272.

Garets, D., & Davis, M. (2006, January). *Electronic medical records vs. electronic health records: Yes, there is a difference.* Chicago: HIMSS Analytics.

Gaudie, T. (2006, June 19). Electronic prescribing gathers steam. *NJBIZ.* Retrieved from: http://www.njbiz.com/article.asp?aid=67585

Gazmararian, J. A., Baker, D. W., & Williams, M. V. (1999). Health literacy among Medicare enrollees in a managed care organization. *Journal of the American Medical Association, 281,* 545–551.

Goldschmidt, P. G. (2005). HIT and MIS: Implications of health information technology and medical information systems. *Communications of the ACM, 48,* 69–74.

Gupta, A., K. (2003). Consumer satisfaction: The arrival of consumer-centric healthcare. *Managed Care Quarterly, 11,* 20–23.

Hillestad, R., Bigelow, J., Bower, A., Girosi, F., Meili, R., & Scoville, R. (2005). Can electronic medical record systems transform health care? Potential health benefits, savings, and costs. *Health Affairs, 24,* 1103–1117.

Horrigan, J. B., & Smith, A. (2007). *Home broadband adoption 2007, data memo, Pew Internet and American Life Project,* Washington, DC: Pew Internet and American Life Project.

Houston, T., Sands, D., Nash, B., & Ford, D. (2003, October). Experiences of physicians who frequently use e-mail with patients. *Health Communication, 15*(4), 515–525.

Jones, D. S. (2008, December). Quality, defensibility, and the electronic medical record. *Journal of Health Care Compliance,* 41–46, 71–72.

Jones, S., & Fox, S. (2009, January 28). Generations online in 2009. *Pew Research Center Publications.* Retrieved July 25, 2009, from http://www.boulderdowntown.com/_files/docs/generations-online-in-2009.pdf

Jonietz, E. (2003, April). Paperless medicine. *Technology Review,* 59–64.

Kahn, J. S., Aulakh, V., & Bosworth, A. (2009). What it takes: Characteristics of the ideal personal health record. *Health Affairs, 28,* 369–376.

Kane, B., & Sands, K. Z. (1998). Guidelines for the clinical use of electronic mail with patients. *American Medical Informatics Association 5,* 104–111.

Ketteridge, G., Delbridge, H., & Delbridge, L. (2005). How effective is email communication for patients requiring elective surgery? *ANZ Journal of Surgery,* 75, 680–683.

Lenhart, A., Horrigan, J., Rainie, L., Allen, K., Boyce, A., & Madden, M. (2003).*The ever-shifting internet population: A new look at internet access and the digital divide.* Washington, DC: Pew Internet and American Life Project.

Lin, C., Wittevrongel, L., Moore, L., Beaty, B., & Ross, S. (2005, July). An internet-based patient-provider communication system: randomized controlled trial. *Journal of Medical Internet Research,* 7(4), e47.

Macias, W., & McMillan, S. (2008). The return of the house call: The role of internet-based interactivity in bringing health information home to older adults. *Health Communication, 23,* 34–44.

Mandl, K. D., Kohane, I. S., & Brandt, A. M. (1998). Electronic patient-physician communication: Problems and promise. *Annals of Internal Medicine, 129,* 495–500.

Mandl, K. D., Szolovits, P., & Kohane, I. S. (2001). Public standards and patients' control: How to keep electronic medical records accessible, but private. *BMJ, 322,* 283–287.

McGrath, J. M., Arar, N. H., & Pugh, J. A. (2007). *The influence of electronic medical record usage on nonverbal communication in the medical interview.* Paper presented at the annual meeting of the International Communication Association, San Francisco, CA.

McLuhan, M. (1964). *Understanding Media: The Extensions of Man.* Cambridge, MA: MIT Press.

Miller, C. C. (2009, January 6). Doctors will make web calls in Hawaii. *The New York Times.* Retrieved July 25, 2009, from http://www.nytimes.com/2009/01/06/technology/internet/06health.html

Mitchell, R. L. (2007, July 14). E-medical records: What seems to be the problem? *Computerworld,* 27–34.

Mo, P. K. H., Malik, S. H., & Coulson, N. S. (2009). Gender differences in computer-mediated communication: A systematic literature review of online health-related support groups. *Patient Education and Counseling, 75,* 16–24.

Moyer, C. A., Stern, D. T., Katz, S. J., & Fendrick, A. M. (1999) "We got mail": Electronic communication between physicians and patients. *American Journal of Managed Care, 5,* 1513–1522.

Neil, R. A., Mainous, A. G., Clark, J. R., & Hagen, M. D. (1994). The utility of electronic mail as a medium for patient-physician communication. *Archives of Family Medicine,* 3, 268–271.

Noonan, D. (2007, October 29) "More information, please." *Newsweek.*

Norman, C. D., & Skinner, H. (2006). eHealth literacy: Essential skills for consumer health in a networked world. *Journal of Medical Internet Research, 8.* Retrieved September 2, 2009, from http://www.pubmedcentral.nih.gov/articlerender.fcgi?artid=1550701

Pelletier, A. L., Sutton, G. R., & Walker, R. R. (2007). Are your patients ready for electronic communication? *Family Practice Medicine 14*(9), 25.

Polack, E. P., Richmond, V. P., & McCroskey, J. C. (2008). *Applied Communication for Health Professionals.* Dubuque, IA: Kendall Hunt.

Rannefeld, L. (2004-2005). The doctor will e-mail you now: Physicians' use of telemedicine to treat patients over the internet. *Journal of Law and Health,* 19, 75–105.

Rideout, V., Neuman, T., Kitchman, M., & Brodie, M. (2005). *E-health and the elderly: how seniors use the internet for health information.* Menlo Park, CA: Kaiser.

Roter, D. L., Larson, S., Sands, D. Z., Ford, D. E., & Houston, T. (2008). Can e-mail messages between patients and physicians be patient-centered? *Health Communication,* 23, 80–86.

Sands, D. Z., Safran, C., Slack, W. V., & Bleich, H. L. (1993). Use of electronic mail in a teaching hospital. *Proceedings of the Annual Symposium on Computer Application in Medical Care,* 17, 306–310.

Shaw, A., Ibrahim, S., Reid, F., Ussher, M., & Rowlands, G. (2009). Patients' perspectives of the doctor-patient relationship and information giving across a range of literacy levels. *Patient Education and Counseling, 75,* 114–120.

Shaw, B., Gutstafson, D. H., Hawkins, R., McTavish, F., McDowell, H., Pingree, S., & Ballard, D. (2006). How underserved breast cancer patients use and benefit from ehealth programs: Implications for closing the digital divide. *American Behavioral Scientist, 49* (6), 823–834.

Silverman, J., Kurtz, S., & Draper, J. (2005). *Skills for communicating with patients.* 2nd ed. Oxford: Radcliffe.

Slack, W. V. (2004). A 67-year-old man who e-mails his physician. *Journal of the American Medical Association, 292* (18), 2255–2261.

Slack, W. V., & Bleich, H. L. (1999). The CCC system in two teaching hospitals: A progress report. *International Journal of Medical Information, 54,* 183–196.

Spielberg, A. R. (1999). Online without a net: Physician-patient communication by electronic mail. *American Journal of Law & Medicine, 25* (2/3), 267.

Stalberg, P., Yeh, M., Ketteridge, G., Delbridge, H., Delbridge, L. (2008). E-mail access and improved communication between patient and surgeon. *Archives of Surgery,* 143 (2), 164–168.

Starr, P. (1984). *The social transformation of American medicine: The rise of a sovereign profession in the making of a vast industry.* New York: Basic.

Suggs, L. S. (2006). A 10-year retrospective of research in new technologies for health communication. *Journal of Health Communication, 11,* 61–74.

Tang, P. C., Ash, J. S., Bates, D. W., Overhage, J. M., & Sands, D. Z. (2006). Personal health records: definitions, benefits, strategies for overcoming barriers to adoption. *Journal of American Medical Informatics Association, 13,* 121–126.

Thompson, T. L., Dorsey, A. M., Miller, L. I., & Parrot, R. (2003). *Handbook of health communication.* Mahwah, NJ: Lawrence Erlbaum.

Tjora, A., Tran, T., & Faxvaag, A. (2005, April). Privacy vs usability: A qualitative exploration of patients' experiences with secure internet communication with their general practitioner. *Journal of Medical Internet Research,* 7(2), e15.

Tu, H. T., & Cohen, G. R. (2008). Striking job in consumer seeking health information. *Tracking Report* (Center for Studying Health System Change), 20.

United States Census Bureau, Population Division, Education & Social Stratification Branch. (October 2007). Computer and internet use in the United States. Retrieved July 22, 2009, from http://www.census.gov/population/www/socdemo/computer/2007.html

University of California, San Francisco (2009, March 10). Barriers to adoption of electronic personal health records outlined. *Science Daily.* Retrieved from http://www.sciencedaily.com/releases/2009/03/090310084727.htm

Williams, M. V., Parker, R. M., & Baker, D. W. (1995). Inadequate functional health literacy among patients at two public hospitals. *Journal of the American Medical Association, 274,* 1677–1682.

Wootton, R. (1996). Information in practice. Telemedicine: A cautious welcome. *BMJ, 313,* 1375–1377.

Wootton, R. (2001). Recent advances: Telemedicine. *BMJ, 323,* 557–560.

PART FOUR

The Dark Side of Computer-Mediated Communication in Personal Relationships

CHAPTER FIFTEEN

Family Imbalance and Adjustment to Information and Communication Technologies

Gustavo S. Mesch

Michal Frenkel

In the last 10 years there has been an increase in academic and research interest on the effects of information and communication technologies (ICTs) on families with children and adolescents. This interest, focused on a specific stage in the family life cycle, arises from two parallel and associated processes. The first is the rapid increase in household adoption of these technologies. Out of all types of families, the percentage of those with children and adolescents having access to the Internet is the highest (Lenhart, Madden, Rankin, & Smith, 2007). So out of all the age groups in the population, adolescents have the highest rate of Internet access and use. The second process is the potential effects of Information and communication technologies on family patterns such as family communication, boundaries, time, and conflicts, all of which may have important consequences for adolescents' developmental processes (Livingstone, 2007; Mesch, 2006).

The purpose of this chapter is to provide an updated and comprehensive review of studies to ascertain the effects of information and communication technologies on families with adolescents. In this review we rely on a family-systems and developmental approach.

Family systems theory holds that families are goal-directed, dynamic and interconnected systems that both affect and are affected by their environment. In that sense technology is part of the family environment, and the adoption of new technologies affects the family system. This conceptualization of the family emphasizes the importance of several dimensions of the system that might be influenced by information and communication technologies. Family boundaries are a central dimension of family life. The boundaries between the family and the external world are important, and necessary to preserve the parental role of socialization; the nuclear family erects clear-cut, often rigid, boundaries between public and private life and between children and adults. Family boundaries are reflected in the need for privacy (Berardo,1998). Strong family relationships evolve through an awareness of boundaries between family members and the rest of the world. In their lives together, parents and children negotiate ideas about how and why they are similar to and different from each other, as well as other families and people, and what activities and information are shared with family members and non-family members (Berardo, 1998).

Two other central dimensions are family cohesion and adaptability, factors that are facilitated by communication and shared time (Olson, 1993). Olson characterized family cohesion as the emotional bonding among family members. It is this dimension that balances the importance of independence or differentiation with the mutuality of membership of the family system. Family cohesion is closely related to family time and family communication.

Through spending time together in shared activities, a collective history is created, facilitating the formation of a shared identity that highlights the system's boundaries. Family communication during shared time supports mutual understanding of roles and rules that facilitate family cohesion. Like any system, the family needs to adapt to new experiences and technologies that can affect one or more of these dimensions. Family adaptability is related to the flexibility in rules of relationships. This is the dimension that refers to the family system's need to change appropriately, to be flexible, or to adapt to and learn from different experiences and situations.

Shared activities may be one of the most salient forces establishing and maintaining boundaries in the contemporary family system. Shared experiences emphasize the uniqueness of families, thus yielding attachments and bonding in family relationships. The collective identity developed through family-shared-time not only strengthens attachments of system members, but constantly offers new sources for increased family cohesion and bonding (Orthner & Mancini, 1991). Research indicates that family-time activities are highly correlated to positive communication. In the context of positive communication family members can clarify family roles, negotiate old and new rules of interaction and behavior and practice ways to enforce them (Zabrieskie & McCormick, 2001).

Families are changing systems and their tasks change according to different stages in the family life cycle. The family goes through six stages: (1) young single people; (2) young couples with no children; (3) young couples with young children; (4) couples with adolescent children; (5) couples with no children at home; (6) older single people (Stolzenberg, Blair-Loy, & Waite, 1995). Technology can exercise different effects at different developmental stages. Here we draw on the conceptual framework developed by Watt and White (1999), who suggests a family-systems and development approach to the study of the effects of ICTs on the family. These effects are assumed to depend on the family's developmental stage (Watt & White, 1999), as the following examples show. The communication aspect of computers can support the process of mate selection, where individuals who have never met before can establish a close and intimate relationship; during pregnancy the couple can rely on online information and forums to acquire health-related information and advice; at the stage of early preschool children, participation in a community bulletin board may provide the family with help in babysitting and finding activities for children (Mesch & Levanon, 2003). For families with young children, connection to the Internet provides access to online information on family-related issues such as parenting, children's education, and family health. Participation in online family discussions facilitates access to social networks that supply social support, advice, and guidance to families (Hughes & Hans, 2001). When the couple has young children, information and communication technologies support regular contact with the extended family, and in many cases help families to reconnect and overcome geographical divides (Langian, Bold, & Chenoweth, 2009). For the empty nest, communication and information technologies support parents' communication with their older children attending college or residing in a different location, between face-to-face meetings (Chen & Katz, 2009).

Regarding the more specific influences of ICT, at any stage of development the family is a social system with permeable borders, and technology is conceived as a major source of change. Technological innovations enter the family, creating changes in norms and social roles. A simple example is the need to accommodate the computer and to find space for the machine. While parents usually prefer to place media in the public spaces of the home, adolescents would rather locate them in their bedroom (Horst, 2008). The bedroom has a special meaning for adolescents as a place where they can conduct private activities. In that sense, arranging their room according to their needs, consuming media of their preference, and decorating the room with symbols of youth culture are exercises in autonomy.

New information is converted into new functions, and one possible result is specialization as children sometimes acquire computer skills rapidly and take on the role of advisors and aides to other family members. This specialization involves changes in the nature of the relationships in the family subsystem, as knowledge is often reflected in doing tasks for the parents and at the same time monopolizing computer time.

Following the systems-developmental approach, rather than discussing the family in general, we now focus on one stage of the family life-cycle: families with adolescents. At this stage, parents' and adolescents' relationships focus on issues of family boundaries, family time, adolescent autonomy and expansion of adolescents' social circles (Fuligni, 1998; Smetana, 1988).

Communication Technologies and Family Boundaries

The introduction of ICTs to the home can have an effect on parents. Technology can be used to bring home to work as well as work to home (Chesley, 2005; Wajcman, Bittman, & Brown, 2008). Information and communication technologies provide additional ways to access individuals anytime, anywhere. Employers use the Internet and cell phones to communicate and update their employees. As a result the boundaries between work and family have become increasingly blurred (Valcour & Hunter, 2005; Wajcman et al., 2008). The potential consequences of the permeability of work-family boundaries are unclear. The spillover hypothesis suggests that home access to information and communication technologies results in overworking, social isolation, and endless interruption of family life, leading to higher levels of stress and a lowering of family satisfaction (Chesley, 2005). Others suggest that working from home and continuous contact with children while at work enhance flexibility, reducing conflicts between family and work (Wajcman et al., 2008). A recent study in the United States exploring the use of the Internet and cell phones for family communication concluded that for some Internet users new communication tools led to an increase in the time they spent working, whether the work was done at the office or at home. One in five employed Internet users (19%) said that using the Internet increased the amount of time they spent working from home, and one in ten (11%) said that it increased the amount of time they spent working from the office. Most respondents did not feel that this change had a negative effect on family time and conflicts (Kennedy & Wellman, 2007). Two recent studies directly tested the work-home and home-work spillover hypothesis. Chesley (2005) conducted a longitudinal study on the effect of the use of Internet at home and of cell phones on parental relations and psychological distress. The study found that the results differed according to the type of technology. Use of cell phones over a two-year period was associated with a rise in negative work-family spillover for both men and women and with a rise in negative family-work spillover for women only. This spillover proved linked to higher distress and lower family satisfaction. However, according to this study Internet use at home was not associated with work-family or family-work spillover (Chesley, 2005). A study in Australia tested the extent that ICT use was perceived as affecting the work-family-life balance. It found that the use of cell phones affected the perception of spillover of family life onto work time but not that of spillover of work activities onto family life (Wacjman et al., 2008). Note that while the cell phone has been associated with spillover, Internet use has not. One possible explanation is that while Internet use is a planned activity, conducted in a different room that can be temporarily separated from family activities, the cell phone can ring anytime and interrupt a conversation or a family meal. Thus, the ability to control technology's invasion of family life mediates the effects of ICT on work-family spillover and its effect on family boundaries. Apparently Internet use is easier to control than cell phone use.

An important source of tension in families with adolescents, with implications for family boundaries, is that ICT use exposes adolescents to larger amounts of information, covering any topic, anytime, anywhere. When adopting the Internet, parents report their wish to use the new technologies to promote desired values. For parents, new technologies represent an investment in their child's future that will support his or her success at school. At the same time, ICTs are a source of anxiety as the new media lay bare information that parents have no control over. A central concern for parents is that adolescents nowadays are more exposed not only to valuable information but to negative and mendacious content as well, and this exposure might have a negative developmental impact. Parents are anxious that Internet content challenges their values. While they try to control this access they know that their ability to control and regulate external information is limited (Horst, 2008; Livingstone & Hellsper, 2008).

A major worry arising from the increase in Internet access from home has become adolescents' exposure to pornography. This content is perceived as more accessible to minors through the Internet than in its traditional forms (Greenfield, 2004; Mesch, 2009). The literature suggests a number of negative effects of frequent and long-term exposure to this material. First, it leads to more liberal sexual attitudes and greater belief that peers are sexually active, which increase the likelihood of first intercourse at an early age (Flood, 2007). Second, adolescents exposed to sexual behaviours outside cultural norms may develop a distorted view of sex as unrelated to love, affection and intimacy, and a desire for emotionally uncommitted sexual involvement (Byrne & Osland, 2000). Third, youth exposed to pornography may develop attitudes supportive of the "rape myth," which ascribes responsibility for sexual assault to the female victim (Flood, 2007; Seto, Maric, & Barbaree, 2001). A study on deliberate exposure of youth to pornography compared frequent Internet users for pornography consumption with frequent users for communication, information, entertainment, and learning purposes. The results showed that the percentage of adolescents who used the Internet for pornography consumption was lower than that of those who used it for other purposes, indicating that they represent a different and particular non-normative group among the adolescent population. The availability and ease of access to Internet pornography provide the opportunity, but taking it seems to be associated with individuals' choices, which have to do with differences in self-control resulting from a low quality of social bonding. This is a highly relevant factor here as adolescents who use the Internet for pornography are less socially integrated and more socially marginal. They express less commitment to their families, fewer prosocial attitudes, and less attachment to school than do their fellows who do not use the Internet for that purpose (Mesch, 2009).

Another tension in family boundaries is the risk of exposure to online hate content (websites and blogs), inciting to violence against, separation from, defamation of, deception about, or hostility towards others based on race, religion, ethnicity, gender, or sexual orientation. Frequent exposure to hate might result in the adoption of stereotypes, attitudes supporting racism, lack of empathy for minorities, aggressive behaviour, and involvement in hate crimes. From the research literature it appears that despite the wide availability of hate sites, only 20% of adolescents report having been exposed to content that advocates hatred of minorities. Of those exposed, more than a third (36%) reacted by ignoring the site and its content. Another third informed a friend or an adult about the site (MNets, 2001).

An additional concern for parents is that youngsters using the web might convey information about themselves and their families to marketers, enabling them to create detailed profiles of a family's lifestyle (Turow, 2001). Accurate or not, such portraits can influence how marketers treat family members: for example, what discounts to offer them, what materials to send them, how much to communicate with them, and even whether to deal with them at all.

In sum, Internet use for the search of information presents a challenge to family boundaries as it increases unsupervised exposure of adolescents to a wide variety of content. Much of this content is important, for example, in supporting homework and searching for answers to existential issues. Yet this increased exposure might blur family boundaries, creating tensions, as teens are exposed to, or provide information that might contradict parental values. Families deal with these tensions and try to adjust to the new challenges that Internet access poses by using technological and social devices to mediate adolescents' exposure.

Controlling Family Boundaries

One of the ways that demonstrate how much parents feel that family boundaries have become permeable is their attempts to control them. Parents use technological devices and social practices to control and regulate Internet content that their children are allowed to see. *Parental mediation* is a concept that has been used in media research to understand the process of television influence on audience attitudes and behaviours. According to this model, individuals are exposed to media content that may affect their attitudes and behaviours (Rothfuss-Buerkel & Buerkel, 2001) The model assumes that this effect is mediated by intervening variables whereby the extent that some viewers may adopt attitudes and behaviours presented in the media depends on parental activities that affect which information is received, and how it is processed and acted on by the audience (Bybee, Robinson, & Turow, 1982). According to the literature there are various types of mediation, but we restrict our discussion to only two techniques, as follows. *Restrictive mediation* involves limiting the child's viewing time and the content accessed. It is restrictive as it does not involve the child's active participation but is the parent's decision. This kind of mediation covers the use of electronic devices that restrict the content and websites that youngsters can access. *Evaluative mediation* represents an open discussion on issues related to Internet use, and the formulation of rules about time allocated to Internet use, websites allowed and not allowed, and placing the computer in a common space that allows parents to co-use the Internet with their children and to be available for questions (Eastin, Bradley, Greenberg, & Hofschire, 2006; Bybee et al., 1982).

There is a sense that parental control of Internet use has intensified over time. A study of teens and parents in the United States found that most parents of teenagers who go online set time limits on their Internet activities. While in 2000 only 41% of parents had installed filtering software on home computers, by 2005 the figure was 54%. In both years the same percentage of parents (62%) reported checking the websites that their children visited. In 2006, 69% of parents reported having rules about how much time the child could spend online, 85% had rules on the kinds of personal information the child could share, and 85% had rules on the kind of websites that their children may or may not visit (Lenhart et al., 2007). Similar findings are reported in a study in Canada comparing findings for 2001 and 2005. In 2005 the percentage of parents reporting rules was higher for all rules, indicating an increase over time in parental monitoring of adolescents' online behaviour (Media Awareness Network, 2005).

There is empirical evidence that parental evaluative mediation is more effective than restrictive mediation. A study of 222 children in Korea investigated the effect of four techniques of parent control of adolescents' Internet content access. Evaluative mediation, measured as parents' recommendation of websites and co-use of the Internet, were related to children's use of the medium for educational purposes. Restrictive mediation, such as time limits and website restrictions, were not related to type of Internet use (Lee & Chae, 2007). Caution is needed in the interpretation of these findings as there is consistent empirical evidence of discrepancies between parents and adolescents in their report of parental mediation. Parents are more likely to report monitoring than their children, a finding that implies disagreement between them regarding the extent of monitoring of children activities. In a study in the United States of 523 parent-adolescent dyads, the two sides were asked about parental monitoring. Internet strategies included placing the computer in an open space, limiting the amount of time that the adolescent could use the computer, using software to block content, and checking websites. Reports of parents and adolescents differed significantly. While 77% of the parents reported that they had limited the amount of time the adolescent spent on the computer, only 59% of the youth reported this; 68% of the parents reported their seeing the sites that their children frequented, but only 55% of the children reported this (Cotrell, Branstetter, Cottrell, Rishel, & Stanton, 2007). A study of 1,124 adolescents in Singapore found that parents reported checking bookmarks or browser history more often than adolescents reported their parents' doing so. Mothers proved more aware of youth Internet activities, apparently because adolescents were more likely to disclose their behaviour to their mothers (Liau, Khoo, & Ang, 2008). Similarly, a study of parents and adolescents in the United Kingdom found a discrepancy between parents and children in their account of the overall level of parental restrictions: three-quarters of the parents, but less than half of the children, reported that there were rules at home regulating at what hours and for how long children may use the computer. Regarding online activities, 77% of the parents but only 54% of the children said that the adolescent was not allowed to buy anything online; 62% of the parents but only 40% of the children said they were not allowed to chat online. Since some parents claimed to regulate media use while their children seemed unaware of this; apparently, either that parents over-claim, being less effective than they would hope, or the children under-claim, being less independent than they would hope (Livingstone, 2007).

Family Time and Communication

Another concern associated with Internet use is that this activity might decrease family time. This is closely associated with family cohesion, defined as the "emotional bonding that family members have toward one another" (Olson, Russell, & Sprenkle, 1983). The significance of this is positive involvement of parents with their children, as reflected in family time devoted to shared activities, supportive behaviour, and affection.

Family time can be seen as a central dimension of both family cohesion and family boundary construction and preservation. If we think of families as social systems having a collective identity, that identity is the result of shared recollections of togetherness created as family members spend time together in shared meals, games, and chatting. Studies have shown that the time a family spends together in activities such as recreation is positively related to family cohesion (Orthner & Mancini, 1991). Shared activities are described as forces contributing to clarifying and strenghtening family boundaries because they create opportunities for interaction, communication, and memories, which contribute to the perception of one's family's identity and uniqueness (Zabrieskie & McCormick, 2001; Hofferth & Sandberg, 2001).

Regarding the effects of the Internet on family time, the argument is that time spent on one activity cannot be spent on another (Nie, Hillygus, & Erbring, 2002). Internet use is a time-consuming activity, and in families connected to its high frequency of use might be negatively associated with family time. This concern has received empirical support in early Internet studies, which, based on family-time diaries, found that Internet use at home was negatively related to time spent with family (Nie et al., 2002). Other studies reported that not only the amount but also the quality of family time seems to be associated with Internet use as studies reported low use was associated with better relationships with parents and friends than was high use (Mesch, 2001; Sanders, Field, Miguel, & Kaplan, 2000). There is some evidence for this argument, as a study conducted on a large adolescent sample found that the more adolescents spent time on Internet activities, the less parents and children spent time together. Adolescents who use the Internet for social purposes are more likely to report spending less time with their parents and experiencing more conflicts with their parents. From the results it appears that Internet use affects family cohesion indirectly through its negative effect on family time and its positive effect on family conflicts (Mesch, 2006). The perception that the amount of daily Internet use might affect family time is expressed by both parents and adolescents. A study in the United States of a sample of youth and parents found that 29% of parents believed that online time interfered with family time, while only 16% of teens believed this (Rosen, Cheever, & Carrier, 2008).

While these studies provide evidence of a relationship between Internet use and a reduction in family time, they treat the Internet as a unified technology. Youth differ in the extent they use the Internet for communication, information search, and entertainment; different uses probably have different effects on family time. A recent study found that Internet use decreased the amount of time youth spent with their families, but different activities had different effects. Playing online games decreased both total time spent with the family and time communicating with family members. Using the Internet for communication with friends resulted in a small decrease in family time. However, using it for educational purposes such as searching for homework or doing homework did not affect family time (Lee & Chae, 2007).

The family as a system attempts to adjust to the technology. One way is to incorporate the Internet as both a shared activity and a channel of communication. Does the Internet serve as a shared family activity, its technology used by children and parents together? A recent large-scale study in the United States on Internet use in families with adolescents found that members of all the families explicitly stated how much they valued spending time together. Many families scheduled time to watch TV shows, movies, and videos together. That is, parents and adolescents conceived the shared consumption of media as facilitating communication and bonding (Horst, 2008). A small but growing proportion of parents with adolescents engages in the production of media together with their teenage children. In this way adolescents worked with the support of parents, learning new skills of multimedia production; parents described these activities as becoming involved in their childrens' interests and culture (Horst, 2008). A study conducted in a Toronto suburb investigated the extent that the Internet was used together with the respondent's spouse and children. The number of hours per week that the Internet was used with children depended on the frequency of its use: the higher the use, the greater the time parents spent using the Internet together with the children (Kennedy & Wellman, 2007). In a recent study in the United States with a representative sample of the Internet population, 27% of the parents reported going online with their children (Kennedy, Smith, Wells, & Wellman, 2008). However, the extent that parents are able to join and participate in Internet-user content production depends on their level of education, occupation, Internet experience, and skills. Many parents of low education and income groups are limited in their ability to transform the Internet into a shared family activity.

As to the incorporation of the new communication technologies as a channel of family communication, it seems that the cell phone is becoming highly integrated and the Internet much less. In a study in Israel conducted among a representative sample of the country's population, we found that the cell phone had been rapidly adopted as an important channel of mediated communication between parents and between parents and children. Regarding spouses, 63% of married respondents called their partner by cell phone every day; 51% of those with children aged 12–18 called them every day. Email proved much less common: only 4% of respondents sent an email to their spouse every day, and 1.7% sent an email to their children every day. In Canada, mobile phone calls are also the leading channel of mediated communication between spouses and between parents and children (Kennedy & Wellman, 2007). In the United States, 47% of couples with children and Internet access communicate daily with their partner by cell phone, and only 8% send an email daily. Regarding communication with children, 42% of the couples with children 7–17 years old communicate daily by cell phone, and only 3% send an email (Kennedy et al., 2008).

These studies in different countries yield a consistent picture and indicate that only a minority of families with children have managed to integrate the Internet into their shared family activities, and far fewer use it for family communication. For the vast majority of families, mobile phones have been better integrated into family communication. As a shared activity, the Internet has been integrated by about a third of families, indicating that this is certainly one form of adjusting family activities to Internet. As a family channel of communication, the Internet is well behind the cell phone, and it is difficult to say whether it will be integrated at all as a frequent family communication channel and activity (Mesch & Frenkel, 2009).

Parent-Adolescent Conflicts

Adolescence is a period replete with conflicts between parents and children. Typically, they increase from pre- to mid-adolescence, reflecting renegotiation of divergent expectations of the roles of either party (Shanahan, McHale, Osgood, & Crouter, 2007; Sillars, Koerner, & Fitzpatrick, 2005). Adolescence is a time when families need to adjust and adapt their relationships to accommodate the increasingly maturing adolescent. Many of their exchanges concern parents' regulation of adolescents' everyday lives, such as media rules, friendship relations, and personal activities like phone, TV, and Internet use (Collins & Russel, 1991). Studies on adolescents show that as they become older they submit ever less to parental authority over aspects of their personal lives. At the same time they demand more and more autonomy and show greater readiness to disagree openly with their parents (Fuligni, 1998). Adolescents' understanding of changes desired by parents is lower than parents' understanding of these changes, and adolescents' grasp of the existence of rules limiting their use of the Internet is less than parents' grasp of the existence of family rules. Gaps in the understanding of rules lead to intergenerational conflicts, associated with parents' and adolescents' low relationship satisfaction (Sillars et al., 2005). Of course, parents and adolescents might hold different perspectives and still understand each other. Yet in most families the conflicts usually amount to little more than mild arguments. By successfully negotiating solutions to issues such as adolescents' demand for greater autonomy, parent-adolescent relationships are restructured to allow adolescents' growth while maintaining close family ties (Reuter & Conger, 1995). At the same time, high levels of conflict are negatively related to family cohesion (Mesch, 2006).

There is some concern that Internet access in the household may negatively affect patterns of interaction between parents and children, increasing intergenerational conflicts and weakening family cohesion (Watt & White, 1999; Livingstone, 2007). Supporting this argument a study in the United Kingdom found that 19% of parents and 9% of adolescents reported frequent arguments over Internet time and type of Internet use (Livingstone, 2007). Another study based on U.S. data found that 40% of parents reported conflicts with their adolescent children over Internet use (Mesch, 2006).

One source of these conflicts is that families are social systems characterized by a hierarchy of authority. The introduction of the computer has the potential to change that hierarchy as the adolescent becomes the family expert on whom other family members rely for technical advice and guidance (Watt & White, 1999; Kiesler, Zdaniuk, Lundmark, & Kraut et al., 2000). Under these conditions the adolescent increases his/her resources relative to the parents', and also his/her ability to dominate the family sphere (Kiesler et al., 2000). Similarly in the United Kingdom, traditional adult–child relations appear to be reversed in many households because children are more technologically competent than their parents (Holloway & Valentine, 2003). In a study based on U.S. data this assumption was confirmed. Conflicts between parents and their adolescent children were more likely as frequency of adolescents' Internet use was higher, and when the adolescent reported being the family Internet expert. The significance of this finding is a reversal of traditional family roles, in which parents provide adolescents with guidance and expertise (Mesch, 2006).

Over time it appears that families are adjusting to this power imbalance. In many households the web has become work for the school-aged and parents assign them "cyberchores." A recent study reports that 38% of youth helped parents to send photos and emails to relatives online, and 36% helped with their parents' information search. Almost half of the teen respondents (47%) said they performed "cyberchores" because their parents lacked online skills; 29% stated that they did so because their parents did not have time to do this job themselves (Canada.com, 2007).

Parents attempt to guide the use of the Internet by creating rules on time and websites permitted. Yet parents and adolescents perceive the rules differently and this discrepancy is another source of intergenerational conflicts. In one study parents were able to articulate the rules but the children forgot to mention them or stated that their parents would mention rules but these were open to negotiation (Horst, 2008; Liau et al., 2008). The pattern of communication existing in the family may be an explanation for the gaps found in the perception of rules regulating the use of the Internet. In families in which parents and adolescents perceived their communication as open, empathetic, encouraging, and trusting, their reports regarding family rules on the use of media were more consistent (Cottrell et al., 2007).

Another source of conflicts is the development of expectation gaps between parents and youth. Parents seem to view the Internet as a positive new force in children's lives, and surveys in different countries report that the chief reason families buy computers and connect their children to the Internet at home is for educational purposes (Lenhart et al., 2007; Livingstone & Hellsper, 2008). Many parents believe that the Internet can help their children to do better at school, do better research for homework, and help them learn worthwhile things (Livingstone & Bober, 2004; Van Rompey, Roe, & Struys, 2002). But not all teens use the Internet in the same way. While some spend most of their Internet time searching for information, acquiring skills, and researching for homework, others mostly use it for social purposes (email, instant messaging, and participation in chat rooms) and entertainment purposes (playing games online) (Lenhart et al., 2007; Livingstone & Bober, 2004). It is plausible to assume that when youth use the Internet for social and entertainment purposes, parental expectations contradict the actual use, aggravating conflict. Conversely, using the Internet for learning and education purposes, a usage highly valued by parents and consistent with parental expectations, will be negatively associated with family conflict (Mesch, 2003; Horst, 2008).

An additional source of potential conflict between parents and adolescents is the social use of the Internet. Online communication expands the size of teens' peer group, involving the formation of ties with friends who are unknown to parents (Mesch, 2006). During adolescence, as teens spend more time with peers and away from their parents, they have increased opportunities to manage information on their online activities, thereby keeping things secret and making choices as to which friends, activities, and behaviour to disclose to their parents. Teenagers' management of information varies according to the different types of activities and how much they believe these activities are part of their domain of autonomy and privacy (Smetana, Metzger, Gettman, & Campione-Barr, 2006). Adolescents tend to reject parents' legitimate authority to regulate personal issues (control over body), privacy, and choices regarding clothes and recreational activities. As they get older, adolescents increasingly assert autonomy over multifaceted issues such as media consumption (Smetana & Asquith, 1994). Smetana et al. (2006) investigated the extent that teens were willing to disclose information to their parents, including multifaceted issues such as websites visited and seeing friends parents do not know or like. The study found that adolescents' and parents' beliefs about adolescents' obligations to disclose to parents were closely associated with their beliefs about parents' legitimate authority to regulate their acts in different domains. Moral and conventional issues were judged to be more legitimately subject to parental authority than multifaceted and personal issues (such as websites watched, R-rated movies watched, or seeing friends parents do not know or like). Multifaceted issues are major sources of conflict in adolescent parent relationships (Smetana et al., 2006; Smetana & Asquith, 1994).

At this point the central question is how these conflicts compare with other prevalent conflicts in adolescence. Mesch (2006), using a representative sample of the Israeli youth population, investigated perceptions of adolescents' conflicts with their parents. He targeted conflicts over mundane issues such as household chores and homework, and the association of Internet use with the frequency of the various types of arguments. In particular, the study found that the frequency of perceived conflicts over the Internet and computers was higher than that of conflicts over school-related issues and equal to that of disagreements over household chores. The most salient finding of the multivariate analysis was that computer and Internet use exerted a generalized effect on perception of conflicts with parents. The amount of Internet use was positively associated with the perception of frequent arguments over school-related and household chores issues. After identifying the association between Internet use and frequency of perceived adolescent-parent arguments, the final question of the study was whether these arguments had consequences for the young people's perception of family closeness. The most salient result of the multivariate analysis was that arguments over the Internet and computer use had a statistically significant negative effect on family closeness. Arguments over household chores and school-related issues did not undermine relations between parents and children, while computer and Internet use did.

Summary and Future Directions for Research

In this chapter a family systems and developmental approach has been used to organize the growing research and cumulative findings on the process of family adjustment to information and communication technologies. The empirical evidence indicates that these technologies are a challenge for the preservation of family boundaries. Information technologies provide new sources of information and may compete with parental values. Youth are exposed to large amount of information, most of it valuable for school and vocation-oriented activities. In this electronic environment youth participate not only as consumers but are active in the production of content in social networking sites. While there is empirical evidence on the patterns of Internet and cell phone use by adolescents, less is known about the characteristics of the families that incorporate these technologies into their families' activities, and what kind of shared activities are replaced thereby. Future research should be directed to understanding the different strategies used in different families.

Communication technologies are used not only for relationship maintenance but also for relationship formation. Parental control of adolescents' associations which for developmental reasons (adolescents' search for autonomy) is less effective during adolescence is further undermined by the use of the Internet and cell phone for communication. Conflicts may arise from parental request for adolescents' information disclosure in areas that adolescents consider not subjected to parental authority. Conflicts may decrease family cohesiveness and be consequential for adolescents' maturation. As social systems, families attempt to adjust and restore the balance using different strategies. Some rely on restrictive mediation, incorporating software that blocks access to certain websites. Others resort to open communications, sharing activities, and guidance of adolescents' web-surfing. Conflicts over the Internet have a generalized effect: the higher the likelihood of conflicts over the Internet, the higher the likelihood of conflicts being reported in other areas as well. It is reasonable to assume that the emergence of family conflicts over the use of information and communication technologies is not universal. Future studies should identify the characteristics of families that develop these conflicts for a better understanding of their nature.

There is some evidence that parents are dealing with the challenges posed by information and communication technologies, as the mobile phone is becoming incorporated and more and more parents and adolescents communicate by cell phone and the Internet. Less is known of the implications of relying on ICTs for family communication on family life. Future studies should investigate the differences in family processes between families that do and do not rely on these channels of mediated communication, and the implications for family life.

As has been shown in the chapter, the entry of the Internet into the home affects various spheres of family functioning. The family becomes exposed to large amounts of information (both positive and negative) that circulates inside and outside the home. Youth become exposed to new acquaintances, and parental control over their friends declines. Family time and cohesion might be reduced. At the same time this is a temporal process, and studies monitoring the process through time are needed to better understand how family time is allocated, how parents and adolescents deal with these conflicts, and what their outcomes are.

Research has been directed to the understanding of the process of family adoption of information and communication technologies. Yet Internet and cell phone use is not universal. Families are still not connected, due to low resources or choice. Future research should be directed to understanding the consequences for families and adolescents of not being connected in the information age. This research can contribute to our understanding of the amplification of social disadvantage in the information society.

In this chapter the process of Internet domestication has been described. We took a developmental and systemic approach. After concluding that the Internet influences the permeability of the family system, our discussion went on to inquire into the extent that the technology affects family boundaries. The study of this process has received empirical attention mainly among families with young and adolescent children. Less research attention has been directed to the understanding of the adoption of these technologies among couples without children and empty-nest families. Empirical evidence from families with children and young adolescents indicates the possibility of a decrease of family cohesion due to reduced family time, and an increase in intergenerational conflicts. However, to the extent that these are stable results, they are limited to one stage in the course of family life. Understanding of family life in the information age will benefit from a life-course approach that studies the adjustment to information and communication technologies, comparing couples without children, families with children, and empty-nest families. Such an approach might provide us with a picture that changes dramatically according to the stage in family life.

References

Berardo, F. (1998) Family privacy: Issues and concepts, *Journal of Family Issues, 118*: 18–41.

Bybee, C., Robinson, D., & Turow, J. (1982). Determinants of parental guidance of children's TV viewing for a special subgroup: Mass media scholars. *Journal of Broadcasting, 26*(3), 697-711.

Byrne, D., & Osland, J.A. (2000) Sexual fantasy and erotica/pornography: Internal and external imagery. In L.T. Szuchman & F. Muscarela (Eds.), *Psychological perspectives on human sexuality* (pp. 283–305), New York: John Wiley.

Canada.com(2007). *Kids help 'clueless' parents with cyberchores.* Retrieved July 10, 2010, from http://www.canada.com/topics/technology/story.html?id=2c81000d-9a03-4675-adaf-9c8e0545fae3&k=48901.

Chen, Y. F., & Katz, J, E. (2009). Extending family to school life: College students' use of the mobile phone. *International Journal of Human-Computer Studies, 67*,179–191.

Chesley, N. (2005). Blurring boundaries? Linking technology use, spillover, individual distress and family satisfaction. *Journal of Marriage and Family, 67*,1237–1248.

Collins, W. A. & Russel, G.J. (1991) Mother-child and father-child relationships in middle-childhood and adolescence: A Developmental Analysis, *Developmental Review, 11*, 99-136.

Cotrell, L., Branstetter, S., Cottrell, S., Rishel, C., & Stanton, B.F. (2007). Comparing adolescent and parent perceptions of current and future disapproved Internet use. *Journal of Children and Media, 1*, 210–226.

Eastin, M., Greenbers B.S., & Hofschire, L. (2006). Parenting the Internet. *Journal of Communication, 56*: 486–504.

Fuligni, A. J. (1998). Authority, autonomy, parent-adolescent conflict and cohesion. *Developmental Psychology, 4*, 782–792.

Flood, M. (2007). Exposure to pornography among youth in Australia, *Journal of Sociology, 43*, 45–60.

Greenfield, P.M. (2004). Inadvertent exposure to pornography in the Internet: Implications for peer to peer file sharing networks for child development and families. *Applied Developmental Psychology, 25*, 741–750.

Hofferth, S. L. and Sandberg, J. F. (2001) How American children spend their time. *Journal of Marriage and Family*, 63, 295-308.

Holloway, S. an& Valentine, G. (2003). *Cyber-kids: Youth Identities and Communities in an On-Line World*, London: Routledge.

Horst, H. (2008). Families. In I. Mizuko, H. Horst, M. Bitanti, d. boyd, S. Herr, P. G. Lange, C. J. Pascoe, & L. Robinson (Eds.), *Hanging out, messing around, geeking out: Living and learning with new media* (pp.123–156), Boston: MIT Press.

Hughes, T. R., & Hans, J. G. (2001). Computers, the Internet and families. *Journal of Family Issues, 22*, 776–790.

Kennedy, T. L. M., Smith, A., Wells, A.T., & Wellman, B. (2008). *Networked families*, Washington, DC: Pew and American Life Project.

Kennedy, T. L.M., & Wellman, B. (2007). The networked household. *Information, Communication and Society, 10(5)*, 645–670.

Kiesler, S., Zdaniuk, B., Lundmark, V., & Kraut, R. (2000). Troubles with the Internet: The dynamics of help at home. *Human-Computer Interaction, 15,* 322–351.

Langian, J. D., Bold, M., & Chenoweth, L. (2009). Computers in the family context: Perceived impact on family time and relationships. *Family Science Review, 14,*16–32.

Lee, S. J. and Chae, Y. G. (2007) Children's Internet use in a family context: Influence on family relationships and parental mediation, *Cyber Psychology and Behavior, 10(2),* 640–644.

Lenhart, A., Madden, M., Rankin. A., & Smith, M. A. (2007). *Teens and social media.* Washington, DC: Pew and American Life Project.

Liau, A.K., Khoo, A., & Ang, P.H. (2008). Parental awareness and monitoring of adolescent internet use. *Current Psychology,* 27, 217–233.

Livingstone, S. (2007). Strategies of parental regulation in the media rich home. *Computers in Human Behavior, 23(3),* 920–941.

Livingstone, S., & Bober, M. (2004). *UK Children Go Online: Surveying the Experiences of Young People and Their Parents.* London: LSE Report.

Livingstone, S., & Helsper, E. J. (2008). Parental mediation of children's Internet use. *Journal of Broadcasting & Electronic Media, 51(4),* 581–600.

Media Awareness Network (2005) *Media Education in Canada–An Overview.* Accessed on June 22, 2010.

Mesch, G. S. (2001). Social relationships and Internet use among adolescents in Israel. *Social Science Quarterly, 82(2),* 329–340.

Mesch G. S. (2006). The family and the Internet: Exploring a social boundaries approach. *Journal of Family Communication, 6(2),* 119-138.

Mesch, G. S. (2009). Social bonds and Internet pornographic exposure among adolescents. *Journal of Adolescence, 32, 601–618.*

Mesch, G. S., & Frenkel, M. (2009). *The networked family in Israel.* Paper presented at the meetings of the Israel Sociological Association, Rishon Le-Tzion, Israel.

Mesch, G. S., & Levanon, Y. (2003). Community networking and locally based social ties in two suburban localities. *City and Community,* 2, 335–351.

Mnet, (2001) *Young Canadians in a wired world-Mnet survey.* Retrieved July 19, 2009, from http://www.mediaawareness/ca/english/special _initiatives/surveys/index.cfm.

Nie, N. H., Hillygus, D. S., & Erbring, L. (2002). Internet use, interpersonal relations and sociability: A time diary study. In B. Wellman & C. Haythornthwaite (Eds.),. *Internet in everyday life* (pp. 215–244), Oxford: Blackwell.

Olson, D. H. (1993). "Circumplex model of marital and family systems: Assessing family functioning." In F. Walsh (Ed.), *Normative Family Processes* (2nd ed.). New York: Guilford.

Olson, D. H., Russel, C. S., & Sprenkle, D. H. (1983). Circumflex model of marital and family systems. *Family Processes, 22,* 69–83.

Orthner, D. K., & Mancini, J. A. (1991). Benefits of leisure experiences for family bonding. In B. L. Driver, P. J. Brown, & G. L. Peterson (Eds.), *Benefits of leisure* (pp. 215–301), State College, PA: Vantage.

Reuter, M. A., & Conger, R. D. (1995). Antecedents of parent-adolescent disagreements. *Journal of Marriage and Family, 57,* 435–448.

Rosen, L. D., Cheever, N. A., & Carrier, L.M. (2008). The association of parenting style and child age with parental limit setting and adolescent MySpace behavior. *Journal of Applied Developmental Psychology, 29,* 459–471.

Rothfuss-Buerkel, N. L. & Buerkel, R.A.(2001) Family mediation. In J. Bryant & A. J. Bryant, (eds). *Television and the American Family* (pp. 335-375). Lawrence Earlbaum.

Sanders, C. E., Field, T. M., Miguel, D., & Kaplan, M. (2000). The relationship of Internet use to depression and social isolation among adolescents, *Adolescence, 35*, 237–242.

Seto, M. C., Maric, A., & Barbaree, H. E. (2001). The role of pornography in the etiologic of sexual aggression, *Aggression and Violent Behaviour, 6*, 35–53.

Shanahan, L., McHale, S. M., Osgood, D.W., & Crouter, A. (2007). Conflict frequency with mothers and fathers from middle childhood to late adolescence: Within and between family comparisons. *Developmental Psychology, 43*(3), 539–550.

Sillars, A., Koerner, A., & Fitzpatrick, M.A. (2005), Communication and understanding in parent-adolescent relationships. *Human Communication Research, 31*, 102–126.

Smetana, J. G. (1988) Adolescents' and Parents' Conceptions of Parental Authority, *Child Development*, 59, 321-335.

Smetana, J. G., & Asquith, P. (1994). Adolescents' and parents' conceptions of parental authority and personal autonomy. *Child Development, 65*, 1147–1162.

Smetana, J. G., Metzger, A., Gettman, D. C., & Campione-Barr, N. (2006). Disclosure and secrecy in adolescent-parent relationships. *Child Development*, 77, 201–217.

Stolzenberg, R. M., Blair-Loy, M., & Waite, L. J. (1995). Religious participation in early adulthood: Age and family life cycle effects on church membership. *American Sociological Review*, 60, 84–103.

Turow, J. (2001). Family boundaries, commercialism, and the Internet: A framework for research. *Applied Developmental Psychology, 22*, 73–86.

Valcour, P. M., & Hunter, L. W. (2005). Technology, organizations, and work-life integration. In E. E. Kossek & S. J. Lambert (Eds.), *Managing work-life integration in organizations: Future directions for research and practice* (pp. 61–84), Mahwah, NJ: Lawrence Erlbaum.

Van Rompaey, V., Roe, K., & Struys, K. (2002). Children's influence on Internet access at home. *Information Communication and Society, 5*, 189–206.

Wajcman, J., Bittman, M., & Brown, J. E. (2008). Families without borders: Mobile phones, connectedness and work home divisions. *Sociology*, 42, 635–652.

Watt, D., & White, J. M. (1994). Computers and the family life: A family developmental perspective. *Journal of Comparative Family Studies*, 30, 1–15.

Zabrieskie, R. B., & McCormick, B. P. (2001). The influences of family leisure patterns on perceptions of family functioning. *Family Relations, 50(3)*, 281–289.

CHAPTER SIXTEEN

Online Performances of Gender: Blogs, Gender-Bending, and Cybersex as Relational Exemplars[1]

Mark L. Hans

Brittney D. Selvidge

Katie A. Tinker

Lynne M. Webb

Computer-mediated-communication (CMC) scholarship examines how partners employ online communication to create and maintain relationships (Stern, 2008). In addition to other concerns, CMC research examines the Internet as a gendered space. Such an interpretation of the Internet suggests that the medium itself might play a role in the construction of online gender identities as well as provide a social context for gender construction in the enactment of online relationships (Ridgeway & Correll, 2004).

Current theorizing distinguishes gender as a *social construct* (Wood, 2009). Given that the Internet has been identified as a tool that *shapes* social reality, a viable relationship between the two constructs appears reasonable. Researchers argue that the relationship between the Internet and gender is fairly complex as the technology of the medium itself is *gendered* (Bailey & Telford, 2007), but it provides a new space for gender to be *performed* (White, 2003).

A *Gendered Medium.* From a historical and cultural perspective, the use of technology has been predominantly male-governed; socialization of women for domestic work limited their need for technical expertise. As recently as 2000, in a sample of 185 users from 84 U.S. families, men reported spending almost twice as many hours per week online at home as women (Kayany & Yelsma, 2000). Because of these inequities, technological advancements can be viewed as masculine tools of power (Bailey & Telford, 2007). Despite steps to increase equity, men continue to dominate in specialized fields of technology such as software design (Anderson & Buzzanell, 2007)—making the Internet a male-occupied space. Liberal feminists argued that because of female exclusion, technology reflects a patri-

1 The authors acknowledge their equal contributions to this essay.

archal hierarchy that produces tools of oppression detrimental to women (i.e., pornography; Podlas, 2000).

However, in the more recent era of technological growth, women comprise an increasing percentage of the Internet population. Since 2001, male versus female access to the Internet has reached parity (U.S. Department of Commerce, 2004). The data regarding relational use of online technologies is especially interesting. The 2000 Pew Internet and American Life project reported that women use the Internet to maintain relationships more than men. Among 713 college students, women were four to five times more likely than men to use social networking websites (Tufekci, 2008). Women report more Facebook "friends" than men and report spending more time on Facebook than men, regardless of the size of their networks (Acar, 2008). Another recent survey of college students documented no differences between male versus female reports of the amount of time spent online communicating with romantic partners (Sidelinger, Ayash, & Tibbles, 2008). Furthermore, women and men spend equal time playing online games (Williams, Consalvo, Caplan, & Yee, 2009). Contemporary feminists view these multiple measures of online equity as indicating that the Internet can provide a space for women's empowerment and agency, given that it provides "unparalleled mechanisms for widespread dissemination and communication" (Bailey & Telford, 2007, p. 244). For a detailed review of the research on women's versus men's use of the Internet, see Royal (2005).

A *Gender Performance Space.* Some scholars have argued that both male and female users communicate online in ways that "replicate and disrupt" established gender practices (Anderson & Buzzanell, 2007, p. 32). Thus, the Internet can be viewed as a space with liberating potential, where gender can be performed in new ways; innovative identities can be imagined by online representation (White, 2003), and gendered scripts can be re-conceptualized through use of language in online discourse (Bruckman, 1993; Kelly, Pomerantz, & Currie, 2006).

Although the "anonymity" argument has been severely critiqued (e.g., Boler, 2007; Kennedy, 2006), scholars have long recognized that the anonymity of the Internet allows for the creation of complex identities (e.g., Kuntsman, 2008) and thus gender can be experienced outside the rigid binary system without facing social repercussion. Furthermore, the ever-changing nature of the Internet and the ability to freely navigate through different online cultures permits the fluidness of gender to be realized and experienced (Bailey & Telford, 2007). Such experimentation typically challenges mainstream conceptions of gender.

The Internet, as a medium of mass communication, provides a space in which gendered social issues can be given widespread attention. Information-giving and interactional sites allow engagement and discussion of ideas outside the hegemonic gendered discourse. Also, it provides a forum for voices that are often overlooked or silenced in society (i.e., lesbian teens). Additionally, the Internet allows for networking and establishing a sense of connection necessary for movements attempting social change. For example, the materialization of modern feminism is "marked by the emergence of networks and contacts which need no centralized organizations and evade its structures of command control" (Bailey & Telford, 2007, p. 259). From this viewpoint, the Internet is a space of liberation for gender to be performed, conceptualized, and theorized about in desirable ways.

Performative Theory and Gender on the Internet

The previous discussion described the Internet as providing a space where gender identities can be enacted and performed in new ways. Based on this depiction, Butler's theory of gender performativity (1990) can provide a framework for interpreting the research on CMC and gender. The theory argues that humans create gender identities through expression and performance (Wood, 2009). Butler makes a clear distinction between biological sex and gender: whereas biological sex is a mere accident of birth, gender is produced and maintained through cultural discourses. Humans enact gender via multiple modes of expression (i.e., communication) within societal inscriptions of gender as performed through words and actions (Menard-Warwick, 2007). Performative theory posits that gender is not specifically something humans *have*, but rather, something they *do* (Menard-Warwick, 2007). Gender is an active expression of identity and an outward performance (Bell, 2006); the central claim of the theory states that without the performance of gender, there is no gender (Wood, 2009).

Purpose and Preview

A large body of research exists on gender and CMC—too large for full review in this chapter. To provide the reader with a reasonable sampling of the research, including topics of both popular and scholarly interest, we review three lines of research and interpret the findings via performative theory: gendered blogging, gender-bending, and cybersex. Our research for this chapter included examining public and scholarly discourses on CMC and gender performance as well as viewing online performances of gender.

Gendered Blogging

Because identity arises from "publicly validated performances," bloggers enact gender identity through blogging (García-Gómez, 2009, p. 613). Blog authors present their performative gendered identities through *both* visual and discursive means (van Doorn, van Zoonen, & Wyatt, 2007) as they form relationships with those who read their blogs.

Weblogs or Blogs

Weblogs, more popularly referred to as blogs, are "frequently updated websites where content (text, pictures, music, etc.) is posted on a regular basis and displayed in a chronological order" (Haferkamp & Krämer, 2008, p. 3). Blogs are interesting spaces, as they incorporate both CMC and human-computer interaction, but are really only elaborated personal homepages, making the blog a result of user/machine dialogue. Most blogs are interactive (Stefanone & Jang, 2008), allowing readers to comment and leave feedback; frequent interaction can facilitate development of relationships between authors and readers via conversations and feedback. Because blogs are (a) created as messages, (b) sent through the medium of the Internet, and (c) designed to be interactive, blogs are an ideal example of CMC technology. Furthermore, blogs are easy to create and maintain, "allowing anyone with access to a computer and the Internet to create and maintain a blog, as little technical knowledge (e.g., HTML) is required" (p. 124).

Two main features of blogs emerge: blogs are updated frequently and updates are short, requiring minimal writing time. Lenhart and Fox (2006) found that 13% of bloggers post at least once a day, 15% post three to five days a week, and 25% post one or two times a week. Additionally, on average, most bloggers spend two hours a week blogging (Lenhart & Fox, 2006). The frequency with which bloggers post influences the depth of relationships they have with readers as well as the strength of the online-created identities. Regular contact between blogger and reader can build close relationships and create a strong online identity by reinforcing identity portrayals.

Because of the popularity of blogs, scholarly attention has focused on the uses and gratifications of blogs. Seven identified motivations for blogging include self-documentation, improving writing, self-expression, medium appeal, information, passing time, and socialization (Li, 2007). Papacharizzi (2002) posits that the most important of these motivations are self-expression and social interaction. Alternately, Kaye (2005) identified six motivating factors for weblog use: information seeking by checking mass media websites, convenience, personal fulfillment, political surveillance, social surveillance, and expression affiliation.

Blog posts blend mass and interpersonal communication as they are written with the intent of being read by an audience but promote one-on-one relationships between frequent blog interactants. Bloggers write to an "ambiguous audience" that may include family, friends, and strangers (Kleman, 2007). Bloggers write to communicate within relationships, as well as to transmit messages to mass audiences. Therefore blogging is the "epitome of masspersonal communication" (p. 2). Bloggers perform gender for their audience of readers via their writing; readers can become familiar and friendly with bloggers after reading their posts regularly and respond with their own performances of gender.

Performing Gender via Blogging

Twelve million Americans report blogging (Lehhart & Fox, 2006); men and women blog approximately equally (Haferkamp & Krämer, 2008). In creating online identities, almost all bloggers reveal their gender on their blogs (Kleeman,

2007). In addition to explicitly stating gender, blog creators employ various forms of nonverbal behaviors (e.g., colors, backgrounds, fonts, and pictures) that perform gendered identity. For example, a self-proclaimed "girly girl" could select a pink background for her blog.

Gendered Use of Blogs

Gender is performed quite distinctively in the context of blogs. Males are more likely to write filter blogs (Wei, 2009; Karlsson, 2007), containing primarily information external to the author such as news and political events where certain items are discussed and others are excluded (Wei, 2009). In a study of British bloggers, Pedersen and Macafee (2007) report that men's blog content focuses on sharing information, providing opinions, and highlighting links. This finding paints a gendered picture for how males share information through blogging—a picture consistent with typical ways males communicate in face-to-face (FtF) interactions. Tannen (1990) argues men engage in report talk, giving information and opinions as a means of gaining or sustaining status. Furthermore, van Doorn et al. (2007) found that male authors carefully avoided being too "emotional," focusing their blogs on information and ideas. Herring and Paolillo (2006) examined gendered communication on blogs, and found that filter blogs had more "male" stylistic features, such as statements and restatements of facts.

Gender can be performed via language choices (Motschenbacher, 2009). Through linguistic practices, gender identities are evoked by men using facts and emotionless language versus women employing expressive and inclusive language. Filter blogs also have masculine content, focusing on concepts and ideas rather than people and processes (Herring & Paolillo, 2006)—what Tannen (1990) described as male report-talk. By engaging in masculine blogging, men enact their gender through their writing on blogs and by posting opinions on others' blogs.

Medical blogs serve as a prime example of how men perform masculinity on blogs. Kovic, Lulic, and Brumini surveyed medical blogs, defined as "a blog whose main topic was related to health or medicine" (2008, p. 2), and discovered that 59% of medical bloggers were male; 74% of the bloggers reported being motivated to post on medical blogs to share knowledge and skills, and 56% by the desire of gaining insights from others. (Respondents could choose more than one motivation; therefore the percent total exceeds 100%.) Two-thirds of medical bloggers received attention from the news media about their blogs. It could be argued that males were more likely to participate in these medical blogs because the nature of these blogs aligns with a masculine communication style, allowing the authors to perform their gender through their blogs. Because men are more likely to write informational blogs (Pedersen & Macafee, 2007; van Doorn et al., 2007), they also are more likely to be seen as credible bloggers (Armstrong & McAdams, 2009). In short, male bloggers are seen as information transmitters and form blogging relationships based on sharing of information and the credibility of that information.

Women, on the other hand, are more likely to write journal blogs, or diary blogs (Attwood, 2009; Karlsson, 2007; Wei, 2009). Such blogs describe personal life with internal content (Wei, 2009). However, unlike traditional diaries, journal blogs do not have the connotation of privacy and instead seek an online audience. Diary blogs are personal and emotion-laden, creating "readerly attachment" (Karlsson, 2007, p. 139). Bloggers invite the readers of such journal blogs to identify with and share in a relationship with the author through responses and comments. Readers who habitually read these blogs are more likely to be female (Karlsson, 2007), and the creation of these support networks on blogs is consistent to the more communal, relational communication characteristic of women.

Women enact their femininity through the interpersonal nature of journal blogs and their accompanying social bonding. Female communication style allows women to share, create, and maintain relationships, bring others into the conversation, and respond to ideas (Wood, 2009). Through journal blogging, reading blogs regularly, and leaving feedback, women engage in rapport-talk, described by Tannen as "negotiations for closeness in which people try to seek and give confirmation and support, and to reach consensus" (1990, p. 25). Blogs can provide a shared emotional connection (Stavrositu & Sundar, 2008), where members of the blog community share life experiences and events. Blogs can functions as a vehicle for bloggers and readers to share personal events and experiences. One blog feature that aids relationship building (van Doorn et al., 2007) is the 'blogroll' (a list of links that allows the user to add others' blogs to their blogroll, creating a network of blogs sometimes called the "blogosphere"). The use of blogrolls "fosters a reciprocal relationship" where people add each other's blogs to their blogrolls (p. 146).

The language of women's blogs plays a central role in the performance of their gendered identities as the "features of 'women's language' are powerful resources to linguistically index female identities" (Motschenbacher, 2009, p. 19). Teen girls' diary blogs provide an obvious example of feminine gender performance via statements such as "I am a woman, not a girl!" and "Since I was a little girl" (García-Gómez, 2009, p. 615). Also, women's language is more inclusive and expressive, passive, cooperative, and accommodating (Herring & Paolillo, 2006) than the language used by male bloggers. Women bloggers construct their gendered identities using sexualized imagery and words, often while talking about domesticity and taking care of the home (van Doorn et al., 2007). This juxtaposed mix creates a unique female gender identity combining traditional views of women such as the mother and sex object (Wood, 2009).

Women's Movements on Blogs

Women in the blogosphere face some inequities (Pedersen & Macafee, 2007), as more men than women maintain blogs (Taylor, 2004). The mass media portrayal of blogging as "adult and masculine" (Herring et al., 2004, as cited in van Doorn et al., 2007), might prompt many women to view blogs as "not for me." Conversely, because blogging is seen as a gendered technology that marginalizes women, women can become empowered through simply using the medium and transforming blogging into a feminine performance. Stavrositu and Sundar (2008) assert that because women are misrepresented and underrepresented in traditional media, women bloggers gain a voice through blogging.

Female sexual liberation can be enacted via blogging. Attwood (2009) studied women's sex blogs, and sex "blooks" (blogs turned into books). Attwood describes "blooks" as "the world's fastest growing new kind of book" (p. 5). Through these blogs and blooks, female authors emphasize sexual openness, empowerment, and pleasure. Through the digital mediums of blogs, women authors redefine their sexuality and femininity by exploring their sex lives through blogging, writing about what many people would consider the most intimate and personal form of communication.

Blogs provide a vital venue for female self-expression, especially in countries that limit freedom of expression (Monteiro, 2008). An examination of the work of female bloggers in Egypt, Saudi Arabia, and Jordan indicates that women are turning to blogs to make sense of their identities. Some of these blogs are forms of political activism, while others are centers of expression, featuring short stories and prose. One female blogger describes blogging as a haven:

> Blogs don't only give you the chance to hide, they give you another valuable thing: a space without a title. But what happens after a while of creating the blog you find yourself in the midst of what you once escaped. The pseudo name is no longer a curtain that hides you, but it becomes the name of the being exposed by the posts, one after the other. You gain an identity among your neighbors in blogging—an identity made more defined and clear by every new post. I want to blog for ten years and remain, to the tenth year, thinking about this place as my own place where my rules apply, and that I could, if I desired, post blank posts. (p. 50)

These female bloggers find refuge, identity, and comfort in their blogs. When the blog quoted above was blocked by the Saudi Arabian government, its author began emailing new posts from her cell phone. Readers wanted to stay connected to the author, so they desired to read her posts however posted. Thus, blogging can give repressed women power and voice.

Female blogging is on the increase. BlogHer, one of the leading networks for women online (BlogHer network, 2007), received more than 4.23 million unique visitors per month in October 2007. The blogs in the BlogHer network cover a wide variety of topics including parenting, health, food, career, money, and politics. The site's mission statement reflects the goal of creating opportunities for women who blog to pursue exposure, education, community, and economic empowerment (BlogHer network, 2007). This blog community for women maintains that women find socialization through blogging; in such an online blog community, women perform their femininity by being social and interpersonal. The "mommy blogs" (Thompson, 2007) are becoming a force in the business world as well. Moms are offering product reviews and shopping tips to readers. Their influence is so strong, that Proctor & Gamble incorporated as many as 15 "mommy bloggers" to their marketing strategy (Neff, 2008). By influencing their readers, mommy bloggers gain power.

Video Blogs or Vlogs

CMC also occurs via vlogs, or video logs, where individuals post in video form (Molyneaux, O'Donnell, & Gibson, 2009) accompanied by text-based comments that allow for ongoing interpersonal communication (Kendall, 2008). Most vlogs focus on personal content, and create social networks by allowing both textual and video comments from viewers. Men and women employ vlogs quite differently. In a study on YouTube vlogs, men posted vlogs more than women (Molyneaux et al., 2009). However, female vloggers were more likely to interact with other vloggers by asking questions and responding. The quality of vlogs also differs along sex lines. Vlogs created by men had better sound quality; women created more interactive vlogs with better image quality. Men vlog about public and technology-related topics; women vlog about personal matters. Despite gender differences in the content and creation of vlogs, both men and women reported feeling a part of the YouTube community. Vlogs might have more impact than blogs on interpersonal relationships, as authors don't just write, but literally talk to viewers. Seeing and hearing the author could forge stronger relational bonds.

Implications of Gendered Blogs for "Real Life" Gender

Weblogs, or blogs, are personal webpages updated frequently (Haferkamp & Krämer, 2008) and written to express views, provide information, and serve as a venue for interaction (Li, 2007). Blogs are interactive (Stefanone & Jang, 2008) and garner communities of followers (van Doorn et al., 2007). Men and women blog in fairly equal numbers (Haferkamp & Krämer, 2008; Lehhart & Fox, 2006), but for different purposes (Kaye, 2005) and thus write different types of blogs (Attwood, 2009; Karlsson, 2007; Wei, 2009). Through performative actions, especially the use of gendered language, both men and women enact gender through the discourse and images on their blogs (van Doorn et al., 2007). These gender performances often follow traditional gender roles and norms that can lead to the development of many close, personal relationships between bloggers and readers. The next section of the chapter discusses a more marginalized form of online gender performance, gender-bending.

Gender-Bending

According to the "cyberspace glossary" online, "gender-bending" is broadly defined as "men posing as women, women posing as men, or either posing as nongendered or 'neutral' characters." Gender-bending can be performed through a variety of expressions, such as cross dressing, that fail to conform to culturally accepted enactment of gender. Gender-bending takes place in many venues including the *Saturday Night Live* skits about "Pat" the androgynous character whose gender was never revealed, websites such as "How to Gender Bend," and Swiss Internet relay chats in which users engage in intentional "gender plays" (Rellstab, 2007).

For those who adhere to traditional gender roles (i.e., biological males enact masculine behaviors, biological females enact feminine behaviors), gender-bending is unfathomable and confusing because gender-benders fail to fit easily or readily into existing cognitive categories. "I log in and now I'm a woman. And I'd log off and I'm a man again" (Bruckman, 1993). Gender-benders display "improper" gender identity in a society that considers gender an important part of human interaction (Bruckman, 1993); thus, gender-bending can be viewed as a form of resistance (Rothman, 1993) that poses a threat to the social structure. Gender-benders can develop identity based on performance of unconventionally gendered representations and accordingly can be socially reprimanded and pathologized by evaluators (Plante, 2006).

The anonymity of the Internet presents a space in which gender-bending can be successfully performed without the usual societal repercussions. Brookey and Cannon discuss how the virtual environment of the Internet allows individuals to feel "empowered with the agency to produce their own sexual world" with seemingly countless options (2009, p. 149). Because of this agency in representation, the Internet is widely perceived to be a space in which users can break free of rigid gender binaries.

Gender-Bending Population

Gaming. Although the size of the population of online gender-benders is difficult to determine, men may gender-bend on the Internet more frequently than women. For example, on the website titled "The Daedalus Gateway: The Psychology of MMORPGs- Gender Bending," a massive multiplayer online role-playing games (MMORPGs), the most common demographic for gender-benders is men over the age of 25, which suggests that over 85% of female avatars are actually played by men. In "The Daedalus Project: The psychology of MMORPGs," the producer suggested that one out of every two female characters is played by a man, and about one out of every one hundred male characters is played by a woman." Evaluating gender-benders within a game helps gamers know how to interact with their opponents. Later research has identified and described additional common gender-bending spaces online.

Online Forums and Chat Rooms. The following year, Jaffe, Lee, Huang, and Oshagan (1999) reported that among 114 college students, more women than men selected user names that masked their gender identity when posting in online forums. More recently, in a survey of 823 chat room users, 28% reported gender-swapping (Samp, Wittenberg, & Gillett, 2003), including choice of user name (52%), explicitly announcing a false biological sex (24%), and intentional language choices to enact the opposite sex (15%). Samp et al. (2003) reported that contrary to popular belief, feminine, masculine, and androgynous individuals were equally likely to engage in gender-bending. Furthermore, low self-monitoring users were more likely than high self-monitors to engage in gender-bending and to question the gender of other users. In sum, it appears that both men and women gender bend in a variety of online venues.

Common Spaces for Gender-Bending on the Internet

Gaming provides a popular venue for online gender-bending. Bruckman (1993) identifies MUD, a text-based multi-user virtual-reality environment, as a place where gender-swapping often occurs. Additionally, games such as World of Warcraft and Second Life allow gamers to build avatars (virtual personalities). (Brookey & Cannon, 2009; Bruckman, 1993). In such games, users build structures, choose clothes and accessories, and "are usually at liberty to shape this virtual world in any way they see fit" (Brookey & Cannon, 2009). The number of gamers has grown rapidly, and an estimated 11 million men and 9 million women had online avatars by 2007 (Remington, 2009).

In addition to game sites, chat forums, including dating chat rooms, provide another common venue for gender-bending (del-Teso-Craviotto, 2006, 2008). Thousands of chat lines are available on the Internet and websites designed specifically for gender-bending individuals. For example, "Gender Bender and the underground chatropolis" is the first website that appears in a Google search for "gender-bending and chat rooms." The site explicitly states that "this room welcomes and encourages gender fucking...the practice of bending stereotypical gender appearances and mannerisms, resulting in a mixture of masculinity and femininity, boy, girl, Dom/me, submissive, whatever." Clearly chat forums provide a venue for gender-benders to meet and interact with others, often forming on-going personal relationships.

Outing Gender-Benders

While gender-bending chat lines are filled with individuals who admittedly gender-bend, general chat lines may or may not openly include gender-benders. To gender-bend without being "outed" by communication partners, individuals must make careful language choices, as users perform gender via language choices (Danet, 1998; Motschenbacher, 2009). Indeed, a line of research has documented that male versus female language differs in multiple online venues including instant messaging (e.g., Baron, 2004; Fox, Bukatko, Hallahan, & Crawford, 2007), online support communities (e.g., Ginossar, 2008), and home pages (e.g., Stern, 2004). Furthermore, research indicates that users can question the gender of a potential relational partner and require subtle linguistic cues as well as obvious declarations of biological sex for authentication of gender (del-Teso-Craviotto, 2008). Stereotypic content of messages also leads to deconstructions and assignments of online gender (Herring & Martinson, 2004). Indeed, users are more likely to use gender stereotypic language when discussing gender stereotypic content than general content (Thomson, 2006). Remaining "sexless" may not help; in a recent experiment, Nowak (2003) reported that almost two-thirds of the college-student participants reported guessing the sex of the online genderless interactants. When discovered, gender-benders can deny their sex; they need to reveal their "true" identity only when they desire to "come-out." Gender-benders might refuse to reveal their true identities because they fear negative evaluations from relational partners (Bruckman, 1993).

Motivations for Gender-Bending on the Internet

Researchers have offered numerous rationales for gender-bend in cyberspace. Motivations themselves can be gendered, as they can reflect deferential treatment of men and women in contemporary American society. Men might gender-bend because they desire the attention garnered by a female identity (Danet, 1998) or because they desire the power achieved in misrepresentation or intentionally deceiving others. Women might gender-bend to become more assertive or enjoy the power typically accorded males. Additionally, adopting a masculine identity allows women to avoid online sexual harassment (Danet, 1998). Furthermore, gender-bending allows individuals to experiment with gendered social norms such as differing levels of self-disclosure. Because it is considered more socially acceptable for a woman to self-disclose at a high level than for a man, a man wishing to self-disclose extensively can assume the identity of a woman to avoid questioning. Thus, gender-bending allows users to enact relationships in ways perceived as unavailable in FtF venues.

Online gamers can gender-bend as a gaming strategy that reflects social standards of gendered treatment of men and women. Bruckman (1993) suggests that male players often obtain a list of the names of the players in advance and then deliver unwanted attention and sexual advances to those perceived as female; because of such treatment, females may adopt a masculine identity. On the other hand, because female characters are perceived as receiving more attention, males might play female characters to obtain such attention. Furthermore, as females are perceived to get better treatment, males might play the role of females to obtain more free gifts or be helped more willingly (Totilo, 2005). Accordingly, Bruckman suggests that often the promiscuous and sexually aggressive female characters are actually played by men. Feminist critics argue that male gender-bending can be viewed as another form of female oppression (Totilo, 2005); in virtual realities, males can design images of female characters with sex appeal from the masculine perspective. Choosing such feminine characteristics allows the male gamer to control the female body.

We offer an additional rationale for gender-bending. Users can gender-bend to "try on" new ways of communicating in their FtF as well as online, personal relationships. For example, a woman might request her male FtF romantic partner to engage in behaviors he considers feminine, such as cuddling and expressing feelings. The man might be more comfortable practicing such behaviors online as a woman, especially if he has witnessed only women performing such behaviors. Conversely, a woman desiring to be more assertive with close, work allies can try out such behaviors as a male game character, to discover if she can say the words necessary to express her opinion in the face of opposition from opponents and high-status relational partners. Thus, gender-bending can serve an important relational function by allowing for rehearsal of new, desired communication behaviors for later use in FtF relationships.

Implications of Gender-Bending for "Real Life" Gender

The motivations behind Internet gender-bending provides insight into how gender identity is culturally prioritized and a consciousness of how its very real effects on social interaction. Online gender-bending allows for the critical examination of social constructions of gender and potentially contributes to the long-term destabilization of the way society currently constructs gender (Danet, 1998). The act of gender-bending allows individuals to gather skills, tools, and data with which to challenge rigid notions of gender and sexuality. The performance of an alternative gender identity through online interactions can "defamiliarize" individuals with their real life gender role (Bruckman, 1993), allowing users to address their sexuality and to interact in ways that he/she would not be comfortable doing in "real life," as well as to understand the way sexual politics work in society (Danet, 1998). In FtF forums, people typically enact gender behaviors condoned by society, but in private people are free to explore alternative visions of self, gender, and personal enactment of gender (Bruckman, 1993). Additionally, as users try on different gendered identities or attempt to perform no gender at all, gender-bending can provide insights into the way gender affects and limits interpersonal relationships. The performance of gender through gender-bending also permits transcendence of gender roles:

> "It is indeed a truly disorienting experience the first time one finds oneself being treated as a member of the opposite sex…my own forays into the realm of virtual masculinity were at first frightening…once deprived of the social tools which I, as female, was used to deploying and relying on, I felt rudderless, unable to negotiate the most simple of social interactions." (Reid, 1994 as cited in Danet, 1998, p. 145)

From this standpoint, the Internet can be seen as a "valuable space in which to study gender and sexuality…because unlike traditional forms of print, film, or television media…users are primarily responsible" for the creation and representation of their gender identity (Brookey & Cannon, 2009, p. 146).

Gender-bending on the Internet can be a constructive tool for reconceptualizing gender; it can deconstruct the power relations of the rigid binary gender structure. However, other research suggests that this might be an optimistic view, as sexual and gender norms can "be reproduced in ways that are retrograde" (Brookey & Cannon, 2009, p. 160). As illustrated in the section on gender-bending in online gaming, the performance of gender can further objectify women and encourage the enactment of traditional gender roles. The next section of the chapter examines the performance of gender in the CMC surrounding sexual conduct and relationships.

Cybersex

Young, Griffin-Shelley, Cooper, O'Mara, and Buchanan (2000) defined cybersex as "two online users engaging in private discourse about sexual fantasies. The dialogue is usually accompanied by self-stimulation" (p. 60). Ben-Ze'ev described cybersex as

> a social interaction between at least two people who are exchanging real-time digital messages in order to become sexually aroused. People send provocative and erotic messages to each other, with the purpose of bringing each other to orgasm as they masturbate together in real time. Cybersex requires the articulation of sexual desire to the extent that would be most unusual in face-to-face encounter. In cyberspace, that which often remains unspoken must be put into words. (2004, p. 5)

According to Millner (2008), sex is one of the most searched topics on the Internet. Given users' intense interest in the subject, it is not surprising that users enage in a number of activities associated with cybersex. In a survey of 760 college students, Boies documented four uses of online sexual content: entertainment, sexual gratification, seeking partners, and in-person exploitation. "About 40% went online to meet new people, and to view sexually explicit material" (2002, p. 77).

In this section, we are concerned with the intersection of gender, cybersex, and relational communication. Thus, it is noteworthy that both married and unmarried men and women participate in cybersex (Millner, 2008), resulting in a wide variety of sexual encounters online. Multiple websites match partners who desire single-single dating, affairs between partners married to others, online sexual encounters, and so on. Cybersex covers a wide range of relationships including but not limited to cyber romances (Gibbs, Ellison, & Heino, 2006) and cyber affairs (Young et al., 2000), yet many cybersex partners never meet FtF. When and if they meet, the relationships are often of very short duration. In contrast, cyber partners who establish emotionally connected relationships often interact online for a very long time (Barta & Kiene, 2005). Emotional intimacy can lead to cybersex and cybersex can lead to emotional intimacy (Young et al., 2000). Thus, while cybersex often is characterized by "illusionary, idealized, sexual relationships" (Limacher & Wright, 2006, p. 314), cybersex conversations obviate very little in the expression of ideas, fantasies, and pleasures—thus allowing for the development of relationships that many users experience as "real." Butkus provides a detailed description of the unique and interesting character of online sexual relationships:

> Online romantic relationships combine features of close and remote relationships. In online relationships, people are neither close, intimate friends nor complete strangers;

> they are also not lust friends. Online relationships constitute a unique kind of relationship—termed "detached attachment." Like direct, face-to-face relationships, online relationships can be spontaneous and casual and show intensive personal involvement. (2004, p. 24)

Furthermore, the web offers opportunities to develop relationships with others who enjoy uncommon relational and sexual practices that users might be hesitant to explore in FtF venues, including various polyamory practices (multiple loving partners) (Ritchie & Baker, 2006), homosexuality (Ashford, 2006; Walker, 2009), and exploitative relationships (Brookey & Cannon, 2009).

Self-Disclosure/Self-Presentation

Rosen, Cheever, Cummings, and Felt's (2008) research indicated that the most obvious difference between online versus FtF relationships is rate of self-disclosure. Online partners reveal more disclosures and reveal them earlier in the relationship than FtF partners. Online partners experience limited channel options for expression, as they typically employ primarily text-based email and chat technologies; extensive and emotional expression via such technologies would be difficult without extensive self-disclosure. Does online self-disclosure have positive relational effects? In a college-student survey, Sidelinger et al. (2008) documented that online openness had no significant impact on relational commitment, communication satisfaction, perceived partner perceptiveness, attentiveness, or responsiveness. Thus, increased openness online appears to have neither positive nor negative impact on on-going dating relationships.

Deception

Users can enact positive self-presentation more easily in the cyberworld than the FtF world, as online interactants readily accept willing participants (Gibbs et al., 2006). Ben-Ze'ev (2004) reveals that lying is easy and usually remains uncontested in the cyberworld. Women typically lie for safety concerns and men lie to boost their socio-economic status, but both sexes believe that lying about such factors allows for openness and honesty regarding the more important matters of their emotional experiences and sexual desires (Ben-Ze'ev, 2004).

Young et al. (2000) posits the ACE model (anonymity, convenience, and escape) as a means of understanding the prevalence of deceptive online practices. *Anonymity* is the catalyst for individuals feeling comfortable disclosing information during cybersex chat sessions; users think anonymity will safeguard FtF partners from discovering the cybersex relationship and that online relational partners will only judge them in the context of the chat session, not applying the same criteria as in FtF relationships (Peter & Valkenburg, 2007). Gibbs et al. (2006) concur with this viewpoint adding that CMC allows for selective self-presentation by precluding FtF nonverbal messages and only allowing for discernment of the individual based on the text he/she provides. Peter and Valkenburg (2007) point out that anonymity facilitates free and uninhibited discussion about intimate desires and/or fantasies. With the absence of visual FtF verbal cues, users have more time to reflect on and interpret the intent and context of chat messages. In a way, the digital mode of communication removes some distractions from the complexity of communication, but at the cost of conveying less complete messages. Gibbs et al. concluded that "the Internet is the medium for identity manipulation" (p. 169), including gender-bending. *Convenience* merely is the ability to meet others more readily online than in person. The constant growth of Internet users gives rise to this phenomenon. *Escape* is the idea that cybersex is not actual infidelity because the relationship is never physically consummated; because cybersex is virtual, it cannot be real.

Infidelity

According to Limacher and Wright (2006), "infidelity can be understood as a breach of trust between a couple, in which the secrecy and lies become the culprit in destroying the relationship, not necessarily the sex" (see Pittman, 1989). Internet infidelity takes on many forms—cyber affairs among them.

Although the Internet can be used for factual sexual education, it also can be used for emotional and sexual maladaptive behaviors associated with cybersex (Millner, 2008). Some internet users spend up to 10 hours per week engaged in cybersexual relationships (Cooper, Boies, Maheu, & Greenfield, 2000). What motivates a person to engage in Internet infidelity or as it is more commonly known as—cyber cheating? Barta and Kiene (2005) posit that people often engage in a cost/benefit determination before engaging in cybercheating. Conversely, Peter and Valkenburg (2007) identified two psychological motivations for cyber cheating that they reference as hypotheses: The recreational hypothesis asserts that such users are sexually permissive sensation-seekers. Alternatively, the compensation hypothesis describes such users as looking to find others to compensate for their own inadequacies. Barta and Kiene assert that people with permissive sexual attitudes and with a high number of past sexual relationships are more likely than their more conservative counterparts to engage in infidelity and to employ sexually assertive behaviors including flirting. Another view asserts that there are basically two motivations for infidelity: emotional and sexual desires (Barta & Kiene, 2005; Toates, 2009). Biological sex "is a good predictor of motivation for infidelity" (Barta & Kiene, 2005, p. 341); women are more likely to engage in infidelity when they experience emotional dissatisfaction in their primary relationship; men are more sexually motivated.

Pornography

While moralists and scholars continue to debate the definition of pornography (Paul, 2009) as well as whether or not viewing pornography constitutes relational infidelity, the pervasiveness of Internet pornography continues to grow. Young (2008) reported that the pornography industry earns $12 billion per year in the United States and $57 billion worldwide. The Internet offers thousands of pornography sites, with a significant portion open to the public, including, for example, websites describing sex tourism cites (Chow-White, 2006) and local escort services (Castle & Lee, 2008). Users can easily stumble across a pornography site accidentally—as easily as mistyping a commonly used website called youtube.com.

Multiple researchers have documented that adolescent and young adult males are more likely than females to view sexually explicit online content (e.g., Boies, 2002; Peter & Valkenburg, 2006). Men are more likely than women to seek visual sexual depictions as a means to experience sexual arousal for masturbation; women are more likely to seek out erotic narratives and chat rooms than men; women are less likely than men to self-stimulate while using online materials (Barta & Kiene, 2005).

Whereas men often think of pornography on the Internet as mere visual stimulation for masturbation with no emotional attachment (Limacher & Wright, 2006), their female FtF romantic partners often hold an alternative viewpoint. "Getting caught" by the partner can transform a safe and loving relationship into one of mistrust and distance. Wives who catch their husbands using Internet pornography typically perceive themselves as unpleasing to their husbands and experience emotional pain by the husband's "involvement" with another woman. Wives typically do not view masturbation as wrong, but view masturbation to pornography as wrong; more precisely, the greater the frequency of the husband's use of pornography, the greater the marital issue in the wife's view.

The Dark Side of Cybersex

Young (2008) described a male respondent who stumbled on a porn site. Formerly an average user of the Internet, he became highly involved with the Internet following the accidental discovery of pornographic material. The man became preoccupied with viewing online pornography, began staying late at work, missing important meetings at work, and spending hours at home by himself. One of Millner's (2008) respondents was so taken by a cyber relationship that he was willing to leave his family to develop a permanent relationship with his cyber-cheating partner. Preoccupation with the Internet or with pornography is one indication that the individual is becoming an addict (Millner, 2008; Paul, 2009; Toates, 2009; Young, 2008). Levert defines Internet pornography addiction/compulsion as a "pathological preoccupation with sexual behaviors in an effort to create a mood-altering experience" (2007, p. 147). Pathological behavior interferes with daily functioning, becomes uncontrollable, and provides the potential for major conflicts in FtF personal relationships. Beyond addiction, users who develop cyber relationships can experience online sexual harassment, cyber stalking, and cyber predators who stalk adults (Kelly et al., 2006; Philips & Morrissey, 2004).

Implications of Cybersex for "Real Life" Gender

Both men and women engage in cybersex, but for different reasons (sexual versus emotional fulfillment). Cybersexual (vs. FtF) relationships typically involve deception and escalated self-disclosure. Although cybersex provides the opportunity for users to engage in gendered, virtual, sexual behavior with online partners, such interactions can lead to negative consequences for their FtF romantic relationships. Cybersexual relationships can involve gender-bending, but they also can follow traditional gender roles and norms, leading to deceptive and pathological behaviors such as women lying for safety reasons and men sexually harassing women online.

Chapter Summary and Conclusion

The Internet can be viewed as a gendered space as well as a space for performing gender. Users can perform gender in many ways including blogging, gender-bending, and engaging in cybersex. The three activities discussed above allow for the development of close, personal, online relationships. However, these three activities have the potential to alter close, personal FtF relationships—for both good and ill. Taken as a whole, this body of research paints a picture of users surfing the Internet for solutions to offline problems, perhaps through enacting "real" identities (giving voice to gendered identities via blogging), perhaps through experimenting with identity (by gender-bending), or perhaps by discovering alternative relationships per se (via developing cybersexual relationships). In all three cases, the Internet offers neither a panacea to offline problems, nor a useless escape, but rather a forum for exploration and discovery that could lead to improved or deteriorated online and offline relationships.

We close this chapter with two optimistic observations: (a) CMC allows relational partners to explore the meanings of gender in unique and insightful ways that can move beyond play to reinvention. (b) When relational partners play at new ways of being and being together, they explore possible metamorphoses. A relational metamorphosis often involves change for the better but inevitably destroys the previous status quo relationship. Thus, we offer a caution from the U.S. novelist Kurt Vonnegut: "We are what we pretend to be, so we must be careful what we pretend to be."

References

Acar, A. (2008). Antecedents and consequences of online social networking behavior: The case of Facebook. *Journal of Website Promotion, 3*, 62–83.

Anderson, W. K. Z., & Buzzanell, P. M. (2007). "Outcast among outcasts": Identity, gender, and leadership in a mac users group. *Women and Language, 30*, 32–45.

Armstrong, C. L., & McAdams, M. J. (2009). Blogs of information: How gender cues and individual motivations influence perceptions of credibility. ***Journal of Computer-Mediated Communication, 14*, 435–456.**

Ashford, C. (2006). The only gay in the village: Sexuality and the net. *Information & Communications Technology Law, 15,* 275–289.

Attwood, F. (2009). Intimate adventures: Sex blogs, sex "blooks" and women's sexual narration. *European Journal of Cultural Studies, 12,* 5–20.

Bailey, J., & Telford, A. (2007). What's so cyber about it: Reflections on cyberfeminism contribution to legal studies. *Canadian Journal of Women & the Law, 19,* 243–272.

Baron, N. S. (2004). See you online: Gender issues in college student use of instant messaging. *Journal of Language and Social Psychology, 23,* 397–423.

Barta W., & Kiene, S. (2005). Motivations for infidelity in heterosexual dating couples: The roles of gender, personality differences, and sociosexual orientation. *Journal of Social and Personal Relationships, 22,* 339–360.

Bell, V. (2006). Performative knowledge. *Theory, Culture, & Society, 23, 214–217.*

Ben-Ze'ev, A. (2004). *Love online: Emotions on the Internet.* Cambridge, UK: Cambridge University Press.

BlogHer network joins ranks of top women's networks online. (2007). *Media Report to Women,* 35(4), 1-3.

Boies, S. C. (2002). University students' uses of and reactions to online sexual information and entertainment: Links to online and offline sexual behavior. *Canadian Journal of Human Sexuality, 11,* 77–89.

Boler, M. (2007). Hypes, hopes and actualities: New digital Cartesianism and bodies in cyberspace. *New Media & Society, 9,* 139–168.

Brookey, R. A., & Cannon, K. L. (2009). Sex lives in second life. *Critical Studies in Media Communication, 26,* 145–164.

Bruckman, A. S. (1993), *Gender-swapping on the Internet.* Retrieved July 2, 2009, from www.inform.umd.edu/EdRes/Topic/WomensStudies/Computing/Articl/gender-swapping

Butkus, C. (2004). Female porn providers and Internet services. *Convergence: The Journal of Research into New Media Technologies, 10,* 24–42.

Butler, J. (1990). Subversive bodily acts. In J. Butler (Ed.), *Gender Trouble: Feminism and the subversion of identity* (pp. 163–180). New York: Routledge.

Castle, T., & Lee, J. (2008). Ordering sex in cyberspace: A content analysis of escort websites. *International Journal of Cultural Studies, 11,* 107–121.

Chow-White, P. A. (2006). Race, gender and sex on the net: Semantic networks of selling and storytelling sex tourism. *Media, Culture, & Society, 28,* 883–905.

Cooper, A., Boies, S., Maheu, M., & Greenfield, D. (2000). Sexuality and the Internet: The next sexual revolution. In F. Muscarella & L. Szuchman (Eds.), *Psychological perspectives on human sexuality* (pp. 519-545). New York: Wiley.

Danet, B. (1998). Text as mask: Gender, play, and performance on the Internet. In S. G. Jones (Ed.), *CyberSociety 2.0: Revising computer-mediated communication and community* (pp. 129–158). Thousand Oaks, CA: Sage.

del-Teso-Craviotto, M. (2006). Language and sexuality in Spanish and English dating chats. *Journal of Sociolinguistics, 10,* 460–480.

del-Teso-Craviotto, M. (2008). Gender and sexual identity authentication in language use: The case of chat rooms. *Discourse Studies, 10,* 251–270.

Fox, A. B., Bukatko, D., Hallahan, M., & Crawford, M. (2007). The medium makes a difference: Gender similarities and differences in instant messaging. *Journal of Language and Social Psychology, 26,* 389–397.

García-Gómez, A. (2009). Teenage girls' personal weblog writing: Truly a new gender discourse? *Information, Communication & Society, 12,* 611–638.

Gibbs, J., Ellison, N., & Heino, R. (2006). Self-presentation in online personals: The role of anticipated future interaction, self-disclosure, and perceived success in Internet dating. *Communication Research, 33*, 152–177.

Ginossar, T. (2008). Online participation: A content analysis of differences in utilization of two online cancer communities by men and women, patients and family members. *Health Communication, 23*, 1–12.

Haferkamp, N., & Krämer, N. (2008, May). *Entering the blogosphere: Motives for reading, writing, and commenting.* Paper presented at the meeting of the International Communication Association, Montreal.

Herring, S. C., & Martinson, A. (2004). Assessing gender authenticity in computer-mediated language use: Evidence from an identity game. *Journal of Language and Social Psychology, 23*, 424–446.

Herring, S. C., & Paolillo, J. C. (2006). Gender and genre variation in weblogs. *Journal of Sociolinguistics, 10*, 439–459.

Jaffe, J. M., Lee, Y. E., Huang, L. N., & Oshagan, H. (1999). Gender identification, interdependence, and pseudonyms in CMC: Language patterns in an electronic conference. *The Information Society, 15*, 221–234.

Karlsson, L. (2007). Desperately seeking sameness: The processes and pleasures of identification in women's diary blog reading. *Feminist Media Studies, 7*, 137–153.

Kayany, J. M., & Yelsma, P. (2000). Displacement effects of online media in the socio-technical contexts of households. *Journal of Broadcasting and Electronic Media, 44*, 215–229.

Kaye, B. (2005). It's a blog, blog, blog, blog world. *Atlantic Journal of Communication, 13*, 73–95.

Kelly, D. M., Pomerantz, S., & Currie, D. H. (2006). "No boundaries"? Girls' interactive, online learning about femininities. *Youth and Society, 38*, 3–28.

Kendall, L. (2008). Beyond media producers and consumers: Online multimedia productions as interpersonal communication. *Information, Communication & Society, 11*, 207–220.

Kennedy, H. (2006). Beyond anonymity, or future directions for Internet identity research. *New Media & Society, 8*, 859–876.

Kleman, E. (2007, November). *Journaling for the world (wide web) to see: A proposed model of self-disclosure intimacy in blogs.* Paper presented at the meeting of the National Communication Association, Chicago.

Kovic, I., Lulic, I., & Brumini, G. (2008). Examining the medical blogosphere: An online survey of medical bloggers. *Journal of Medical Internet Research 10*(3), e28. Retrieved July 2, 2009, from http://www.ncbi.nlm.nih.gov/pmc/articles/PMC2626433/

Kuntsman, A. (2008). Written in blood: Contested borders and the politics of passing in Israel/Palestine and in cyberspace. *Feminist Media Studies, 8*, 268–283.

Lenhart, A., & Fox, S. (2006). Bloggers: A portrait of the Internet's new storytellers. *Pew Internet & American Life Project.* Retrieved July 2, 2009, from http://www.pewinternet.org/~/media//
Files/Reports/2006/PIP%20Bloggers%20Report%20July%2019%202006.pdf.pdf

Levert, N. (2007). A comparison of Christian and non-Christian males, authoritarianism, and their relationship to internet pornography addiction/compulsion. *Sexual Addiction & Compulsivity, 14*, 145–166.

Li, D. (2007, May). *Why do you blog: A uses-and-gratifications inquiry into bloggers' motivations.* Paper presented at the meeting of the International Communication Association, San Francisco.

Limacher, L., & Wright, L. (2006). Exploring the therapeutic family intervention of commendations. *Journal of Family Nursing, 12*, 307–331.

Menard-Warwick, J. (2007). "My little sister had a disaster, she had a baby": Gendered performance, relational identities, and dialogic voicing. *Narrative Inquiry, 17,* 279–297.

Millner, V. (2008). Internet infidelity: A case of intimacy with detachment. *The Family Journal: Counseling and Therapy for Couples and Families, 16,* 78–82.

Molyneaux, H., O'Donnell, S., & Gibson, K. (2009). YouTube vlogs: An analysis of the gender divide. *Media Report to Women, 37,* 6–11.

Monteiro, B. (2008). Blogs and female expression in the Middle East. *Media Development, 55,* 47–53.

Motschenbacher, H. (2009). Speaking of the gendered body: The performative construction of commercial femininities and masculinities via body-part vocabulary. *Language in Society, 38,* 1–22.

Neff, J. (2008). P&G relies on power of mommy bloggers. *Advertising Age, 79,* 4–24.

Nowak, K. L. (2003). Sex categorization in computer-mediated communication (CMC): Exploring the utopian promise. *Media Psychology, 5,* 83–103.

Papacharizzi, Z. (2002). Presentation of the self in virtual life: Characteristics of personal home pages. *Journalism and Mass Communication Quarterly, 79,* 643–660.

Paul, B. (2009). Predicting Internet pornography use and arousal: The role of individual difference variables. *Journal of Sex Research, 46,* 344–357.

Pedersen, S., & Macafee, C. (2007). Gender differences in British blogging. *Journal of Computer-Mediated Communication, 12,* 1472–1492.

Peter, J., & Valkenburg, P. M. (2006). Adolescents' exposure to sexually explicit online material and recreational attitudes toward sex. *Journal of Communication, 56,* 639–660.

Peter, J., & Valkenburg, P. M. (2007). Who looks for casual dates on the Internet? A test of the compensation and the recreation hypotheses. *New Media & Society,* 9 (3), 455–474.

Pew Internet and American Life Project (2000, May). *Tracking online life: How women use the Internet to cultivate relationships with family and friends.* Retrieved July 2, 2009, from Pew Internet and American Life website: http://www.pewinternet.org

Philips, F., & Morrissey, G. (2004). Cyberstalking and cyberpredators: A threat to safe sexuality on the internet. *Convergence: The Journal of Research into New Media Technologies, 10,* 66–79.

Pittman, F. (1989). *Private lies: Infidelity and the betrayal of intimacy.* New York: Norton.

Plante, R. F. (2006). Sexual spanking, the self, and the construction of deviance. *Journal of Homosexuality,* 50(2/3), 59—79.

Podlas, K. (2000). Mistresses of their domain: How female entrepreneurs in cyberporn are initiating a gender power shift. *Cyberpsychology and Behavior, 3,* 847–854.

Rellstab, D. H. (2007). Staging gender online: Gender plays in Swiss Internet relay chats. *Discourse and Society, 18,* 765–787.

Remington, A. (2009).Gender-bending gamers dress for success. In The book of odds: The odds of everyday life. Retrieved July 2, 2009, from http://www.bookofodds.com/Daily-Life-Activities/Entertainment-Media/Articles/A0008-Gender-Bending-Gamers-Dress-for-Success

Ridgeway, C. L., & Correll, S. J. (2004). Unpacking the gender system: A theoretical perspective on gender beliefs and social relations. *Gender & Society, 18,* 510–531.

Ritchie, A., & Barker, M. (2006). "There aren't words for what we do or how we feel so we have to make them up": Constructing polyamorous languages in a culture of compulsory monogamy. *Sexualities, 9,* 584–601.

Rosen, L. D., Cheever, N. A., Cummings, C., & Felt, J. (2008). The impact of emotionality and self-disclosure on online dating versus traditional dating. *Computers in Human Behavior, 24,* 2124–2157.

Rothman, R. A. (1993). *Inequity and stratification: Class, color, and gender.* Englewood Cliffs, NJ: Prentice Hall.

Royal, C. (2005). A meta-analysis of journal articles intersecting issues of Internet and gender. *Journal of Technical Writing and Communication, 35*, 403–429.

Samp, J. A., Wittenberg, E. M., & Gillett, D. L. (2003). Presenting and monitoring a gender-defined self on the Internet. *Communication Research Reports, 20*, 1–12.

Sidelinger, R. J., Ayash, G., & Tibbles, D. (2008). Couples go online: Relational maintenance behaviors and relational characteristics use in dating relationships. *Human Communication, 11*, 341–356.

Stavrositu, C., & Sundar, S. S. (2008). Can blogs empower women? Designing agency-enhancing and community-building interfaces. In CHI 2008 Proceedings of the Conference on Human Factors in Computing Systems: Works in progress (pp. 2781–2786. Retrieved July 2, 2009, from http://delivery.acm.org/10.1145/1360000/1358761/p2781-stavrositu.pdf?key1= 1358761&key2=5505519621&coll=GUIDE&dl=GUIDE&CFID=80978056&CFTOKEN=37792974

Stefanone, M., & Jang, C. (2008). Writing for friends and family: The interpersonal nature of blogs. *Journal of Computer-Mediated Communication, 13*, 123–140.

Stern, M. J. (2008). How locality, frequency of communication, and Internet usage affect modes of communication within core social networks. *Information, Communication & Society, 11*, 591–616.

Stern, S. R. (2004). Expressions of identity online: Prominent features and gender differences in adolescents' world wide web home pages. *Journal of Broadcasting & Electronic Media, 48*, 218–243.

Tannen, D. (1990). *You just don't understand: Women and men in conversation.* New York: William Morrow.

Taylor, R. (2004, May 11). Is blog a masculine noun? *The Guardian.* Retrieved July 2, 2009, from http://politics.guardian.co.uk/comment/story/0,9115,1214393,00.html

Thomson, R. (2006). The effect of topic of discussion on gendered language in computer-mediated communication discussion. *Journal of Language and Social Psychology, 25*, 167–178.

Thompson, S. (2007). Mommy blogs: A marketer's dream. *Advertising Age, 78*, 6–11.

Toates, F. (2009). An integrative theoretical framework for understanding sexual motivation, arousal, and behavior. *Journal of Sex Research, 46*, 168–193.

Totilo, S. (2005, December 5). First film about French riots comes courtesy of a videogame. *MTV News.* Retrieved July 2, 2009, from http://www.mtv.com/news/articles/1517481/20051205/index.jhtml

Tufekci, Z. (2008). Grooming, gossip, Facebook and Myspace: What can we learn about these sites from those who won't assimilate? *Information, Communication & Society, 11*, 544–564.

U.S. Department of Commerce. (2004). *A nation online: Entering the broadband age.* Retrieved July 2, 2009, from http://www.ntia.doc.gov/reports/anol/NationOnlineBroadband04.htm

van Doorn, N., van Zoonen, L., & Wyatt, S. (2007). Writing from experience: Presentations of gender identity on weblogs. *European Journal of Women's Studies, 14*, 143–159.

Walker, B. (2009). Imagining the future of lesbian community: The case of online lesbian communities and the issue of trans. *Continuum: Journal of Media & Cultural Studies, 23*, 921–935.

Wei, L. (2009). Filter blogs vs. personal journals: Understanding the knowledge production gap on the Internet. *Journal of Computer-Mediated Communication, 14*, 532–558.

White, M. (2003). Too close to see: Men, women and webcams. *New Media & Society, 5*, 105–121.

Williams, D., Consalvo, M., Caplan, S., & Yee, N. (2009). Looking for gender: Gender roles and behaviors among online gamers. *Journal of Communication, 59*, 700–725.

Wood, J. (2009). *Gendered lives: Communication, gender, and culture.* Boston: Wadsworth.

Young, K. (2008). Internet sex addiction: Risk factors, stages of development, and treatment. *American Behavioral Scientist, 52*, 21–37.

Young, K., Griffin-Shelley, E., Cooper, A., O'Mara, J., & Buchanan, J. (2000). Online infidelity: A new dimension in couple relationships with implications for evaluation and treatment. *Sexual Addiction and Compulsivity, 7*, 59-74.

CHAPTER SEVENTEEN

Digital Deception in Personal Relationships

Norah E. Dunbar

Matthew Jensen

Trust and sincerity are the foundation on which close relationships are built. Paradoxically, it is that trust which allows deception to flourish because it prevents us from fully scrutinizing the messages from our relationship partners (Knapp, 2008; Guerrero & Floyd, 2006). Individuals lie to their closest relationship partners for a variety of reasons and the discovery of deception can have a profound impact on both the partners and the relationship itself. New technologies present an interesting challenge for deception researchers because the number of verbal and nonverbal cues available is more limited than when we are speaking in person. Whether it is a college student emailing her parents about her latest calculus test, a forward-deployed military officer using Skype to speak with his wife, or a salesperson in the field using her iPhone to communicate with her boss at the head office, deception in personal relationships is affected by the modality of the communication.

The use of computer-mediated communication (CMC) to talk with friends and family has changed the way we communicate in our personal relationships somewhat but not entirely. Although the advent of the Internet was both heralded as a revolutionizing force that would bring people together around the globe and a dystopian force that would segment and isolate people hampering face-to-face (FtF) relationships, the past few decades of research have not revealed CMC to meet either of these expectations (Boase & Wellman, 2006). Internet use is not associated with declines in contact with friends or family and often is used to enhance FtF relationships. In a summary of research on mobile phones, for example, users said their phone was a relationship-enhancing tool (Katz, Rice, Acord, Dasgupta, & David, 2004). It is a way for them to stay connected to their social networks, secure new relationships, and enhance sociality. Only one in seven users in a national survey said their mobile phone caused problems in their relationships (Katz et al., 2004). In fact, CMC can strengthen time spent with friends because it serves as a reminder of the relationship and can be used to schedule FtF meetings (Boase & Wellman, 2006).

In examining the effects of CMC on deception in personal relationships, three critical questions need to be addressed. Does communicating via computer mediation with your relationship partners make it more difficult to detect deception than communicating FtF? Are the motives behind deception different when communicating via CMC than FtF? What are the features of CMC that might be relevant to the discovery of deception in CMC? In answering these questions, we will review the available literature that examines the use of deception within CMC with an emphasis on personal relationships. Although there is relatively little literature on the use of deception in personal relationships via CMC, the chapter will marry the extensive literature on deception in personal relationships with the literature on deception in CMC to propose some future avenues for research in this area.

Deception in Context

Deception Defined

Scholars have defined *deception* in many different ways but generally agree that it entails deliberately inducing in another a belief the deceiver knows is false (DePaulo & DePaulo, 1989; Knapp, 2008; Masip, Garrido, & Herrero, 2004; Vrij, 2008). The communicator's *intention* is important because an act can be classified as "deceptive" if the person tries to give a false impression, regardless of whether the recipient of the communication is actually fooled (DePaulo & DePaulo, 1989). The deception may be successful or unsuccessful and may be delivered via verbal or nonverbal means. Although some scholars have distinguished *lies* from the more general term deception (Knapp, 2006), we follow the lead of other scholars and use the terms interchangeably (Vrij, 2008; Massip et al., 2004).

Theories of Deception Detection

In conceptualizing deception as a persuasive activity, Stiff (1995) argued that many theories of persuasion could be applied to the study of deception as well. For example, Anderson's (Anderson, 1973, 1981; Anderson & Graesser, 1976; Anderson, Lindner, & Lopes, 1973) information integration theory could be used to explain the fact that perceivers are generally fooled by deceptive messages when they are accompanied by sincere-looking demeanor because we give more weight to the easily manipulated visual cues than verbal or vocal ones. Similarly, Chaiken's (Chaiken, 1980, 1987; Chaiken, Liberman, & Eagly, 1989) heuristic-systematic processing model can be helpful in explaining deception because quick decision-rules (or heuristics) like "experts are usually correct" or "my partner is trustworthy" can short-circuit our ability to fully and systematically process a deceptive message. Vrij (2006) argues that we use several different heuristics to judge the messages of others and these heuristics lead to systematic errors and biases. For example, the *truth bias* is the general assumption that others are telling the truth and it is especially strong among familiar others. As the relationship between two people intensifies, they become more convinced that they can accurately detect their partners' lies and become more trusting that their partners do not or would not lie to them (Vrij, 2006). The *probing heuristic* is the tendency to believe a source is truthful if it has been probed. In fact, probing is not effective because deceivers often use probing as a way to gauge the suspicion of the receiver and adapt accordingly (Buller & Burgoon, 1996; Levine & McCornack, 2001).

Other theories have been developed specifically for the study of deception. The four-factor model of deception (Zuckerman, DePaulo, & Rosenthal, 1981; Zuckerman & Driver, 1985) asserts that deception is meant to be concealed and thus, unlike emotional expressions, should not have any distinctive verbal or nonverbal pattern (Zuckerman & Driver, 1985). On the contrary, deception detectors should look for unintentional sources of "leakage" that might give away the deceiver (Ekman & Friesen, 1969). They assert that four factors influence behavior and might contribute to leakage: attempted control, felt emotion, physiological arousal, and cognitive processing (Zuckerman et al., 1981). In order to avoid giving themselves away, deceivers typically attempt to control their verbal and nonverbal messages and thus often avoid gesturing naturally and sometimes appear stiff (Vrij, 2008). They also experience greater fear or anxiety (and sometimes excitement or delight) while deceiving and thus might embrace their lies to a lesser extent or show more stress or tension than truth-tellers (Ekman, 2001). The physiological arousal associated with deception results in changes in respiration, palmar sweating, and blood pressure which is what the polygraph is designed to test although the reliability of these cues are questionable because they are often signs of nervousness rather than deception per se (Iacono & Patrick, 2006; Vrij, 2008). Lying may be more cognitively demanding than telling the truth and certain types of lies, like complete fabrications, might be more cognitively demanding than other types, like omissions or half-truths (Vrij, 2006).

McCornack's (1992) information manipulation theory suggests that messages which are commonly thought of as deceptive derive from *covert* violations of Grice's (1989, 2002) conversational maxims or rules: quantity, quality, manner, and relation. Deceptive communication violates the quantity maxim when not all the available information is given to the receiver. The quality maxim is violated when the information is not truthful. The manner maxim is violated if the information is related indirectly or ambiguously and the relation maxim is violated if the information is not relevant. Although McCornack found that deceptive communicators violate the maxims more than truth-tellers, Stiff (1995) found that truthful people violate some of the maxims even more than deceivers. He questions whether we really do adhere to all of Grice's maxims in truthful interactions at all.

Finally, Buller and Burgoon's (1996) interpersonal deception theory offers a unique communication-based perspective of deception because it suggests that deception is interactive. Not only do deceivers experience the emotional and cognitive factors posited by Zuckerman et al.'s (Zuckerman et al., 1981; Zuckerman & Driver, 1985) four-factor theory, which might cause them to be more detectable, but the receiver's behavior in exhibiting suspicion is also manifest. The receiver's suspicion triggers behavioral adjustments on the part of the deceiver which makes them more difficult to detect. In fact, many of Burgoon et al.'s studies (Burgoon, Buller, White, Afifi, & Buslig, 1999; Burgoon & Qin, 2006; White & Burgoon, 2001) demonstrate that deception becomes more difficult to detect over the course of a lengthy interaction because deceivers adapt over time and take on the appearance of truth-tellers.

The Difficulties of Detecting Deception Accurately

A substantial body of literature has explored issues related to deception detection accuracy and the identification of specific cues that differentiate liars from truth-tellers (Bond & DePaulo, 2006; DePaulo, Lindsay, Malone, Muhlenbruck, Charlton, & Cooper, 2003; Vrij, Edward, & Bull, 2001). Despite the depth of this literature, extant research provided inconsistent findings about the ability of individuals to detect the deceiver (O'Hair, Cody, & McLaughlin, 1981; Vrij et al., 2001). Specific nonverbal cues, such as a lack of eye contact or foot tapping, are often thought to be associated with deception; however, few cues are reliable indicators of deception (Zuckerman, Driver, & Guadagno, 1985). Bond and DePaulo's (2006) meta-analysis examined 206 studies and found the average detection accuracy reported is only 54% (not far off from chance although statistically significant). In mediated interactions, research has found detection accuracy can be even lower, ranging from 8–20% (George & Marett, 2005).

Verbal and nonverbal cues. Recent meta-analyses (DePaulo et al., 2003; Sporer & Schwandt, 2006, 2007) and an exhaustive review of the literature by Vrij (2008) suggest there are few reliable verbal and nonverbal cues to deception. Although individual nonverbal cues have not been found to be reliable indicators on their own, a pattern of deceivers' communication style has emerged. It encompasses three groupings of behaviors. First of all, liars are more tense than truth-tellers which is revealed in behaviors like higher pitch, more pupil dilation, nervousness, vocal tension, lip pressing, and overall facial unpleasantness. Second, lying is more cognitively complex than telling the truth and so deceivers typically pause longer, wait longer to answer, use fewer illustrator gestures that accompany speech, use fewer hand/finger movements, fewer leg/foot movements, and more repetitions. Third, liars tend to endorse their lies less than truth tellers and so they appear more ambivalent, less involved, and more uncertain (Vrij, 2008). Again, it is the constellation of behaviors that make up each pattern that appears to be diagnostic rather than individual cues themselves.

There are fewer studies which analyze the verbal messages themselves of deceivers and truth-tellers in FtF communication. In DePaulo et al.'s (2003) meta-analysis, they found over 100 nonverbal cues that have been examined in the extant research but only 30 verbal cues. However, trends have emerged for the use of more negative statements such as using denials or communication that indicate depressed affect or negative mood, more generalizing terms such as "always," "never," or "nobody," fewer self-references such as using the terms "I," "me," or "mine," a lack of reasonableness or plausibility, and a lack of directness or immediacy (DePaulo et al., 2003; Vrij, 2008). Overall, the verbal and vocal cues tend to be more diagnostic of deception than the nonverbal cues because the nonverbal cues are more easily manipulated by deceivers which is consistent with the leakage hypothesis that posits that the least controlled channels are the best indicators of deceit (Ekman & Friesen, 1969).

Humans as poor lie detectors. Vrij (2006; 2008) outlines several reasons why we are poor lie detectors: (a) behaviors have different meanings in different contexts and in association with other behaviors, (b) competent liars alter their behavior and adapt to suspicion when it is apparent, and (c) liar behavior changes when the situation changes so the type of lie, the motivation behind the lie, the context in which the lie is told, the relationship of the deceiver to the lie target, and the severity of the lie can all change the liar's behavior. Often, we fall prey to myths and stereotypes and are duped by innocent-looking demeanor when the real clues to deception go unnoticed. Bond and his Global Deception Research Team (2006) found that around the world, the stereotype of the gaze avoidant, nervous, and fidgeting liar persists across many cultures despite the fact that research has disputed this stereotype for decades. In a comparison between the objective cues known to be diagnostic of deception and the subjective beliefs of laypersons and professional lie detectors, Vrij (2008) found that of the ten diagnostic cues, people use only three of them and of the eight cues people say they don't rely on, five are diagnostic.

Furthermore, in our close relationships, it might often be the case that when we suspect deception is occurring we choose to overlook it for the sake of the relationship or because our partner's altruistic lies are preferable to the truth. Do we really want our partners to be brutally honest about our appearance or cooking skills? Sometimes we prefer not to detect deception. McCornack and Levine (1990a) suggest that even in the case of serious lies, such as infidelity, a partner's suspicion might be counteracted by the fervent hope that their suspicions are unfounded. They will wait until they have definitive proof of the infidelity before confronting the partner. Even in the face of an accusation, though, deceivers may opt to add to their story or continue denying they were lying if they have high confidence they can succeed in the deception (Boon & McLeod, 2001) which raises more doubt and prevents the deception from being detected.

How deception in CMC differs from other channels

When considering the influence of CMC on deceptive communication in close relationships, one might assume that restricting the cues available might make it more difficult to detect deception. But, perhaps the overwhelming sincerity of the physical demeanor makes it more difficult to detect lies than if you were using audio-only or perhaps even simply text. The effects of CMC on deceptive communication are complex and at times competing. For example, individuals are more honest and disclose more about themselves in an anonymous online setting than when they are identified (Whitty & Joinson, 2009), yet anonymity has also made possible a host of fraudulent schemes on the Internet (Grazioli & Jarvenpaa, 2003). As the world moves towards increasingly digital forms of communication, the effects of CMC on deception must be dissected. However, before discussing the effects of media on deceptive communication, it is helpful to carefully define types of deception that can take place in CMC.

Deception that occurs in CMC has been termed *digital deception* and takes place in a technologically mediated interaction, meaning that the messages must be exchanged in a medium other than FtF (Hancock, 2007). Although most individuals may think digital deception occurs solely in text-based, web environments, it also incorporates media such as telephone and video conferencing.

Types of digital deception. Digital deception has been recently characterized by two general categories: identity-based deception and message-based deception (Hancock, 2007). First, identity-based deception generally concerns qualities and characteristics of the deceiver. Just as Tom Hanks concealed his true identity from his online friend in the film *You've Got Mail*, it is relatively easy to disguise oneself in an environment where you may only be known by a handle or a screen name. In digital media such as text-based messages, individuals have a tremendous amount of control over how they represent themselves (Walther, 1996). For example, lies about gender, age, and race (Berman & Bruckman, 2001) and height, weight, and socio-economic status (Hancock, Toma, & Ellison, 2007; Whitty, 2002) are easily accommodated. Another example of identity deception is the never ending waves of phishing attacks and other scams where the sender tries to represent a legitimate source but is, in fact, attempting to defraud the receiver.

An early investigation of identity-based deception was performed by Donath (1999) who investigated online Usenet groups. She observed that a person could have many virtual personas and that others determine one's true identity by examining assessment signals (things you are; e.g., physical characteristics) and conventional signals (things you have, do, or say; e.g., online personal profile). Assessment signals are costly to attain and relate directly to a trait or characteristic. For example, muscular arms on a man may signal his strength. Conventional signals are commonly associated with assessment signals, but do not require the same high cost. In a virtual environment, the majority of information that is exchanged falls under the category of conventional signals and this opens the door to deception. For example, a man who desires to appear strong to others may use the user name "StrongMan1." The user name is a conventional signal of strength that costs little and is not evidence of actual strength. As few assessment signals are available in Usenet groups, one could manipulate the conventional signals to give a false impression. Donath (1999) describes several types of identity-based deception including trolling (posing as a group member and sparking unwanted debate), identity concealment, and category deception (attempts to influence Usenet members' perceptions of certain social groups).

The separation of physical presence from text-based assertions permits deception about a range of topics as there is often no way for the receivers to verify the assertions. Examples of identity-based deception abound on the Internet. In the online dating environment, individuals have been shown to frequently be deceptive about their physical and socio-economic characteristics (Hancock et al., 2007; Whitty, 2002; Whitty & Carville, 2008). In anonymous chat rooms, participants can choose who they want to be. In email phishing scams, email recipients are informed that they have just won the lottery or that they will be handsomely compensated if they assist a foreign dignitary in distress (Whitty & Joinson, 2009). Even skilled Internet users fall prey to identity deception when distinguishing between valid websites and fraudulent websites (Grazioli & Jarvenpaa, 2000).

Identity-based deception typically occurs when the deceiver is unknown to the receiver and the receiver has little opportunity to verify the deceiver's claims. However, a great many of the interactions that occur in CMC involve people who know each other. For example, an employee emailing an employer, two friends instant messaging, and a husband calling his wife are all examples where identity-based deception is unlikely. Rather in these cases, the second form of digital deception may take place: message-based deception. In these cases, deceptive messages do not concern who is sending the message, rather the deception is in the content of the message (Hancock, 2007). So, the employee may be telling her boss that she likes the new marketing plan her boss came up with when she really thinks it will fail. These messages are more analogous to deceptive messages in a FtF interaction and yet, modality of communication affects these messages too.

Digital Deception in Personal Relationships

The Incidence of Deception

Most people believe honesty is an essential ingredient for a successful relationship. In studies of attitudes towards honesty, trustworthiness or honesty are often rated as the most desirable traits in a potential romantic partners (see Knapp, 2006, for a review). Although we generally espouse the belief that trust and honesty are valuable in our close relationship partners, the fact remains that lying is an unavoidable part of everyday life. In an informal study conducted by O'Hair and Cody (1994), only 20.8% of their survey respondents agreed with the statement that "one should never lie" in close relationships whereas 59% of respondents indicated that deception was a prevalent communication strategy or that it had its place in interpersonal communication. Boon and McLeod's (2001) survey respondents pointed out that although they believe in honesty, there are many situations in dealing with close relationships where honesty is not the best policy—that in some circumstances deception is not only necessary but the ethical choice. In a diary study of both students and community members, lies were told in all sorts of close relationships including spouses, best friends, family, children, and nonspouse romantic partners (DePaulo & Kashy, 1998). Lies were told in 1 out of every 10 interactions but the closest relationships (i.e., spouses) had the fewest lies. Other research suggests that spouses use outright fabrications less than other close relationships (Metts, 1989).

The increased usage of communication technologies has prompted several groups of researchers to investigate the incidence of digital deception as well. DePaulo et al. (1996) conducted one of the first studies that compared the prevalence of deception in writing, telephone, and FtF messages in student and community populations. They discovered that across modalities 31% of social interactions involving students contained deception and 20% of social interactions of community members contained deception. The largest number of lies was told in person; however, they also noted that the FtF modality also had the most overall messages. When messages containing lies were examined as a proportion of total messages, the telephone modality had the largest proportion of messages containing lies.

In a diary study that was intended to replicate the study conducted by DePaulo et al. (1996), Hancock, Thom-Santelli, and Ritchie (2004) found that across modalities, 26% of social interactions contained a lie. Participants told the greatest percentage of lies during telephone conversations (37%). This was followed by FtF (27%), instant messaging (21%), and email (14%). In further replications, George and Robb (2008) conducted two additional diary studies. In the first study, 25.2% of all interactions contained lies and in the second, 21.9% of interactions contained lies. In both studies, no significant differences were reported in the proportions of lies told in the FtF, telephone, instant messaging, and email modalities.

The Motives for Deception

Typically, lies are considered to be either self-serving or other-serving (altruistic). Self-serving lies are told to avoid punishment and blame, make ourselves more desirable, protect our privacy, or attain instrumental goals (Anderson, Ansfield, & DePaulo, 1999). Deception is also used as a method of conflict avoidance which might benefit both the self or other (Peterson, 1996). Altruistic lies or "white lies" are seen as less serious because they are told for the benefit of another person or to follow politeness norms (Peterson, 1996). When asked why they lie to those they are closest to, most people report that their lies are altruistic (DePaulo & Kashy, 1998). They say their lies are meant to protect their partners' feelings, build or maintain their partners' self-esteem, show concern for their partners, and to help their partners attain their goals (Knapp, 2006). People also lie to avoid confrontation or conflict especially on sensitive topics such as sexual desirability (Marelich, Lundquist, Painter, & Mechanic, 2008; Williams, 2001). In diary studies of everyday lies (DePaulo & Kashy, 1998; DePaulo, Kashy, Kirkendol, Wyer, & Epstein, 1996), less than 20% of the lies were told for personal advantage.

However, those who are the recipients of those lies do not necessarily describe them as altruistic. In a study comparing the narratives of lie-tellers to lie-receivers,

Kaplar and Gordon (2004) found that lie-tellers were 11 times more likely to reference altruistic motivations than lie-receivers. Lie-receivers saw the lies as more serious and more self-motivated than lie-tellers. O'Sullivan (2008) argues that the fact that liars experience guilt about telling lies, despite labeling them altruistic, represents a form of self-deception on the part of deceivers.

These "everyday" lies should be distinguished from the more "serious" lies which are told relatively rarely but can have catastrophic implications for a close relationship. In a study where community members and college students were asked to write narratives about serious lies they had told or been told, most serious lies were told to cover up bad behavior and relatively few were told to altruistically protect the partner (DePaulo, Ansfield, Kirkendol, & Boden, 2004). "In the community sample, 94.4% of the lies were self-serving and 5.6% were other-oriented, and in the college sample 85.9% of the lies were self-serving and 14.1% of the lies were other-oriented" (DePaulo et al., 2004, p. 157).

Deception in CMC mirrors many of the same motives for deception in FtF interactions. People lie online for altruistic and self-serving purposes; however, self-serving lies have received more attention from researchers. In a rare study involving altruistic lies in CMC, Whitty and Carville (2008) found that altruistic lies are generally told to familiar individuals, while self-serving lies are told more frequently to unfamiliar individuals. Interestingly, Whitty and Carville also note that individuals in their study felt more compelled to tell altruistic lies than self-serving lies, suggesting an elevated desire to protect those with whom they are familiar from any potential pain the truth might cause.

Of the self-serving lies in CMC, one of the more accessible categories involves identity deception. Whitty (2002) showed that in online chat rooms, deception was pervasive and that men were more likely to lie than women. Men were more likely to lie about their occupation, education, and income. In Internet dating sites, deception was also frequent with individuals attempting to appear more attractive by deceiving about their physical description (Hancock et al., 2007; Whitty & Joinson, 2009). In her exploration of possible motives behind common forms of digital deception, Utz (2005) found that deception about attractiveness was associated most with idealized self-presentation, category deception (gender switching) was associated with playing with new roles of the self, and identity concealment was associated most with privacy concerns. Of the three types of deception examined by Utz, attractiveness deception was judged to be the most serious. Similar to privacy concerns, Whitty (2002) also found that women were more likely to lie for safety reasons. An additional motivation for digital deception is financial enrichment through fraud. Grazioli and Jarvenpaa (2003) noted that from the years 1995–2000, the incidence of fraud online is keeping pace with the number of U.S. and worldwide users of the Internet.

Deception Detection in Personal Relationships and CMC

Several researchers have investigated whether it is easier or more difficult to detect deception in our close relationships than with strangers but the answer is somewhat muddy. On the one hand, when we are close to someone, we know what their "normal" baseline behavior looks like and so we can recognize when our partners deviate from that norm. The more we know about our partners, the less we rely on stereotypical cues and instead rely on deviations from our partners' normative baseline through inconsistency or pattern violations (Helme, 2002; Henningsen, Cruz, & Morr, 2000). Additionally, it may be more difficult for our partners to lie because we are privy to their social networks and their personal history that might make it more difficult for them to concoct a plausible lie (Guerrero & Floyd, 2006). This corresponds with the finding that most deception in close relationships is not discovered via nonverbal cues in the heat-of-the-moment but rather through third-party information, confessions, or other means sometimes weeks or months later (Park, Levine, McCornack, Morrison, & Ferrera, 2002).

On the other hand, our truth bias is much stronger in our close relationships because we have already established trust with them and so we are not expecting them to lie to us. In fact, we may be so confident that our partners would never dare lie to us because we would surely detect it, that we may overlook telltale signs of deception (Vrij, 2006). This is in direct contrast to the finding that deceivers also have high confidence that they can lie to their partners and get away with it and that people believe they are more successful at deceiving their partners than their partners are at deceiving them (Boon & McLeod, 2001; Cole, 2001). Peterson (1996) reports that since most people claim they are telling white lies to their partners, they assume their partners are doing the same in return. Cole (2001) argues that this reciprocity effect is illusory because partners rarely are matched in their deceptiveness. Some research has suggested that a moderate amount of suspicion or heightened vigilance when evaluating messages improves our ability to detect deception and counteracts the truth bias (Levine, Feeley, McCornack, Hughes, & Harms, 2005; McCornack & Levine, 1990b) but the link between familiarity and deception detection ability remains elusive.

Some scholars (Guerrero & Floyd, 2006; Miller & Stiff, 1992) have suggested that perhaps the relationship between familiarity and deception detection ability is curvilinear. It is possible that those with moderate familiarity (such as acquaintances) will be more accurate than those who are strangers or those who are very familiar (such as spouses). The rationale behind this perspective is that those who are very close suffer from information overload, have a heightened truth bias, or have become fatigued or bored with their interactions and so lose their ability to remain vigilant to potential deception cues. Those who are unfamiliar with the deceiver are lacking the baseline of normal behavior and so do not have relevant comparable truths to measure the deceptive statement against. Comparably, those who are moderately familiar have the baseline of normal behavior but are not overly trusting and do not have so much behavioral or informational knowledge about the deceiver that they are overloaded. This curvilinear hypothesis requires more testing in future empirical investigation.

When modality of the interaction is added to the equation, the implications for the detection of digital deception are also not very clear. One might conclude from the four-factor model of deception formulated by Zuckerman et al. (1981), that deception would be more difficult to detect in CMC as the amount of nonverbal leakage would be curtailed. Although, in message-based deception, detection performance has been generally poor (George & Marett, 2005), there are mixed results concerning detection in richer and leaner modalities. Heinrich and Borkenau (1998) argued and demonstrated that deception will be more apparent in richer modalities as inconsistencies between modalities become evident during deception. However, Bond and DePaulo (2006) in their meta-analysis examining detection accuracy find that receivers are more accurate at detecting audible than visible lies. When comparing deception detection accuracy rates of interactive chat and FtF groups, George and Marett (2004) uncovered no differences across modalities. Still others have demonstrated that receivers find deceivers most credible when receivers have access to nonverbal cues as in a recorded video, as opposed to only audio or a transcription (Burgoon, Blair, & Strom, 2008).

Burgoon, Chen, and Twitchell (2005) also showed that receivers found deceivers more credible when the deceivers used synchronous CMC as opposed to asynchronous CMC. Consistent with interpersonal deception theory, Burgoon, Bonito, Ramirez, Dunbar, Kam, and Fischer (2002) proposed the principle of interactivity that suggests that human communication processes like credibility assessment vary systematically with the degree of interactivity that is afforded by a communication modality. Interactive media that promote mutuality and involvement lead to higher perception of a partner's credibility (Burgoon, Bonito, Bengtsson, Ramirez, Dunbar, & Miczo, 1999). Interestingly, receivers tend to prefer richer media (Carlson & George, 2004; Kahai & Cooper, 2003) and a familiar partner (Carlson & George, 2004) when detecting potential deception.

To explain effects of modality on message-based deception, several competing hypotheses have been forwarded. The first is media richness theory (Daft & Lengel, 1986; Daft, Lengel, & Trevino, 1987) which was developed in managerial settings and predicts that for complex, equivocal tasks such as deception, richer media will be preferred. Richer media provide instant feedback, multiple sets of cues, language variety, and the capability to personalize the message. Daft and colleagues (1987) supported this claim with numerous interviews and reviews of interactions from actual managerial work where managers preferred FtF interaction for sharing complex messages.

A competing hypothesis was forwarded by DePaulo et al. (1996), which suggested that deceivers were more likely to use distant modalities because deception can make deceivers feel uneasy. Therefore distance between the deceiver and receiver should be preferred by deceiver. This hypothesis has been termed the social distance hypothesis (Hancock, 2007; Hancock et al., 2004) and predicts that deceivers will favor leaner media when deceiving. Evidence supporting this hypothesis was provided by DePaulo et al. (1996) who conducted a diary study where participants recorded their deceptions for one week. They showed that deceivers used the telephone more often than FtF to deceive others. An exception to their prediction was that written communication, which also allows distance between the deceiver and receiver, did not contain more deception than FtF.

Hancock and colleagues note that both of these early attempts to explain modality differences during deception explore only a single feature (i.e., richness or distance), while CMC provides a large number of capabilities that have significant implications for deceivers (Hancock, 2007; Hancock et al., 2004). For example, a deceptive email leaves a record that the deceiver may not want and therefore may choose a medium that is recordless.

There have been several attempts to examine mediated deception from a feature-based perspective. Hancock et al. (2004) identified three media features that may affect the incidence of deception in different media: synchronicity, recordability, and physical distribution of interactants. They point to research demonstrating that deception occurs spontaneously (c.f. DePaulo et al., 1996) and posited that as a result deception would be more prevalent in more synchronous media. They also suggested that deceivers will be more hesitant to deceive in media which are easily recorded such as email or instant messaging. Finally, they suggest that deceivers who are distributed may more easily deceive about topics such as where they are and what they are doing. Hancock et al. suggested that based on their feature-based model, the telephone would be preferred the most by deceivers, followed by FtF, then instant messaging and finally email. Using a diary study patterned after DePaulo et al.'s (1996) work, this was shown to be the case (Hancock et al., 2004).

By building on media synchronicity theory (c.f. Dennis & Valacich, 1999), Carlson, George, Burgoon, Adkins, and White (2004) also developed a feature-based model of deception in different media; however, their primary focus was on deception effectiveness. Consistent with the principle of interactivity and interpersonal deception theory, synchronicity was hypothesized to improve deception success as deceivers would garner greater involvement and mutuality during their interactions. Consistent with Hancock et al.'s (2004) notion of recordability, Carlson et al.'s (2004) reprocessability was hypothesized to hamper deception success as receivers would be able to review past messages a deceiver had sent and discover discrepancies. Carlson et al. (2004) suggested that higher levels of symbol variety, tailorability, and rehearsability also increase the likelihood of deception success. They also posited that deception will be most successful if the deceiver uses media that afford high levels of social presence. Further, they suggested that a high level of communication ability within a particular medium and that experience with message, medium, or context will also make deception more difficult to detect. This view offers deceivers a complex choice of media where each medium has benefits and drawbacks and numerous factors external to the structural affordances of the medium (e.g., motivation, experience in the context of the deception) come into play.

Research comparing these views of deception in CMC has uncovered mixed results. The results of the DePaulo et al. (1996) and Hancock et al. (2004) diary studies are consistent with the feature-based view of deception as the highest proportion of deception occurred when participants used the telephone. Further, in survey responses, university employees reported they preferred highly synchronous media for deception (Carlson & George, 2004). However, in further diary study replications involving more media (e.g., texting), George and Robb (2008) found no significant differences in the proportion of lies told between the media. Other researchers have shown support for the social distance hypothesis by demonstrating that individuals prefer to send self-serving lies over email and have found no effect for other-serving lies (Whitty & Carville, 2008).

In an attempt to more effectively identify deception when it occurs in CMC, researchers have begun to investigate the linguistic properties of CMC messages. Using linguistic cues (such as lexical diversity, word counts, affective words, etc.), researchers have demonstrated that deceivers can be identified by textual features in email (Zhou, Burgoon, Nunamaker, & Twitchell, 2004) and chat (Hancock & Dunham, 2001; Zhou, 2005). These features are automatically extracted from text-based media using computer-based text mining and the features are not consistently diagnostic across studies. However, within a given context, such computer-based tools may offer hope of improving detection accuracy. Likewise, in richer modalities such as FtF, other computer-based tools that consider multiple channels (e.g., kinesic, linguistic, and vocalic behavior) may potentially assist receivers in detecting deception when it is present (Jensen, Meservy, Burgoon, & Nunamaker, in press; Meservy, Jensen, Kruse, Twitchell, Tsechpenakis, Burgoon, 2005).

The Effects of Discovered Deception on Close Relationships

Once we discover the deception, how do we react? What are the implications of discovered deception for the future of our close relationships? The consequences often depend on an appraisal of the motivation of the deceiver by the recipient, the level of commitment in the relationship, and also the severity of the deception. DePaulo et al. (2004) summarize three consequences for relationships that emerged from their diary studies: a lack of trust, a reduction in closeness, and an increase in guardedness. Negative consequences can be found for the target of the lie (hurt, suspicion, lowered self-esteem, desire for revenge), the liar (loss of credibility, trust, respect), or the relationship itself (lowered satisfaction and commitment, tension) (Knapp, 2006). Even undiscovered deception can have negative consequences for relationships because the deceiver views the recipient as less honest as a result of the liar's own deception. This fuels a downward spiral of suspicion that affects the trust and intimacy in the relationship as a whole (Sagarin, Rhoads, & Cialdini, 1998). Peterson (1996) found that those who self-reported using the most lies (other than white lies) were the least satisfied with their relationships. In cases where the deception is especially severe such as in the case of sexual betrayal, the relationship is often terminated (Feldman & Cauffman, 1999; McCornack & Levine, 1990a).

However, positive consequences can emerge if the result of the discovered deception is that the couple addresses a problem that had long been ignored. Relational partners are most likely to see improvements in their relationship quality when prosocial repair strategies are used such as apologizing, soothing the partner, invoking the importance of the relationship, or working on strengthening the relationship (Aune, Metts, & Ebesu Hubbard, 1998). The recipients of deception or other betrayals are most likely to forgive the deceiver when they are committed to continuing the relationship or are willing to think about the betrayal in benevolent terms by giving their partner the benefit of the doubt (Finkel, Rusbult, Kumashiro, & Hannon, 2002). Knapp (2006) suggests that close relationships can survive even the most serious lies when there is (a) a positive history with the liar and the couple has a history of trust in their relationship, (b) the deceiver makes sincere pleas for forgiveness and makes attempts to rebuild the trust that has been compromised, (c) the couple communicates openly about why the deception happened, (d) the deception serves as a reminder to couple that the relationship is worth saving, and (e) the couple engages in mutual problem solving that can strengthen the relationship bonds. These are strategies that can be used no matter what medium is being used to communicate.

Conclusions

Although there is very little research on digital deception in close relationships, we believe that by marrying the vast literature on deception in close relationships and the burgeoning literature on mediated deception, we are able to draw some tentative conclusions that may spur future researchers in continuing in this line of research. We began this chapter by asking three questions to guide our inquiry and we will attempt to provide some answers to them with the understanding that more research is needed in this area.

First, we asked whether communicating via CMC would make it more difficult to detect deception than communicating FtF in personal relationships. We believe it does not even though there may be evidence on both sides of this question. Our personal relationships are based on trust and our ability to make ourselves vulnerable to our partners regardless of what medium we use to communicate with them. Thus, the truth bias appears to be just as applicable in CMC modalities as in FtF when we are speaking to our partners whom we know and love. However, there are four reasons why it might be *easier* to detect deception in mediated environments. First, the reduced attention on nonverbal cues in CMC makes it more likely that receivers will not be misled by honest-looking demeanor and may pay attention to the more diagnostic verbal cues. Second, the fact that a record of interactions and the asynchronous nature of some modalities (such as email) means that receivers can compare messages over time and look for inconsistencies between messages or between senders. Third, the ability of the receiver to hide their suspicion may be increased due to the lack of nonverbal cues which should reduce the deceivers' ability to adapt and enhance their credibility as IDT would predict (Buller & Burgoon, 1996). Although Burgoon and Qin (2006) found that the diagnostic ability of text-based cues fades over time, if skilled receivers can successfully gather information without giving away their suspicions, it may make deception easier to detect. Fourth, channel constriction might reduce the cognitive load of the receiver which allows them to spend more time scrutinizing the behavior of the deceiver. This is consistent with previous research that has suggested that third-party observers are better able to detect deception than participants who are immersed in a conversation (Dunbar, Ramirez, & Burgoon, 2003).

Our second question asked whether the motives behind deception were different when communicating via CMC than FtF. We believe the answer to this question is "no." In close relationships, most deceivers report the reason for their deception is altruistic. They report that their lies are primarily "white lies" told for the sake of their partners—to protect their partners from hurt, to save their partners' face or self-esteem, or to make their partners happy. The exception is the relatively rare "serious" lies which are told primarily for self-interested reasons such as to protect oneself from punishment. In CMC, the motives appear to be the same. We generally tell self-interested lies to strangers and other-oriented lies to our friends and romantic partners. Like Boase and Wellman (2006) who conclude that our relationships have not changed much since we began communicating via CMC, we agree that our motives for deception have not changed much as a result of our mode of communication either.

Our final question asked what features of CMC might be relevant to the discovery of deception in CMC. There seem to be features of CMC that are relevant during digital deception. According to the feature-based view of digital deception, deceivers should select media that are synchronous, recordless, and allow interactants to be physically distributed. Synchronicity increases the number of messages between sender and receiver and thus the interactivity of synchronous media should make messages seem more believable to the receivers (Burgoon, Bonito, et al., 1999). Recordless media should prevent receivers from reviewing past messages from deceivers, thus hampering detection of past deception (Carlson et al., 2004; Hancock et al., 2004). Physical separation of receiver and deceiver allows lies to be told about the deceiver's location and activities (Hancock et al., 2004) and reduces the inhibitions to sending self-oriented lies (Whitty & Carville, 2008). All of these features favor the deceiver during digital deception. However, as listed above, there are also several CMC features that may favor the receiver in detecting deception. The net result of these opposing effects of CMC is unclear and, as Carlson et al. (2004) suggest, a host of factors beside the structural capabilities of the communication medium are likely to be influential in affecting a receiver's ability to detect deception in CMC. This area of inquiry is in great need of additional systematic research.

References

Anderson, D. E., Ansfield, M. E., & DePaulo, B. M. (1999). Love's best habit: Deception in the context of relationships. In P. Philippot, R. S. Feldman, & E. J. Coats (Eds.), *The social context of nonverbal behavior.* (pp. 372–409). New York: Cambridge University Press.

Anderson, N. H. (1973). Information integration theory applied to attitudes about US presidents. *Journal of Educational Psychology, 64*(1), 1–8.

Anderson, N. H. (1981). *Foundations of information integration theory*. San Diego, CA: Academic.

Anderson, N. H., & Graesser, C. C. (1976). An information integration analysis of attitude change in group discussion. *Journal of Personality and Social Psychology, 34*(2), 210–222.

Anderson, N. H., Lindner, R., & Lopes, L. L. (1973). Integration theory applied to judgments of group attractiveness. *Journal of Personality and Social Psychology, 26*(3), 400–408.

Aune, R. K., Metts, S., & Ebesu Hubbard, A. S. (1998). Managing the outcomes of discovered deception. *Journal of Social Psychology, 138*(6), 677–689.

Berman, J., & Bruckman, A. S. (2001). The Turing game: Exploring identity in an online environment. *Convergence, 7*(3), 83.

Boase, J., & Wellman, B. (2006). Personal relationships: On and off the Internet. In A. L. Vangelisti & D. Perlman (Eds.), *The Cambridge handbook of personal relationships* (pp. 709–723). New York: Cambridge University Press.

Bond, C. F., & DePaulo, B. M. (2006). Accuracy of deception judgments. *Personality and Social Psychology Review, 10*(3), 214–234.

Boon, S. D., & McLeod, B. A. (2001). Deception in romantic relationships: Subjective estimates of success at deceiving and attitudes toward deception. *Journal of Social and Personal Relationships, 18*(4), 463–476.

Buller, D. B., & Burgoon, J. K. (1996). Interpersonal deception theory. *Communication Theory, 6*, 203–242.

Burgoon, J. K., Blair, J. P., & Strom, R. E. (2008). Cognitive biases and nonverbal cue availability in detecting deception. *Human Communication Research, 34*(4), 572–599.

Burgoon, J. K., Bonito, J. A., Bengtsson, B., Ramirez, A., Dunbar, N. E., & Miczo, N. (1999). Testing the interactivity model: Communication processes, partner assessments, and the quality of collaborative work. *Journal of Management Information Systems, 16*(3), 33–56.

Burgoon, J. K., Bonito, J. A., Ramirez, A., Dunbar, N. E., Kam, K., & Fischer, J. (2002). Testing the interactivity principle: Effects of mediation, propinquity, and verbal and nonverbal modalities in interpersonal interaction. *The Journal of Communication, 52*(3), 657–677.

Burgoon, J. K., Buller, D. B., White, C. H., Afifi, W., & Buslig, A. L. S. (1999). The role of conversational involvement in deceptive interpersonal interactions. *Personality and Social Psychology Bulletin, 25*(6), 669–685.

Burgoon, J. K., Chen, F., & Twitchell, D. P. (2005). *Deception and Its Detection under Synchronous and Asynchronous Computer-Mediated Communication.* Paper presented at the International Communication Association Conference, New York.

Burgoon, J. K., & Qin, T. (2006). The dynamic nature of deceptive verbal communication. *Journal of Language and Social Psychology, 25*(1), 76–96.

Carlson, J. R., & George, J. F. (2004). Media appropriateness in the conduct and discovery of deceptive communication: The relative influence of richness and synchronicity. *Group Decision and Negotiation, 13*(2), 191-210.

Carlson, J. R., George, J. F., Burgoon, J. K., Adkins, M., & White, C. H. (2004). Deception in computer-mediated communication. *Group Decision and Negotiation, 13*(1), 5-28.

Chaiken, S. (1980). Heuristic versus systematic information processing and the use of source versus message cues in persuasion. *Journal of Personality and Social Psychology, 39*(5), 752-766.

Chaiken, S. (1987). The heuristic model of persuasion. In M. P. Zanna, J. M. Olson, & C. P. Herman (Eds.), *Social Influence: The Ontario Symposium* (Vol. 5, pp. 3-39). Hillsdale, NJ: Lawrence Erlbaum.

Chaiken, S., Liberman, A., & Eagly, A. H. (1989). Heuristic and systematic information processing within and beyond the persuasion context. In J. S. Uleman & J. A. Bargh (Eds.), *Unintended thought* (pp. 212-252). New York: Guilford.

Cole, T. (2001). Lying to the one you love: The use of deception in romantic relationships. *Journal of Social and Personal Relationships, 18*(1), 107-129.

Daft, R. L., & Lengel, R. H. (1986). Organizational information requirements, media richness and structural design. *Management Science, 32*(5), 554-571.

Daft, R. L., Lengel, R. H., & Trevino, L. K. (1987). Message equivocality, media selection, and manager performance: Implications for information systems. *MIS Quarterly, 11*(3), 355-366.

Dennis, A. R., & Valacich, J. S. (1999). *Rethinking Media Richness: Towards a Theory of Media Synchronicity*. Paper presented at the Proceedings of the 32nd Hawaii International Conference on System Sciences.

DePaulo, B. M., Ansfield, M. E., Kirkendol, S. E., & Boden, J. M. (2004). Serious lies. *Basic and Applied Social Psychology, 26*(2), 147-167.

DePaulo, B. M., & Kashy, D. A. (1998). Everyday lies in close and casual relationships. *Journal of Personality and Social Psychology, 74*(1), 63-79.

DePaulo, B. M., Kashy, D. A., Kirkendol, S. E., Wyer, M. M., & Epstein, J. A. (1996). Lying in everyday life. *Journal of Personality and Social Psychology, 70*(5), 979-995.

DePaulo, B. M., Lindsay, J. J., Malone, B. E., Muhlenbruck, L., Charlton, K., & Cooper, H. (2003). Cues to deception. *Psychological Bulletin, 129*(1), 74-118.

DePaulo, P. J., & DePaulo, B. M. (1989). Can deception by salespersons and customers be detected through nonverbal behavioral cues? *Journal of Applied Social Psychology, 19*(18), 1552-1577.

Donath, J. S. (1999). Identity and deception in the virtual community. In M. A. Smith, P. Kollack, & I. Heywood (Eds.), *Communities in Cyberspace* (pp. 29-59). London: Taylor & Francis.

Dunbar, N. E., Ramirez, A., Jr., & Burgoon, J. K. (2003). The effects of participation on the ability to judge deceit. *Communication Reports, 16*(1), 23-34.

Ekman, P. (2001). *Telling lies: Clues to deceit in the marketplace, politics, and marriage*. New York: W.W. Norton.

Ekman, P., & Friesen, W. V. (1969). Nonverbal leakage and clues to deception. *Psychiatry: Journal for the Study of Interpersonal Processes, 32*(1), 88-106.

Feldman, S. S., & Cauffman, E. (1999). Sexual betrayal among late adolescents: Perspectives of the perpetrator and the aggrieved. *Journal of Youth and Adolescence, 28*(2), 235-258.

Finkel, E. J., Rusbult, C. E., Kumashiro, M., & Hannon, P. A. (2002). Dealing with betrayal in close relationships: Does commitment promote forgiveness? *Journal of Personality and Social Psychology, 82*(6), 956-974.

George, J. F., & Marett, K. (2004). *Inhibiting deception and its detection*. Paper presented at the Proceedings of the 37th Hawaii International Conference on System Sciences.

George, J. F., & Marett, K. (2005). Deception: The dark side of e-collaboration. *International Journal of e-Collaboration, 1*(4), 24–37.

George, J. F., & Robb, A. (2008). Deception and computer-mediated communication in daily life. *Communication Reports, 21*(2), 92–103.

Grazioli, S., & Jarvenpaa, S. L. (2000). Perils of Internet fraud: An empirical investigation of deception and trust with experienced Internet consumers. *IEEE Transactions on Systems, Man and Cybernetics, Part A, 30*(4), 395–410.

Grazioli, S., & Jarvenpaa, S. L. (2003). Deceived: Under target online. *Communications of the ACM, 46*(12), 196–205.

Grice, H. P. (1989). *Studies in the ways of words*. Cambridge, MA: Harvard University Press.

Grice, H. P. (2002). Logic and conversation. In D. J. Levitin (Ed.), *Foundations of cognitive psychology: Core readings* (pp. 719–732). Cambridge, MA: MIT Press.

Guerrero, L. K., & Floyd, K. (2006). *Nonverbal communciation in close relationships*. Mahwah, NJ: Lawrence Erlbaum.

Hancock, J. (2007). Digital deception: When, where and how people lie online. In A. N. Joinson, K. Y. A. McKenna, T. Postmes & U. Reips (Eds.), *Oxford handbook of internet psychology* (pp. 287-301). Oxford: Oxford University Press.

Hancock, J. T., & Dunham, P. J. (2001). Language use in computer-mediated communication: The role of coordination devices. *Discourse Processes, 31*(1), 91–110.

Hancock, J. T., Thom-Santelli, J., & Ritchie, T. (2004). *Deception and design: The impact of communication technology on lying behavior*. Paper presented at the Proceedings of the SIGCHI conference on Human factors in computing systems, New York.

Hancock, J. T., Toma, C., & Ellison, N. (2007). *The truth about lying in online dating profiles*. Paper presented at the Proceedings of the SIGCHI conference on Human factors in computing systems.

Heinrich, C. U., & Borkenau, P. (1998). Deception and deception detection: The role of cross-modal inconsistency. *Journal of personality, 66*(5), 687–712.

Helme, D. W. (2002). Indicators of deception in marriages: Why do we still rely on nonverbal cues? *Communication & Cognition, 35*(3), 139–171.

Henningsen, D. D., Cruz, M. G., & Morr, M. C. (2000). Pattern violations and perceptions of deception. *Communication Reports, 13*(1), 1-9.

Iacono, W. G., & Patrick, C. J. (2006). Polygraph (lie detector) testing: Current status and emerging trends. In A. K. Hess & I. B. Weiner (Eds.), *The handbook of forensic psychology* (3rd ed., pp. 552–588). Hoboken, NJ: John Wiley.

Jensen, M. L., Meservy, T. O., Burgoon, J. K., & Nunamaker, J. F. (in press). Automatic, multimodal evaluation of human interaction. *Group Decision and Negotiation*.

Kahai, S. S., & Cooper, R. B. (2003). Exploring the core concepts of media richness theory: The impact of cue multiplicity and feedback immediacy on decision quality. *Journal of Management Information Systems, 20*(1), 263–299.

Kaplar, M. E., & Gordon, A. K. (2004). The enigma of altruistic lying: Perspective differences in what motivates and justifies lie telling within romantic relationships. *Personal Relationships, 11*(4), 489–507.

Katz, J. E., Rice, R. E., Acord, S., Dasgupta, K., & David, K. (2004). Personal mediated communication and the concept of community in theory and practice. In P. J. Kalbfleisch (Ed.), *Communication yearbook 28* (pp. 315–371). Mahwah, NJ: Lawrence Erlbaum.

Knapp, M. L. (2006). Lying and deception in close relationships. In A. L. Vangelisti & D. Perlman (Eds.), *The Cambridge handbook of personal relationships* (pp. 517–532). New York: Cambridge University Press.

Knapp, M. L. (2008). *Lying and deception in human interaction*. New York: Allyn & Bacon.

Levine, T. R., Feeley, T. H., McCornack, S. A., Hughes, M., & Harms, C. M. (2005). Testing the effects of nonverbal behavior training on accuracy in deception detection with the inclusion of a bogus training control group. *Western Journal of Communication*, 69(3), 203–217.

Levine, T. R., & McCornack, S. A. (2001). Behavioral adaptation, confidence, and heuristic-based explanations of the probing effect. *Human Communication Research*, 27(4), 471–502.

Marelich, W. D., Lundquist, J., Painter, K., & Mechanic, M. B. (2008). Sexual deception as a social-exchange process: Development of a behavior-based sexual deception scale. *Journal of Sex Research*, 45(1), 27–35.

Masip, J., Garrido, E., & Herrero, C. (2004). Defining deception. *Anales de Psicología*, 20(1), 147–171.

McCornack, S. A. (1992). Information manipulation theory. *Communication Monographs*, 59, 203–242.

McCornack, S. A., & Levine, T. R. (1990a). When lies are uncovered: Emotional and relational outcomes of discovered deception. *Communication Monographs*, 57, 119–135.

McCornack, S. A., & Levine, T. R. (1990b). When lovers become leery: The relationship between suspicion and accuracy in detecting deception. *Communication Monographs*, 57(3), 219–230.

Meservy, T. O., Jensen, M. L., Kruse, J., Twitchell, D. P., Tsechpenakis, G., Burgoon, J. K., Metaxas, D. N., & Nunamaker, J. F. (2005). Deception detection through automatic, unobtrusive analysis of nonverbal behavior. *IEEE Intelligent Systems*, 20(5).

Metts, S. (1989). An exploratory investigation of deception in close relationships. *Journal of Social and Personal Relationships*, 6(2), 159–179.

Miller, G. R., & Stiff, J. B. (1992). Applied issues in studying deceptive communication. In R. S. Feldman (Ed.), *Applications of nonverbal behavioral theories and research* (pp. 217–237). Hillsdale, NJ: Lawrence Erlbaum.

O'Hair, H. D., & Cody, M. J. (1994). Deception. In W. R. Cupach & B. H. Spitzberg (Eds.), *The dark side of interpersonal communication* (pp. 181–213). Hillsdale, NJ: Lawrence Erlbaum.

O'Hair, H. D., Cody, M. J., & McLaughlin, M. L. (1981). Prepared lies, spontaneous lies, Machiavellianism, and nonverbal communication. *Human Communication Research*, 7(4), 325–339.

O'Sullivan, M. (2008). Deception and self-deception as strategies in short- and long-term mating. In G. Geher & G. Miller (Eds.), *Mating intelligence: Sex, relationships, and the mind's reproductive system* (pp. 135–157). Mahwah, NJ: Lawrence Erlbaum.

Park, H. S., Levine, T. R., McCornack, S. A., Morrison, K., & Ferrera, M. (2002). How people really detect lies. *Communication Monographs*, 69, 144–157.

Peterson, C. (1996). Deception in intimate relationships. *International Journal of Psychology*, 31(6), 279–288.

Sagarin, B. J., Rhoads, K. v. L., & Cialdini, R. B. (1998). Deceiver's distrust: Denigration as a consequence of undiscovered deception. *Personality and Social Psychology Bulletin*, 24(11), 1167–1176.

Sporer, S. L., & Schwandt, B. (2006). Paraverbal indicators of deception: A meta-analytic synthesis. *Applied Cognitive Psychology*, 20(4), 421–446.

Sporer, S. L., & Schwandt, B. (2007). Moderators of nonverbal indicators of deception: A meta-analytic synthesis. *Psychology, Public Policy, and Law*, 13(1), 1–34.

Stiff, J. (1995). Conceptualizing deception as a persuasive activity. In C. R. Berger & M. Burgoon (Eds.), *Communication and social influence processes* (pp. 73–90). East Lansing, MI: Michigan State University Press.

The Global Deception Research Team. (2006). A world of lies. *Journal of Cross-Cultural Psychology, 37*(1), 60–74.

Utz, S. (2005). Types of deception and underlying motivation: What people think. *Social Science Computer Review, 23*(1), 49.

Vrij, A. (2006). Nonverbal communication and deception. In V. Manusov & M. L. Patterson (Eds.), *The Sage handbook of nonverbal communication* (pp. 341–359). Thousand Oaks, CA: Sage.

Vrij, A. (2008). *Detecting lies and deceit: Pitfalls and opportunities* (2nd ed.). Chichester, UK: John Wiley.

Vrij, A., Edward, K., & Bull, R. (2001). Stereotypical verbal and nonverbal responses while deceiving others. *Personality and Social Psychology Bulletin, 27*(7), 899–909.

Walther, J. B. (1996). Computer-mediated communication: Impersonal, interpersonal, and hyperpersonal interaction. *Communication Research, 23*(1), 3.

White, C. H., & Burgoon, J. K. (2001). Adaptation and communicative design: Patterns of interaction in truthful and deceptive conversations. *Human Communication Research, 27*(1), 9–37.

Whitty, M. T. (2002). Liar, liar! An examination of how open, supportive and honest people are in chat rooms. *Computers in Human Behavior, 18*(4), 343–352.

Whitty, M. T., & Carville, S. E. (2008). Would I lie to you? Self-serving lies and other-oriented lies told across different media. *Computers in Human Behavior, 24*(3), 1021–1031.

Whitty, M. T., & Joinson, A. N. (2009). *Truth, lies and trust on the Internet*. New York: Routledge.

Williams, S. S. (2001). Sexual lying among college students in close and casual relationships. *Journal of Applied Social Psychology, 31*(11), 2322–2338.

Zhou, L. (2005). An empirical investigation of deception behavior in instant messaging. *IEEE Transactions on Professional Communication, 48*(2), 147–160.

Zhou, L., Burgoon, J. K., Nunamaker, J. F., & Twitchell, D. P. (2004). Automated linguistics-based cues for detecting deception in text-based asynchronous computer-mediated communication: An empirical investigation. *Group Decision and Negotiation, 13*(1), 81–106.

Zuckerman, M., DePaulo, B. M., & Rosenthal, R. (1981). Verbal and nonverbal communication of deception. In L. Berkowitz (Ed.), *Advances in experimental social psychology* (Vol. 14, pp. 1–57). New York: Academic.

Zuckerman, M., & Driver, R. (1985). Telling lies: Verbal and nonverbal correlates of deception. In A. W. Seigman & S. Feldstein (Eds.), *Multichannel integrations of nonverbal behavior* (pp. 129–148). Hillsdale, NJ: Lawrence Erlbaum.

Zuckerman, M., Driver, R., & Guadagno, N. S. (1985). Effects of segmentation patterns on the perception of deception. *Journal of Nonverbal Behavior, 9*(3), 160–168.

CHAPTER EIGHTEEN

Speculating about Spying on MySpace and Beyond: Social Network Surveillance and Obsessive Relational Intrusion

Makenzie Phillips

Brian H. Spitzberg

Every technological innovation provides an opportunity for insight into the process of human evolution. The reliable and efficient production of fire opened the doors of night, provided for the safer consumption of foods, and enabled various pursuits of protection from enemies large and small. The invention of the printing press significantly collapsed the geography and accumulation of knowledge distribution, and helped spark the renaissance and rediscovery of ancient wisdom. The invention of telegraphy and telephony in turn collapsed both space and time by promoting the approach to synchronous interchange across distances. The invention of television vastly expanded the communicative spectrum to nonverbal and empathic dimensions of possibility, as well as facilitating the exportation of culture at all levels of imagination. As the pace of technological change continues to accelerate, there is little doubt that communicative technologies will continue to play a pivotal role in pushing, changing, and transforming the boundaries, and perhaps even the nature, of human relationships.

Some technological innovations play unanticipated roles in the directions of human change. Einstein peered into the theoretical realm of the atom and the galaxies, but as scientists increasingly developed the technologies of exploring the microcosms of nature, they envisioned the possibilities that would lead both to nuclear power, and nuclear war. As medicine could increasingly formulate the technologies that would cure humans from disease, they also enabled the technologies for chemical and biological warfare. Technologies tend not to stand still, and the roles they play in culture often span ambivalent applications. In biological evolution, sometimes form or structure emerges to enhance adaptive survival, yet gets grafted onto new uses. The primitive forebears of the panda lost the thumb to a basic four-toed paw, and over time increasingly "adapted" the uses of a lower part of the forearm (Gould, 1978). In evolution such function-spanning innovations are referred to as spandrels.

Technologies often become spandrels as well. There may have been relatively little initial *function* for having cameras inserted into cellphones, but once available to the masses, varieties of uses were soon imagined. The ambivalence of such uses include the promotion of social relationships by stimulating activities of a more immediate or co-present sense of connection, and at the same time promote the corruption of social relationships, from spoiling the identity of celebrities to "sexting" images to a mass audience capable of intruding in unwanted ways into individual domains of privacy. This chapter examines one such technology, and speculates on the various social and relational spandrels that can currently be anticipated. Specifically, this chapter investigates what is currently known, and what is reasonable to predict, about the development of social network sites as a software-based technological extension of existing technologies, with special attention to some of the functionally ambivalent and darker sides of their use. Given the nature of evolution, however, there is no particular or predictable destination, and therefore it is a virtual certainty that decades from now, the technologies of social network sites will have negotiated their paths into realms that are now impossible to imagine.

With each new technological innovation in communication there are both opportunities for the actualization of affordances and abuses. The advent of social networking sites as a societal phenomenon has ushered in an extensive reformulation of access to personal information and relational uncertainty management. More than ever before, people can "wear their hearts on their technological sleeves." Social networking sites permit the calculated management of personal identity by permitting the author to select which images, whether submitted by self or site visitors, remain as part of the "advertised" and public-personal pastiche. In addition, site authors can determine what kinds of gate-keeping barriers are established in order to control who can, and who cannot, gain access to a person's more personal social networking site information. Given increasing information (Pew, 2006) and theory (Walther & Tidwell, 2000) indicating that new communication technologies can facilitate relationship formation, it is possible that societal norms of privacy are evolving rapidly to accommodate to, and from, these new technological capabilities. Information that people only a generation ago might have considered highly "personal," and not made available for public consumption, much less socially "marketed," may now be considered normatively accessible and uncontroversial. Such an evolutionary trend, to the extent it is real and continues, presages an increasing shift of what was previously considered "private" information to the public sphere. With such normative shifts comes both an increase in relational openness, as well as new opportunities for spying, surveillance, and informational exploitation. Specifically, people who have particular "designs" on pursuing a relational objective with a target person now have access to considerably greater real-time information about this target than ever before. This chapter examines the emerging research and theory relevant to the use of social networking sites as a tool of relational surveillance.

The chapter will proceed as follows. First, the trends of social network site adoption and their uses will be reviewed to demonstrate the importance of this emerging phenomenon. Related technological trends, such as the increasing convergence of social networking sites and online dating/courtship activity, will also be examined.

Particular emphasis will be placed on evidence of the current uses and gratifications sought from social network sites by their users. Second, theories relevant to understanding such trends and their implications for relationship development (identity formation, online disclosure and intimacy development, dialectical theory, privacy management theory, etc.) will be examined. Third, the "dark side" of CMC uses will be briefly conceptualized, and the socially undesirable and more paradoxical sides of social networking sites will be explored. The authors' own original research relevant to the use of CMC in general, and social networking sites in particular, will be prominently featured, along with other relevant emerging studies. The relational goal theory of obsessive relational pursuit and cyber-stalking will be reviewed with attention to its potential insights into social network site surveillance. Fourth, a critique of existing research and theory will be formulated with an eye towards identification of specific research agendas that will need to be pursued to illumine the more valuable extensions of these lines of inquiry. Specifically, the need for a functional theory that extends across specific technologies to accommodate the increasing convergence of technologies will be needed, so that theories and studies do not become archaic within a couple of generations of product development and innovation. Furthermore, issues surrounding measurement and construct validation in the area of CMC relationship development will be critically examined. Finally, some research-based speculations will be offered for the practical management of surveillance issues, with attention to dialectical and privacy management theories as promising directions for the integration of programs of theory-based research in the area of social network site surveillance.

Technological Trends

The technological landscape of our society is evolving rapidly, and with this evolution comes changes in language, interaction, and even in definitions of what exactly characterizes an interpersonal relationship. The use of technology or computer-mediated communication (CMC) in personal relationships has revolutionized the ways in which interaction is framed, occurs, and is evaluated. Specifically, the use of one type of CMC, social networking sites (SNSs), has had an impressive impact on the availability, efficiency, and affordability of interacting with individuals both far and near.

Social Networking Sites

Social networking sites can be understood to be interactive web-based applications in which individuals have the opportunity to present themselves, establish or maintain connections and relationships with other contacts, and define and/or control their own personal social networks (Ellison, Steinfield, & Lampe, 2007). Social networking sites are now utilized by most age groups, but were originally intended for use by individuals in the 18–24 year-old age bracket. This group is heavily dependent on participation in these sites, with slightly more than 85% of college-aged students in the United States belonging to at least one social networking site (Salaway & Caruso, 2008), and at least 50% of those individuals with SNS memberships logging on at least once a day (Peluchette & Karl, 2008). This appeal of the interaction that can occur over the Internet, especially to younger generations, is based on the speed, convenience, inexpensive nature, and user-friendly features available through social networking sites (Lenhart, Madden, & Hitlin, 2005; Lenhart, Rainie, & Lewis, 2001; Subramanyam, Greenfield, Kraut, & Gross, 2002; Wang, 2004). These features have popularized social networking sites to the extent that many individuals now consider their web-based user profiles to be accurate reflections of their real-space lives.

Trends and Usage

Evidence of the extraordinary social networking site adoption and curves of increasing usage among youth clearly presage the potential importance such media may play in the culture of interpersonal relationships in society (PEW 2005; 2006). MySpace alone has as many as 8 million new users every month, and shows no signs of slowing (PEW 2007). These sites are so prevalent in today's communication and interaction, in fact, that the use of social networking sites has created a new forum, and even vocabulary, for interpersonal relationships (Hinduja & Patchin, 2008; Phillips, 2009a). This vocabulary (i.e., "sexting," "friending," "tagging," etc.) has become vernacular for site users and nonusers alike. While originally used for connecting with old friends and meeting new contacts, the vast popularity and marketability offered by and for users has made social network sites places for advertising, networking, romancing, and, unfortunately, for more devious activities ranging from scamming and personal deception to identity fraud and cyber-surveillance. That said, one of the most significant impacts of social network sites is that which occurs in personal relationships.

Social Network Sites and Personal Relationships

Historically, people maintained relationships with those individuals who lived and worked within the same town or village through time spent together, shared activities, work relationships, and the bonds inherent in family relationships. While these aspects are still valuable ways in which to initiate and maintain contact with those in their lives, the modern individual is likely to also have many relationships that are characterized by something our ancestors never had to worry about: distance. The ease with which proximal relationships were enacted has now been confounded by friends and family moving away for educational, occupational, or personal reasons. Technological advances and computer-mediated communication have facilitated the relational maintenance process for those with distant loved ones in dynamic ways, from long-distance telephone calls to Instant Messaging (IM) functions on the Internet. One of the most significant technological trends for personal relationships, however, has been the use of social networking sites. These sites allow members to view photos of distant friends, communicate with them regardless of time-zones or miles, and create personalized profiles, "status updates," and contact groups that allow members to broadcast their activities, interests, and lifestyles to those in their personal networks. Clearly, social network sites have had a dramatic impact on the ways in which individuals initiate, manage, and maintain their close relationships.

Darkness Emerging Out of the Light: Explaining SNS Uses

Although there are many models and theories that have been developed to account for differences between mediated, unmediated (RS, or "real space") and mixed mode relationships (see Walther, 1996, for review), most of these were formulated prior to the development of social network sites, and may or may not be as relevant to the particularly social uses of this medium. It is not the intent here to formulate a full theory of social network site use, but some framework for discussion is necessary for providing some vocabulary for analysis. Most existing models of CMC use and relationships have focused in some way or another on the *richness* (i.e., the extent to which nonverbal and synchrony of interaction are available in the medium), *presence* (i.e., the extent to which the medium recreates a sense of co-occupation of similar space and time—feeling "there" with the other interactants), or *competence* (i.e., the extent to which the users are capable of conveniently or expertly using the technologies). These are clearly important dimensions of CMC communication, and will necessarily figure prominently in any complete theory of SNS uses. As technologies are increasingly adopted, however, and given that the curve of innovation will almost certainly continue to decrease the individual differences in richness, presence and competence, it will become important to examine new variables to help account for variations in the uses of given media. The possibility that SNSs provide certain individual and social affordances is suggested as one theoretical window into such uses.

Technological Affordances of SNSs

Understanding any technology, at any level, generally needs to begin by asking "what needs does it, or can it, serve?" The study of human motivation has traditionally theorized many different models of human needs. In contemporary psychology, theories of human needs range from highly individual (e.g., Maslow, 1943) to more interactional (e.g., Leary, 1957) and dialectical (Altman, Vinsel, & Brown, 1981) in nature. Rather than attempt to explain human nature by selecting a particular model of needs, it is more to the point to explore the affordances that social network sites might serve. An *affordance* is defined here as a feature or facet of a technology that enhances the pursuit of a particular end, function, or objective of human activity. The invention of fire afforded greater activity at night, and the invention of the telephone afforded more rapid coordination of human activity. As social network sites primarily are currently embedded in the technologies of electronic Internet and computer-based technologies (e.g., desktop, laptop, and multiply-enabled cellphones), many of such affordances have already been explored regarding these technologies (e.g., Rice, 1993; Turkle, 1995; Walther & Parks, 2002). For example, there is little surprise in the affordances of improved *efficiency* in message delivery. The question is what the efficiency is *for*. The advent of social network sites has to date been progressing at such a rapid adoption curve that theory and research have not been able to keep up with their expanding uses and affordances.

There may be an infinite number of potential functions served by social network sites, but for the sake of theoretical parsimony, three dimensions help describe the main affordances of this technology: objective, valence, and surveillance (see Figure 18.1). The *objective* is the type of purpose to which an SNS is put by the user, whether a site author or site viewer. There are many objectives to which communication may be put, but one way of classifying them is as instrumental, identity, or relational. *Instrumental objectives* ask whether the communication is functioning to achieve a particular, objective, discrete outcome. For example, an author may want to get a job, and a viewer may want to decide whether or not to hire a site author, and the SNS may provide information for both users in the process of obtaining a job. *Identity objectives* ask whether the purpose of the communication is to express or construct a person's face or sense of self. For example, a site author may decide to get a new hairstyle to suit an intent to change one's look, and in turn, viewers may be interested in seeing how the author has attempted to redefine his or her "look." Social network sites can also be used to serve a *relational objective*, in which the function is to alter the relationship between author and viewers. An author may seek to increase intimacy with a particular user, or seek interested "singles" through an SNS, and both actual and prospective or potential users may locate the author's SNS in the pursuit of a "match."

These objectives of SNS use are discussed in terms of uses, purposes, and functions, but the word *intent* was intentionally avoided. In everyday language use, the concept of use, purpose, and even function often carries with it implications of user intentionality. The problem is that people are not always aware of the many functions their apparent intentions may serve. An author may intend to express relationship status as "single" as a relatively simple expression of personal independence, but such a designation clearly has relational implications for anyone this author is dating. Thus, although the objectives dimension may suggest user-intentionality, many uses of SNSs will have unanticipated effects. Furthermore, although the dimension is suggested as a continuum as a limitation of representing it in two-dimensional space, it is not a true continuum. That is, instrumental objectives do not lead into identity objectives any more or less than they lead into relational objectives. Most uses of SNSs are likely to be undertaken to achieve more of one of these objectives than another, but are likely to affect the other two at some level as well.

A second basic dimension of SNS affordances is *valence*, which refers to the basic evaluative stance a user intends, whether antagonistic or affiliative, negative or positive. All human relations tend to be oriented along this dimension (Leary, 1957), and all symbolic interpretations, regardless of referent, culture, or language (Osgood, May, & Miron, 1975), also reveal this as a fundamental structure of perception. From the deepest evolutionary past of human evolution, it has been vital to evaluate rapidly and with some degree of accuracy whether another person's behavior carries with it potential for good or potential for danger. Furthermore, given the importance of the evaluative dimensions, communicators often invest considerable effort to promote the impression of being a positive, competent (i.e., appropriate and effective), and attractive interactants (Metts & Cupach, 1994; Goffman, 1967) , although interactants often pervert, corrupt, or otherwise "mess up" such performances (Gardner & Martinko, 1998; Shepperd & Kwavnick, 1999; Vonk, 2001). So, for example, a site author may reveal a harrowing personal experience on a personal wall, and a potential stalker might engage in ingratiating, highly supportive message interchanges to curry favor from the author. The author, however, may see through the shallow manipulative nature of the ingratiation, especially if the support seemed excessively needy or suggestive of reasons for face-to-face contact. The manipulator intends the author to make a positive evaluation of the messages, but the author attributes negative intentions to the messages. It is in such disjunctions of interpretation that relationships often go astray.

The third dimension underlying SNS affordances is *surveillance*, which refers to the extent to which users, both authors and reviewers, of author sites use such sites for the purpose of uncertainty-reduction and ongoing awareness of the author's life. Having information about a person is often key to ascertaining whether and how to pursue ongoing interaction or relationship with this person. Research and theory reveal that interactants invest extensively in uncertainty-reduction, reflecting a significant motivation in both initial and ongoing interaction (Afifi & Burgoon, 2000; Berger, 1979, 1987). In surveilling an author's SNS, all manner of information may be gleaned, and all manner of insights into the author's world may be interpreted, but for the purposes of examining the dark side of SNSs, this dimension is defined here by the poles of *geospatial activities* and *communication activities*. By attending to an author's SNS, it is often (but not always) possible to obtain information about the author's travels, routines, intended trips, daily hobbies, and so forth. Such monitoring may provide significant insight into the author's *activity map*, or the social, physical, and cultural spaces through which this author travels in a given time span. Such information can provide valuable information for the purposes of illicit monitoring, arranging "accidental" coincident of contact and meeting, as well as an understanding of areas of common interest. In contrast, surveillance of the author's communications, through imagery, photos, gifts, group memberships, wall postings, expression of status and self-descriptions, musical selections, postings, and of course, direct interaction with the author by the monitor, a virtual (i.e., interacting only as avatars or only online, but not in real space), parasocial (i.e., a monitor or lurker imagining a relationship with a site author, but not actually interacting with the author) or actual relationship can emerge.

These three dimensions are likely to interact in various ways in SNS environments. A relationship may begin innocently enough, with a classmate expressing an interest to be "friends." By allowing this friend in, an SNS author makes an initial, if perhaps innocent or naïve, positive evaluation that the objective is merely to maintain casual friendship. Over time, however, the friend may become hyper-vigilant by commenting on every new posting by the author, revealing an excessive or demanding interest in the author's posted items or experiences, and suggest an objective of pursuing a different kind of intimacy than the author was ever interested in or willing to abide. Indeed, in the process of interpreting the intent of the monitor, the author may in turn engage in surveillance of the monitor's site to reduce uncertainty and make an evaluation of whether or not an increase in intimacy would be a desirable or undesirable use of the SNSs. Both the author and monitor are enacting behaviors across all three dimensions, but as the relationship develops, their perceptual and behavioral movements on these dimensions are likely to become increasingly divergent as some dark side emerges. This divergence along these dimensions is one marker that the dark side of SNSs is near.

There are other affordances that computer-mediated communication provides, especially for some who might use it for more nefarious or harassing purposes. These other affordances might not fit as easily into the three-dimensional space of objectives, valence, and surveillance. For example, SNS and CMC permit a degree of user compensation, in the sense that socially unskilled or shy persons can better control the impressions and messages they construct (Parks & Archey-Ladas, 2003;Walther & Tidwell, 2000). For the truly manipulative pursuer, CMC in general permits greater anonymity of activity. So, for example, many people take on the role of *lurker* by monitoring but not contributing to or interacting in chatrooms or other SNS interaction spaces. Both anonymity and compensation, however, are likely to be subsidiary to uses of SNSs for the purpose of developing a relationship, establishing a positive identity, and maintaining surveillance on the target.

The Marketplace for Mating

The first "personals ads" did not appear until near the early decades of the last century. For a long time, meeting mates was primarily a product of chance encounter, routine processes of everyday interaction, matchmaking individuals or liaisons, and of course, all the powerful mate selection processes involved in attraction, impression management, and interaction skills. Thus, the use of the Internet as a means through which mate selection can occur is a relatively recent phenomenon. A nationwide representative poll by Harris Interactive (2006, n = 2,985) estimated that 3% of committed relationships began in some form of online dating service, with another 3% having begun in chat room contexts, although most of these respondents indicated that they had originally met in more typical or real space contexts (e.g., job, school, social group, etc.). These estimates illustrate both the relatively small overall role that such sites seem to play at present in mate selection, but are clearly (a) only a tip of an iceberg, given that for each current "partner" there may well be the detritus of many previous partnering attempts, and (b) for a technological medium that has only been available for about a decade or two at most as a widely adoptable selection procedure, 6% of the population, reporting significant online influences over their current mateship, is a substantial sea change portending a potentially rapid change towards such means in the future.

At least two theoretical approaches provide heuristic insights into SNSs as such sites would apply to mating processes. Impression management theory posits that humans are inherently social creatures who are constantly cognizant of their social audiences with whom they interact, and in front of whom they enact their behavior. Initially forged by Goffman's extensive insights into everyday dramaturgical routines and foibles (Goffman, 1967, 1969), the concept of *face* was conceptualized as the image a given actor (i.e., person) wants others to have of the actor. Thus, all action is undertaken in part with the idea of an audience and therefore how actors perform their various roles in everyday social interaction is always in part understandable as a performance on the stage of the ongoing human drama. There are at least two basic types of face with which social actors are typically concerned: positive and negative (Brown & Levinson, 1978). *Positive face* is a concern with being liked, being seen as attractive or pleasing. In contrast, *negative face* is a concern for being effective and achieving preferred individual ends or outcomes. Thus, some encounters or performances, such as interpersonal conflict, are inherently problematic for meeting face needs, because one actor's pursuit of negative face (i.e., goal achievement) are resisted in what are perceived as inappropriate ways by another co-actor, thereby threatening the co-actor's positive face (Spitzberg, Canary, & Cupach, 1994). The enactments and negotiations of face are referred to as the process of impression management, especially when such negotiations can be analyzed as a set of tactics in the actors' repertoires performed with varying competence and with identifiable effects on peoples' impressions. Such performances are considered forms of *self-presentation*, which is defined as "the process by which individuals, more or less intentionally, construct a public self that is likely to elicit certain types of attributions from others, attributions that would facilitate the achievement of some goal" (Metts & Grohskopf, 2003, p. 360). Thus, interactants engage in self-presentations as a way of managing their impressions with others so as to achieve personal and social goals.

So what impressions are people likely to present in SNS contexts? One theory that is suggestive is socioevolutionary theory (see Buss, 1994; Shackelford, Goetz, LaMunyon, Quintus, & Weekes-Shackelford, 2004). Socioevolutionary theory argues that like all biological organisms, humans evolved over thousands of generations, both as primates and then as hominids, with certain standard evolutionary forces at work. First, a primary motivation of organisms is to propagate, and pass down their own genes into future generations—that is, organisms seek to survive and mate. Second, there are random variations in genetic code, which produce individual differences in an organism's capabilities, some of which are more advantageous in a given environment than others. The more advantageous variations tend to be *selected* by survival and mating, and therefore passed down increasingly to the species over time. Third, in a sexually dimorphic species such as humans, there are mating pressures that select for particular tendencies. In particular, women invest heavily in carrying and then caring for offspring, and therefore will tend to be selective in preferring mates who will be strong, higher in status, and stable or likely to commit investments in a long-term relationship. In contrast, males are capable of mating extensively and with relatively little investment in the actual process of carrying the product of their mating to term. Thus, males will tend to prefer mates who are sexually fertile. Certain characteristics that indicate genetic fitness would include youth, physical attractiveness, and certain body ratios suggestive of fertility.

Fourth, to the extent such sexual differences have been adapted and selected, then other corollary adaptive mechanisms are likely to have evolved. One extensively studied corollary is a sexual difference in jealousy. Jealousy is an emotional, cognitive, and behavioral tendency to respond to perceived threats to a valued relationship or partner (Guerrero & Andersen, 1995). If socioevolutionary theory is correct, then males and females should differ in the nature or extent of their jealousy experiences. Specifically, both males and females should experience jealousy when it appears that another actor may be threatening to poach their partner, but males should be relatively *more* reactive to their female partner's sexual infidelity, whereas females should be relatively *more* reactive to their male partner's emotional commitment or investment in another person. For the male, it is the potential that he may invest relationally in someone else's progeny that is most threatening, whereas for the female it is the prospect that the male may leave and take his investments with him that is most threatening. Research largely has supported this difference, indicating that males and females are differentially reactive to these infidelity scenarios (see analysis in recent Edlund & Sagarin, 2009).

Fifth, research has largely also found male and female differences in what each sex seeks to *advertise* to others in the mating marketplace (see Spitzberg & Cupach, 2007; Sprecher, Schwartz, Harvey, & Hatfield, 2008). Specifically, in the context of personal ads and online matchmaking sites, males tend to advertise signs of their status (e.g., professional achievements, education), income and possessions of wealth, and physical prowess. Presumably, they advertise these signs in some knowledge that this is what females are seeking. In turn, females tend to advertise their physical features and beauty, their youth, and directly or indirectly, their potential for fidelity (e.g., selectivity in choosing mates). Presumably, they advertise these features because they believe this is what males are seeking. Indeed, when such ads and sites are examined for what males and females claim to be seeking, it is not surprising that males tend to seek youth and beauty, and females tend to seek mates who are taller than themselves, physically fit, and professionally and economically successful. Features like education and professional success are disproportionately likely to be sought by males, and likewise, females are disproportionately unlikely to indicate choosiness regarding physical attractiveness and more accepting of a mate older than themselves (see Spitzberg & Cupach, 2007).

Social networking sites are not explicitly matchmaking sites, although they often both permit such uses and are heavily co-sponsored in advertising budgets by such sites. It seems likely, however, that SNSs can be considered impression management frames (Goffman, 1974), and that males and females might construct their SNS self-presentations consistent with the gender presentations anticipated by such matchmaking sites. Thus, first and foremost, people will seek to emphasize both their positive and negative face on their sites. This basic prediction is the almost taken-for-granted notion that people tend to select out "incriminating" information from their sites, or information that would spoil their identity (e.g., a person might be unlikely to advertise getting caught plagiarizing or having a sexually transmitted disease), and more likely to promote attractive information (e.g., sociality, adventure, fun, professional activity, etc.). Further, socioevolutionary theory would anticipate that these self-presentations might be gendered. Specifically, information in a given site would tend to be selected for promotion of gendered tendencies—females would have more photos advertising skin, youth, slender proportions, attractive dress, vitality, and capability of demonstrating commitment to relational partners, whether friend or partner. In contrast, males would tend to emphasize adventuresome activities, personal and professional achievements, and general status among various contexts and groups.

In sum, social network sites are, among other things, personal advertisements to the social world. They are not necessarily intended for mating, and obviously are often used by those who are already mated (although infidelity is one of the dark sides of SNSs, to be discussed later). Such sites, however, are efforts at impression management, and as such, it is important to understand them in terms of the ways in which people tend to promote their self-presentations to others. One of the ways in which people promote their self-presentations is through gendered extensions of their socioevolutionary preferences and propensities.

Dystopic Potentials for SNSs

There are many potential utopian visions for social network sites. They bring people together, bridge social groups and networks that in real space might never otherwise have met, and they provide an ability to maintain relationships when space and time otherwise might militate against such maintenance. People can bathe in the warmth of knowing that they have dozens, hundreds, or thousands of "friends," and they can communicate at virtually any time by entering the stream of individual action and social interaction that comprises a person's site. Such potentials may indeed define the future of social network sites. There are, however, already signs that the stream is getting polluted, and that entering the stream should at least have warning signs prominently posted. There are at least two significant tributaries of trouble anticipated in this chapter, although others have been examined elsewhere (e.g., online dating: Sprecher, et al. 2008; deception and infidelity: Whitty, 2003): surveillance and cyber-stalking.

One of the ironies of the title *Brave New World* is that it envisions a world likely to evoke far more fear than bravery. Technologies are inherently ambivalent in their functional value—a hammer can build a shelter from the elements, and can be used as a blunt instrument for bludgeoning a person. Likewise, contemporary communication technologies permit both unprecedented access to people as well as those people's personal information. That is, the technological surveillance affordance promotes contact and communication, both consensual and coercive, both invited and intrusive. One of the darker sides of this brave new world is the unprecedented possibility of unwanted surveillance, loss of privacy, intrusion, and cyberstalking.

Surveillance and Relational Monitoring

With so many people turning to social networking sites to create and maintain relationships, it is important to consider the impact these sites are having on the quality and satisfaction of personal relationships. As a means of frequent communication, relational maintenance, and the exchange of information in relationships, social networking sites are likely to serve a significant surveillance function among network members. Surveillance can range in severity from occasionally checking the profiles of others to obsessively tracking the electronic movements of a target individual. Obviously, monitoring within close relationships can create problems and even instigate feelings of jealousy, suspicion, and distrust. In a world where employers access the profiles of potential new hires, school officials revoke scholarships after discovering proof of misconduct or illegal activities, and potential romantic partners Google each other before first dates, it is important to assess to what level these actions are viewed as normative, and to what extent they may be violations of personal privacy.

Social networking sites have evolved over time from small memberships in sites geared towards professional opportunities to global infrastructures linking friends, strangers, celebrities, and just about anyone (or anything) else one could imagine. Members are becoming increasingly dependent on participation in these sites as a means of keeping in touch with, and keeping track of, "friends" or contacts, as well as a means for self-presentation via individual profiles. This increased reliance on social network sites has also spurred an increase in the amount of surveillance individuals engage in, regardless of their propensity to monitor others in "real-space." In fact, one study indicated that as many as 65% of college-aged participants (n = 508) regularly engage in surveillance via social network sites, and that this surveillance exists on covert, obsessive, and problematic levels regardless of if the individuals are monitoring romantic or platonic targets (Phillips, 2009b). Covert surveillance consists of activities ranging from asking friends to "tag" photos of others so that they can be viewed by contacts to creating fake profiles in order to gain access to the profile of a specific target. This type of surveillance is usually indicative of an inclination to monitor without the knowledge of a given target or targets. Usually, covert surveillance is enacted to avoid confrontation, or in order to "check up on" a suspicion of the monitoring individual. Obsessive surveillance is typified by ruminative feelings that prompt an individual to incessantly monitor a specific profile, or to become preoccupied with monitoring the comments, pictures, and other online interactions a target has with others through a social networking site. Problematic surveillance is represented by actions that result in discovering undesirable information from participation in social network sites, engaging in conflicts based on information discovered or contained in social network sites, and questioning individuals about specific comments or messages posted by others on the target's profile. These actions are generally directly confrontational and often jeopardize the existence or quality of relationships between members of social network sites. Such experiences indicate that social network sites are penetrating the fabric of contemporary relationships and are now becoming part of the bases for relational conflicts.

Obsessive Relational Intrusion (ORI) and Cyber-Stalking

In 1990 the State of California passed the first explicit anti-stalking law, in part in response to some dramatic cases of celebrity attacks by fans, and women being attacked by their former partners against whom they had protection orders. Within a decade, all 50 states, the Federal government, and numerous other countries had passed anti-stalking legislation, a pace of political change that is virtually revolutionary in a domain known for more glacial rates of progress. A large part of this rate of change is based in the unique features of the crime of stalking. Even though threats, harassment, theft, fraud, and assault may be involved in stalking, these can all be singular events, and are considered crimes of relatively tangible damage to a person or a person's property. In contrast, stalking represents a pattern of behaviors, any one of which may not amount to an obvious crime, but that collectively present a threatening context of unwanted intrusion. To the extent that technologies provide new affordances for the conduct of stalking activity, it represents an important subject for concern.

Stalking legislation varies from one jurisdiction to another, but most stalking laws identify several of the following components (Miller, 2001; Modena Group on Stalking & Emilia, 2007). First, stalking is a pattern of conduct—not a single act or episode. Stalking occurs over multiple contexts and times. Second, stalking is explicitly unwanted. In some way, it must be clear that the pattern of conduct is undesired, dispreferred, or otherwise unwanted by the target of harassment. Third, stalking serves no other "legitimate" purpose—that is, just because you get a lot of unwanted email ads for political parties, sexual products, or online dating promotions does not make such attempts at sales or retail marketing or political advertising forms of stalking. Fourth, and perhaps most importantly, the pattern of conduct must, as an accumulated set of activities, constitute a fear-inducing or threatening experience. Because different individuals vary in the degree to which they find things fearful, there are two relatively common standards of fear in stalking legislation—either a victim experiences fear, or a "reasonable person" in the victim's place would be likely to experience fear as a result of the pattern of conduct.

Stalking can occur for any number of reasons, but it sometimes represents an attempt to establish a relationship, escalate the level of intimacy in a relationship, or to try and exact punishment against someone for not permitting an escalation or renewal of intimacy in a relationship. In such cases, it is a form of *obsessive relational intrusion* (Cupach & Spitzberg, 2004, 2008a, 2008b; Spitzberg & Cupach, 2002, 2003). Obsessive relational intrusion (ORI) is a pattern of unwanted pursuit of intimacy. Unlike stalking, it does not necessarily evoke fear or anxiety, although it often does (Cupach & Spitzberg, 2000). Indeed, the fact that approximately three-quarters of stalking cases emerge from preexisting relationships, and approximately half emerge from previously romantic relationships, indicates several important features of these processes. First, most stalking is not the product of someone so deranged or mentally disturbed that they were unable to "fly in under the relational radar" of a normal person. Indeed, mental disorder in a stalker actually decreases the likelihood of violence in a stalking case—it is the previous romantic partner who is most dangerous (Mohandie, Meloy, McGowan, & Williams, 2006; Rosenfeld, 2004). Second, despite the anonymity affordance of CMC technologies, most unwanted pursuit and harassment reflects the detritus of a failed relationship between people who know one another. Third, as the availability, efficiency, and capability of CMC technologies have increased in the development of everyday relationships, it seems likely that people will make more use of these means to pursue relationships in ways that are unwanted by the objects of pursuit.

Stalking and obsessive relational intrusion are enacted in a variety of ways. Some pursuers engage in *hyper-intimacy* behaviors—actions that reflect inappropriate or "creepy" forms of courtship. Why send a dozen roses when six dozen can be sent? Why show up on the person's doorstep when a pursuer could show up also at the person's classroom, gym, and workplace? Why send a note of affection when additional notes can be left on the person's car windshield, door and as a text message? Second, pursuers use *mediated contacts* by employing any number of CMC technologies. This category will be elaborated below. Third, pursuers may attempt *interactional contact* by attempting face-to-face or proximal opportunities for communication. This distinction between interactional and mediated contacts is becoming an increasingly important issue in understanding the nature of cyberstalking and social network surveillance. Fourth, some pursuers engage in forms of *surveillance* by seeking to monitor or obtain information about the object of pursuit. This can take the form of following, driving by the person's home, lurking in online environments to observe (but not interact with) the person, or engaging in ongoing monitoring of the person's social network site. Fifth, unwanted pursuit often takes the form of *invasion and intrusion*, which ranges from trespassing to illicit entry into a person's diary or computer sites or files, to the more extreme examples of "B&E" (i.e., breaking and entering into the person's residence). Sixth, many pursuers resort to a campaign of *harassment and intimidation*, which often consists of overly persistent attempts at contact that are unresponsive to pleas to stop, and in more extreme examples, subscribing the person to unwanted publications, harassing legal accusations, or sheer administrative strangulation of the person. Seventh, pursuit can become threatening by implying the possibility that harm will or might occur to the person's self, pets, property, reputation, friends, family, or even to the pursuer (e.g., suicide threats). Eighth, pursuit can become aggressive and violent in the forms of actual property damage or physical assault. Finally, a pursuer can enlist other persons or parties to engage in *proxy pursuit* through any of these previous eight modes of harassment (see Cupach & Spitzberg, 2004; Spitzberg, 2002; Spitzberg & Cupach, 2007).

To a large extent, each of these forms of pursuit or stalking could occur *through* CMC technologies. Gifts and e-cards can be sent in hyper-intimate forms, attempts at facilitating or seeking face-to-face interaction (or video-cam) can be pursued. Further, electronic media can be used to intrude or engage in surveillance into the person's life, using means ranging from simple Google searches to using online "find someone" location services and the illicit use of GPS (global positioning services) devices. So-called Trojan horse software can be hidden in attachments that invade a host's computer and permit electronic monitoring of every keystroke and every address and file that computer possesses. Such media can also be used to harass and intimidate someone by email-bombing, inappropriately subscribing the person, or otherwise jamming their modes of CMC. CMC can, of course, also be used to send and make threats—emailing a woman a digital photo of her, with a rifle-scope target lens image with her head in the crosshairs is obviously a threatening image. Sending a disguised virus that opens up a screen saying "You have just opened a virus that will permanently erase your hard drive in 5 seconds" is obviously a threatening form of interaction. In contrast, if there is an actual virus that destroys the person's computer or blocks the person's computer access, or if CMC, Internet or GPS technologies are employed in the process of planning, coordinating, or enacting physical assault or worse on a person, it represents a form of aggression and violence. Finally, CMC and the Internet can greatly facilitate the recruitment of other parties, or even unwitting accomplices in enacting a campaign of pursuit or harassment.

Stalking has been around at least since ancient times (Kamir, 2001), but the legal concept evolved to some degree concurrently with the massive diffusion of CMC and SNS technologies. It is therefore little surprise that neither research nor legislation has kept up with the rapid pace of technological change and adoption. Cyberstalking can be defined as "threatening behavior or unwanted advances directed at another using the Internet and other forms of ... online electronic communications technology" (Chik, 2008, p. 16). There are numerous legal characteristics that are recognized in US legislation (see Table 18.1). Table 18.2 summarizes a number of findings from studies of stalking and unwanted relationship pursuit. Early studies of stalking tend to reveal relatively limited uses of electronic media for such purposes, but a large-scale study by Baum, Catalano, Rand, and Rose (2009) indicates that such technologies are becoming relatively common both for monitoring (i.e., surveillance) and for stalking purposes (see Table 18.3). These snapshots of cyberstalking prevalence are likely to be the "tips of the iceberg" of the entire domain of cyberstalking and surveillance. For example, such estimates do not reflect the use of such technologies for the sexual seduction of children by adults (Finkelhor, Mitchel, & Wolak, 2000; Mitchell, Finkelhor, & Wolak, 2001, 2007; Wolak, Finkelhor, & Mitchell, 2004). It does not include such facets as industrial espionage, illicit wiretapping by the government, or attempts to locate celebrities for the purpose of photographing them. For example, research thus far suggests that cyberstalking is less likely than realspace stalking to involve prior acquaintances (i.e., more likely to involve a stranger than a prior romantic partner), more likely to target organizations as targets, and less likely to involve sexual or physical violence (Bocij, 2003; D'Ovidio & Doyle, 2003; Moriarty & Freiberger, 2008; Sheridan & Grant, 2007). Some research, however, does suggest that cyberstalking is perceived as relatively threatening and fearful, despite the fact that by itself, it is action "at a distance." Nevertheless, the data of Tables 18.2 and 18.3 certainly presages immense potential for the use of technologies for the purpose of unwanted surveillance, pursuit, harassment, and aggression. With each new technology arise new affordances for the pursuit of such instrumental and relational ends.

Just having the technologies available, or considering how often they are used, reveals relatively little about the nature of cyberstalking. Little is known about cyberstalkers, in part because the very medium often makes it difficult for them to be caught; those who are surveyed often do not view what they are doing as stalking, and the phenomenon is still very understudied.

McFarlane and Bocij (2003) examined a number of US cases of cyberstalkers and concluded there are four basic types. *Vindictive* cyberstalkers tend to be ferocious and threatening in their communication, and pursue the target in realspace as well as cyberspace. *Composed* cyberstalkers tend to use the media primarily for engaging in a campaign of harassment and threats, and are less likely to use realspace for such purposes. *Intimate* cyberstalkers are pursuing greater intimacy with their target, but reflect two distinct sub types: ex-intimate (i.e., expressing ambivalence about renewing the relationship) and infatuates (i.e., no prior relationship but an obsessive or fantasy-based desire to initiate a relationship). *Collective* cyberstalkers represent group-based pursuit or harassment, typically either for the purpose of discrediting an organization or seeking revenge and punishment for a perceived wrong by some group.

A contrasting typology was developed by Sheridan and Grant (2007), who were more explicitly interested in the use of cyberspace for the purpose of relationship pursuit. They identified three types of cyberstalkers. *Cyberstalking-only* pursuers relied exclusively on mediated harassment, representing only 4% of their sample. *Cyberspace-to-Realspace* pursuers began their harassment online but increasingly moved their campaign into the realm of the proximal world. This group also represented a small percentage of the sample (5%). Another group was labeled *cyberspace-and-realspace*, which consists of people who from the beginning employ both means of pursuit (38.5%). Most of the stalking, however, still consisted of *realspace-only* pursuit (52.5%).

Figure 18.1. A model of SNS affordance dimensions

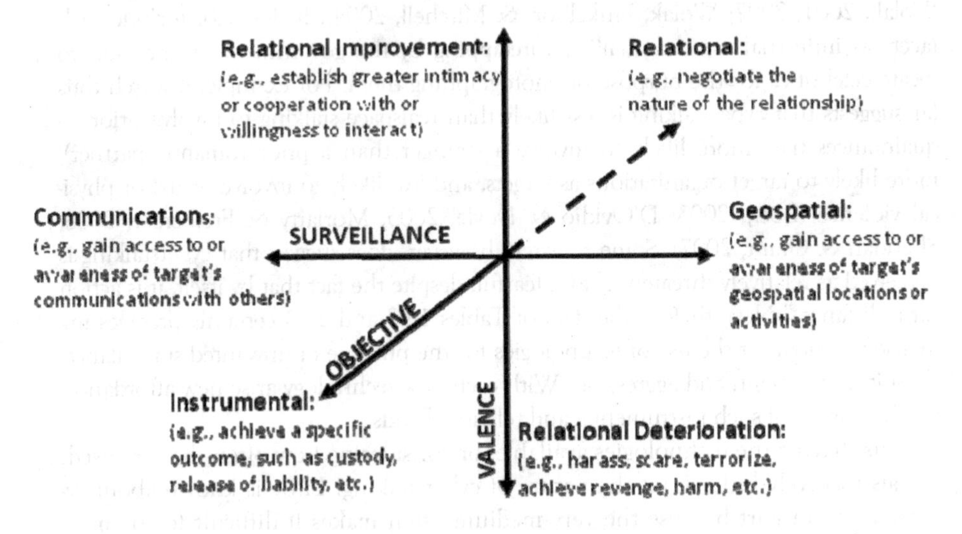

Table 18.1. Legal parameters of cyberstalking in U.S. legislation, broadly adapted from: § 14-196.3.

- Electronic communication, language or symbols
- threatening to...
  - inflict bodily harm,
  - physical injury to property
  - or to extort money or things of value.
- Electronic communication that...
  - repeatedly
  - intends to
  - abuse, annoy, threaten, terrify, harass, or embarrass.
- Electronic communication that...
  - knowingly
  - makes false statements concerning death, injury, illness, disfigurement, indecent conduct, or criminal conduct of recipient
  - intending to
  - abuse, annoy, threaten, terrify, harass, or embarrass.
- Knowingly permits electronic communication device(s) under the person's control to be used for any purpose enumerated above,
- But that excludes constitutionally protected speech or rights of assembly.

Sources: Adapted from http://www.ncsl.org/programs/lis/cip/stalk99.htm; http://www.ncga.state.nc.us/enactedlegislation/statutes/pdf/bysection/chapter_14/gs_14-196.3.pdf

Table 18.2. Prevalence estimates of cyberstalking and media of cyberstalking across studies

- 0% web pages (LeBlanc et al., 2001)
- 1% altering your electronic identity or persona (Spitzberg & Hoobler, 2002)
- 1% first meeting you online and then stalking you (Spitzberg & Hoobler, 2002)
- 1% **stalked by means of the Internet (Meloy et al., 2000)**
- 1% meeting first online and then following you (Spitzberg & Hoobler, 2002)
- 1% meeting first online and then harming you (Spitzberg & Hoobler, 2002)
- 2% directing others to you in threatening ways (Spitzberg & Hoobler, 2002)
- 2% emailed victims (n=519; Morewitz, 2003)
- 2% fax (Oddie, 2000)
- **2% stalked by means of the Internet (n=201; Kamphuis & Emmelkamp, 2001)**
- 3% (31.5% of those stalked) (n=756; Alexy et al., 2005)
- 3% attempting to disable your computer (Spitzberg & Hoobler, 2002)
- 3% cyber-attempting to disable your computer (Spitzberg & Hoobler, 2002)
- 3% meeting first online and then intruding in life (Spitzberg & Hoobler, 2002)
- 3% meeting first online and then threatening (Spitzberg & Hoobler, 2002)
- 3% ordered goods/services in your name (Bocij, 2003)
- 3% taking over your electronic identity or persona (Spitzberg & Hoobler, 2002)
- 4% **stalked by email (Ashmore et al., 2006)**
- 5% cyber-sending threatening pictures/images (Spitzberg & Hoobler, 2002)
- 5% send/leave unwanted email/faxes/pages (n=85; Palarea, 2004)
- 5% sending threatening pictures/images (Spitzberg & Hoobler, 2002)
- **6% cyber-stalked (Oddie, 2000)**
- 6% sending excessively disclosive messages (Spitzberg & Hoobler, 2002)
- 7% using your computer to get information on others (Spitzberg & Hoobler, 2002)
- 7% bugging your car, home, or office (n=235; Spitzberg & Hoobler, 2002)
- 7% using your computer to get information on others (Spitzberg & Hoobler, 2002)
- 8% telephoned, sent mail, or contacted electronically (McLennan, 1996)
- 9% sending threatening written messages (Spitzberg & Hoobler, 2002)
- 9% cyber-sending threatening written messages (Spitzberg & Hoobler, 2002)
- 9% Impersonated you in email messages to your friends/family/colleagues (Bocij, 2003)
- 9% text messages (n=45; Jones & Sheridan, 2009)
- 10% obtaining private information without permission (Spitzberg & Hoobler, 2002)

Continued on next page

Table 18.2. Prevalence estimates of cyberstalking and media of cyberstalking across studies *(continued)*

- **11% cyber-stalking (n=875; Purcell et al., 2009)**
- 11% repeated email (n=103; Morrison, 2008)
- 12% sabotaging your private reputation (Spitzberg & Hoobler, 2002)
- 17% attempted to access confidential information stored on your computer (Bocij, 2003)
- 17% exposing private information about you to others (Spitzberg & Hoobler, 2002)
- 18% sending sexually harassing messages (Spitzberg & Hoobler, 2002)
- 19% sending pornographic/obscene messages (Spitzberg & Hoobler, 2002)
- 20% pretending to be someone she or he wasn't (Spitzberg & Hoobler, 2002)
- **20% stalked "about 1–2 times per year, and 10% about every month" on Facebook (n=358; Stern & Taylor, 2007)**
- **22% "using stricter criteria…could be considered…genuine cyber-stalking" (n=167; Bocij, 2003)**
- 24% encouraged other users to harass, threaten, or insult you (Bocij, 2003)
- 24% posted false information (rumors) about you to a bulletin/chat room (Bocij, 2003)
- 25% "indirect means" (letter, package, cyberspace) (n=1005; Mohandie et al., 2006)
- **25% stalking cases involved email, 2% exclusively (Fisher et al., 2000)**
- 25% sending excessively "needy" or demanding messages (Spitzberg & Hoobler, 2002)
- 31% sending exaggerated messages of affection (Spitzberg & Hoobler, 2002)
- 31% sending tokens of affection (Spitzberg & Hoobler, 2002)
- 27% attempted to monitor your actions with Trojan horse software (Bocij, 2003)
- **27% cyberstalking (n=210; Rosay et al., 2007)**
- **30% "have been stalked by other people on SNSs (social network sites) at least once" (n=643; Shmeleva, 2009)**
- 31% "written missives (emails, letters, or faxes)" (n=200; McEwan et al., in press)
- 39% made threats or abusive comments via IM software (Bocij, 2003)
- 40% sent threatening or abusive email messages (Bocij, 2003)
- 41% attempted to damage your computer system with malicious programs (Bocij, 2003)
- 43% of computer investigation cases of NYPD (n=201; D'Ovidio & Doyle, 2003)
- 48% made threats or abusive comments in chat rooms (Bocij, 2003)
- 50% "unwanted letters, emails, or faxes" (n=400; Stieger et al., 2008)
- 58% "repeated emails," 13% information obtained through the web of those stalked (LeBlanc et al., 2000)
- 67% repeated phone/email (n=100; Morrison, 2001)

Table 18.3. Prevalence estimates of cyberstalking and cyber-surveillance from large scale U.S. nationally representative study

Cyber-stalking prevalence (Baum et al., 2009). Overall survey *N* = 65,000	**All**	**Stalking**	**Harassment**
% of cyber-stalking or monitoring			
• Cyber-stalking	26.6	26.6	27.4
• Electronic monitoring	23.4	21.5	26.4
% of cyber-stalking involving:[1]			
• Email	82.6	82.5	82.7
• Instant messaging	28.7	35.1	20.7
• Blogs or bulletin boards	12.5	12.3	12.8
• Internet sites re: victim	8.8	9.4	8.1
• Chat rooms	4.0	4.4	3.4
% of monitoring involving: [b]			
• Computer spyware	44.1	33.6	81.0
• Video/digital cameras	40.3	46.3	19.3
• Listening devices	35.8	41.8	14.8
• GPS	9.7	10.9	5.2

a. Based on 1,217,680 total victims, 77,870 stalking victims, & 539,820 harassment victims

b. Based on 314,400 total victims, 244,880 stalking victims, & 69,530 harassment victims

References

Afifi, W. A., & Burgoon, J. K. (2000). The impact of violations on uncertainty and the consequences for attractiveness. *Human Communication Research, 26*, 203–233.

Altman, L., Vinsel, A., & Brown, B. (1981). Dialectic conceptions in social psychology: An application to social penetration and privacy regulation. In L. Berkowitz (Ed.), *Advances in experimental social psychology* (Vol. 14, pp. 107-160). New York: Academic.

Baum, K., Catalano, S., Rand, M., & Rose, K. (2009, January). *Stalking victimization in the United States* (NCJ 224527). Washington, DC: Bureau of Justice Programs, U.S. Department of Justice.

Berger, C. R. (1979). Beyond initial interaction: Uncertainty, understanding, and the development of interpersonal relationships. In H. Giles & R. St. Clair (Eds.), *Language and social psychology* (pp. 122–144). Oxford, UK: Basil Blackwell.

Berger, C. R. (1987). Communicating under uncertainty. In M. E. Roloff & G. R. Miller (Eds.), *Interpersonal processes* (pp. 39–62). Newbury Park, CA: Sage.

Bocij, P. (2003). Victims of cyberstalking: An exploratory study of harassment perpetrated via the Internet. *First Monday, 8* (10–6) 1–17.

Brown, P., & Levinson, S. C. (1978). *Politeness: Some universals in language usage*. Cambridge, UK: Cambridge University Press.

Buss, D. M. (1994). *The evolution of desire*. New York: Basic.

Chik, W. (2008). Harassment through the digital medium: A cross-jurisdictional comparative analysis on the law on cyberstalking. *Journal of International Commercial Law and Technology, 3*, 13–44.

Cupach, W. R., & Spitzberg, B. H. (2000). Obsessive relational intrusion: Incidence, perceived severity, and coping. *Violence and Victims, 15*, 357–372.

Cupach, W. R., & Spitzberg, B. H. (2004). *The dark side of relationship pursuit: From attraction to obsession to stalking*. Mahwah, NJ: Lawrence Erlbaum.

Cupach, W. R., & Spitzberg, B. H. (2008a). "Thanks but no thanks...": The occurrence and management of unwanted relationship pursuit. In S. Sprecher, A. Wenzel, & J. Harvey (Eds.), *Handbook of relationship initiation* (pp. 409–424). New York: Taylor & Francis.

Cupach, W. R., & Spitzberg, B. H. (2008b). Unwanted relationship initiation. In J. Harvey & B. Wenzel (Eds.), *Handbook of relationship beginnings*. Mahwah, NJ: Lawrence Erlbaum.

D'Ovidio, R., & Doyle, J. (2003). A study on cyberstalking: Understanding investigative hurdles. *FBI Law Enforcement Bulletin, 72*, 10–17.

Edlund, J. E., & Sagarin, B. J. (2009). Sex differences in jealousy: Misinterpretation of nonsignificant results as refuting the theory. *Personal Relationships, 16*, 67-78.

Ellison, N. B., Steinfield, C., & Lampe, C. (2007). The benefits of Facebook "friends": Social capital and college students' use of online social network sites. *Journal of Computer-Mediated Communication, 12*(4), article 1.

Finkelhor, D., Mitchell, K. J., & Wolak, J. (2000). *Online victimization: A report on the nation's youth*. Alexandria, VA: Crimes Against Children Research Center.

Gardner, W. L., III, & Martinko, M. J. (1998). An organizational perspective of the effects of dysfunctional impression management. *Dysfunctional behavior in organizations: Violent and deviant behavior* (Monographs in Organizational Behavior and Industrial Relations, Vol. 23, Part B, pp. 69–125). Stamford, CT: JAI.

Goffman, E. (1967). On facework: An analysis of ritual elements in social interaction. In A. Jawrski & N. Coupland (Eds.), *The discourse reader* (pp. 306–321). London: Routledge.

Goffman, E. (1969). *Strategic Interaction*. Philadelphia: University of Pennsylvania Press.

Goffman, E. (1974). *Frame analysis: An essay on the organization of experience*. New York: Harper and Row.

Gould, S. J. (1978). The panda's peculiar thumb. *Natural History, 87*, 20–30.

Guerrero, L. K., & Andersen, P. A. (1995). Coping with the green-eyed monster: Conceptualizing and measuring communicative responses to romantic jealousy. *Western Journal of Communication, 59*, 270–304.

Harris Interactive Press Release (2006, February 9). *More think it is important to give than to receive on Valentine's Day according to new survey*. Retrieved July 18, 2009, from http://www.harrisinteractive. com/news/allnewsbydate.asp?NewsID=1018

Hinduja, S., & Patchin, J. W. (2008). Personal information of adolescents on the Internet: A quantitative content analysis of MySpace. *Journal of Adolescence, 31*, 125-146.

Kamir, O. (2001). *Every breath you take: Stalking narratives and the law*. Ann Arbor: University of Michigan Press.

Leary, T. (1957). *Interpersonal diagnosis of personality: A functional theory and methodology for personality evaluation*. New York: Ronald.

Lenhart, A., Madden, M., & Hitlin, P. (2005). Teens and technology. Retrieved July 18, 2009, from http://www.pewinternet.org/pdfs/PIP_Teens_Tech_July2005web.pdf

Lenhart, A., Rainie, L., & Lewis, O. (2001). *Teenage life online: The rise of the instant message generation and the Internet's impact on friendships and family relationships.* Retrieved July 18, 2009, from http://www.pewinternet.org/reports/pdfs/PIP_Teens_Report.pdf

Maslow, A. H. (1943). A theory of human motivation. *Psychological Review, 50*, 370–396.

McFarlane, L., & Bocij, P. (2003). An exploration of predatory behavior in cyberspace: Towards a typology of cyberstalkers. *First Monday*, 8 (9–1), 1–14.

Metts, S., & Cupach, W. R. (1994). *Facework*. Newbury Park, CA: Sage.

Metts, S., & Grohskopf, E. (2003). Impression management: Goals, strategies, and skills. In J. O. Greene & B. R. Burleson (Eds.), *Handbook of communication and social interaction skills* (pp. 357–399). Mahwah, NJ: Lawrence Erlbaum.

Miller, N. (2001). Stalking investigation, law, public policy, and criminal prosecution as problem solver. In J. A. Davis (Ed.), *Stalking crimes and victim protection: Prevention, intervention, threat assessment, and case management* (pp. 387–426). Boca Raton, FL: CRC.

Mitchell, K. J., Finkelhor, D., & Wolak, J. (2001). Risk factors for and impact of online sexual solicitation of youth. *Journal of the American Medical Association, 285*, 3011–3014.

Mitchell, K. J., Finkelhor, D., & Wolak, J. (2007). Online requests for sexual pictures from youth: Risk factors and incident characteristics. *Journal of Adolescent Health, 41*, 196–203.

Modena Group on Stalking, & Emilia, R. (2007, April). *Protecting Women from the New Crime of Stalking: A Comparison of Legislative Approaches within the European Union.* Modena Group on Stalking/Daphne Project Final Report 05–1/125/W. Universita Degli Studi Di Modena and Reggio Emilia. Retrieved July 18, 2009, from http://stalking.medlegmo.unimo.it/RAPPORTO_ versione_finale_011007.pdf

Mohandie, K., Meloy, R., McGowan, M. G., & Williams, J. (2006). The RECON typology of stalking: Reliability and validity based upon a large sample of North American stalkers. *Journal of Forensic Sciences, 51*, 147–155.

Moriarty, L. J., & Freiberger, K. (2008). Cyberstalking: Utilizing newspaper accounts to establish victimization patterns. *Victims and Offenders*, 3, 131–141.

Osgood, C. E., May, W. H., & Miron, M. S. (1975). *Cross-cultural universals of affective meaning.* Urbana: University of Illinois Press.

Parks, M., & Archey-Ladas, T. (2003). *Communicating self through personal homepages: Is identity more than screen deep?* Paper presented at the International Communication Association convention, San Diego, CA.

Peluchette, J., & Karl, K. (2008). Social networking profiles: An examination of student attitudes regarding use and appropriateness of content. *CyberPsychology & Behavior, 11*, 95-97.

Pew Internet & American Life Project (PEW). (2005, July 27). *Teens and technology: Youth are leading the transition to a fully wired and mobile nation.* Washington, DC: Pew Internet and American Life Project. Retrieved July 18, 2009, from http://www.pewinternet.org

Pew Internet & American Life Project. (2006, June 22). *Friends, family and community: Social networking sites.* Washington, DC: Pew Internet and American Life Project. Retrieved July 18, 2009, from http://www.pewinternet.org

Pew Internet & American Life Project (PEW). (2007, January 1). *Social networking and teens.* Washington, DC: Pew Internet and American Life Project. Retrieved July 18, 2009, from http://www.pewinternet.org

Phillips, M. A. (2009a, February). *Myspace or yours?: Connections between the use of social networking sites and cyber-surveillance in romantic relationships.* Paper presented at the Western States Communication Association conference, Mesa, AZ.

Phillips, M. A. (2009b). *You're invading MySpace!: Predicting the use of social networking sites for surveillance in romantic relationships.* Unpublished master's thesis, San Diego State University, San Diego, CA.

Rice, R. E. (1993). Media appropriateness: Using social presence theory to compare traditional and new organizational media. *Human Communication Research, 19*, 451–484.

Rosenfeld, B. (2004). Violence risk factors in stalking and obsessional harassment: A review and preliminary meta-analysis. *Criminal Justice and Behavior, 31*, 9–36.

Salaway, G., & Caruso, J, B., with Nelson, M. R. (2008). *The ECAR Study of Undergraduate Students and Information Technology*, Vol. 8. Boulder, CO: EDUCAUSE Center for Applied Research. Retrieved July 18, 2009, from http://net.educause.edu/ir/library/pdf/ERS0808/RS/ ERS0808w.pdf

Shackelford, T. K., Goetz, A. T., LaMunyon, C. W., Quintas, B. J., & Weekes-Shackelford, V. A. (2004). Sex differences in sexual psychology produce sex-similar preferences for a short-term mate. *Archives of Sexual Behavior, 33*, 405–412.

Shepperd, J. A., & Kwavnick, K. D. (1999). Maladaptive image maintenance. In R. M. Kowalski & M. R. Leary (Eds.), *The social psychology of emotional and behavioral problems* (pp. 249–277). Washington, DC: American Psychological Association.

Sheridan, L. P., & Grant, T. (2007). Is cyberstalking different? *Psychology, Crime & Law, 13*, 627–640.

Spitzberg, B. H. (2002). The tactical topography of stalking, victimization, and management. *Trauma, Violence, & Abuse, 3*, 261–288.

Spitzberg, B. H., Canary, D. J., & Cupach, W. R. (1994). A competence-based approach to the study of interpersonal conflict. In D. D. Cahn (Ed.), *Conflict in personal relationships* (pp. 183–202). Hillsdale, NJ: Lawrence Erlbaum.

Spitzberg, B. H., & Cupach, W. R. (2002). The inappropriateness of relational intrusion. In R. Goodwin & D. Cramer (Eds.), *Inappropriate relationships: The unconventional, the disapproved, and the forbidden* (pp. 191–219). Mahwah, NJ: Lawrence Erlbaum.

Spitzberg, B. H., & Cupach, W. R. (2003). What mad pursuit? Conceptualization and assessment of obsessive relational intrusion and stalking-related phenomena. *Aggression and Violent Behavior: A Review Journal, 8*, 345–375.

Spitzberg, B. H., & Cupach, W. R. (2007). Cyber-stalking as (mis)matchmaking. In M. T. Whitty, A. Baker, & J. Inman (Eds.), *Online matchmaking* (pp. 127–146). Basingstoke, UK: Palgrave Macmillan.

Spitzberg, B. H., & Cupach, W. R. (2007). The state of the art of stalking: Taking stock of the emerging literature. *Aggression and Violent Behavior: A Review Journal, 12*, 64–86.

Spitzberg, B. H., & Hoobler, G. (2002). Cyberstalking and the technologies of interpersonal terrorism. *New Media & Society, 14*, 67-88.

Sprecher, S., Schwartz, P., Harvey, J., & Hatfield, E. (2008). Thebusinessoflove.com: Relationship initiation at Internet matchmaking services. In S. Sprecher, A. Wenzel, & J. Harvey (Eds.), *Handbook of relationship initiation* (pp. 249-265). New York: Psychology Press.

Subramanyam, K., Greenfield, P. M., Kraut, R., & Gross, E. (2002). The impact of computer use on children and adolescents' development. In S. L. Calvert, A. B. Jordan, & R. R. Cocking (Eds.), *Children in the digital age: Influences of electronic media on development* (pp. 3-33). Westport, CT: Praeger.

Turkle, S. (1995). *Life on the screen: Identity in the age of the internet.* New York: Simon & Schuster.

Vonk, R. (2001). Aversive self-presentations. In R. M Kowalski (Ed.), *Behaving badly: Aversive behaviors in interpersonal relationships* (pp. 79–115). Washington, DC: American Psychological Association.

Walther, J. B. (1996). Computer-mediated communication: Impersonal, interpersonal, and hyperpersonal interaction. *Communication Research, 23*, 3–43.

Walther, J. B., & Parks, M. R. (2002). Cues filtered out, cues filtered in: Computer-mediated communication and relationships. In M. L. Knapp & J. A. Daly (Eds.), *Handbook of interpersonal communication* (3rd ed., pp. 529–563). Thousand Oaks, CA: Sage.

Walther, J. B., & Tidwell, L. C. (2000). Computer-mediated communication: Interpersonal interaction online. In K. M. Galvin & P. J. Cooper (Eds.), *Making connections: Readings in relational communication* (2nd ed., pp. 322-329). Los Angeles, CA: Roxbury.

Wang, H. (2004). *Self-disclosure in long-distance friendships: A comparison between face-to-face and computer-mediated communication.* Unpublished master's thesis, San Diego State University, San Diego, CA.

Whitty, M. T. (2003). Cyber-flirting: Playing at love on the Internet. *Theory & Psychology, 13*, 330–357.

Wolak, J., Finkelhor, D., & Mitchell, K. (2004). Internet-initiated sex crimes against minors: Implications for prevention based on findings from a national study. *Journal of Adolescent Health, 35*, 424–433.

CHAPTER NINETEEN

Problematic Youth Interactions Online: Solicitation, Harassment, and Cyberbullying

Andrew R. Schrock

danah boyd

The adoption of the Internet by American youth (Center for the Digital Future, 2008; Madden, 2006) and the recent rise of social media have provided youth with a powerful space for socializing, learning, and engaging in public life (boyd, 2007; Gross, 2004; Ito, Baumer, Bittanti, boyd, Cody, Herr-Stephenson, Horst, Lange, Mahendran, Martinez, Pascoe, Perkel, Robinson, Sims, & Tripp, 2009; Palfrey & Gasser, 2008). Most American youths navigate an online environment from childhood through adolescence, where they explore their identity, interact with peers, and develop relationships through social network sites (SNSs), online chats, massively multiplayer online games (MMOGs), message boards, and blogs. While the majority of parents (59%) say the Internet is a "positive influence" in their children's lives (Rideout, 2007), there is also growing concern about the risks of online interactions. Parents, teachers, and law enforcement have raised concerns about the dangers posed by new forms of online communication, particularly online predators, social network sites (Cassell & Cramer, 2007; Marwick, 2008), anonymous contact, and "sexting" (multimedia messaging on mobile devices). This chapter summarizes and interprets the character and scope of research on two types of problematic interpersonal communication that are central to these fears: online solicitation and cyberbullying (or online harassment). Based on an emerging body of research, conclusions can be drawn on the prevalence of these problematic forms of online communication to young Americans, risk factors, how technologies mediate risks, and areas for future research.

The Role of the Legal System

Laws (local, state, and federal) regulating these communications are constantly shifting in response to new data, media coverage, public demands, and perceptions of lawmakers. It is the authors' contention that social science generally, and communication research specifically, should lead the discussion of the practical and moral questions inherent in risky online interactions. Therefore, we periodically use legal references for context and clarity, but a legalistic analysis is outside of the scope of this document. Beyond the quickly changing nature of laws, are several reasons why solicitation and harassment are particularly difficult to discuss from a legalistic standpoint. The illegal status of these crimes may rest primarily on laws that do not directly address the perceived crime. For instance, in the case of Megan Meier, who committed suicide when an adult neighbor harassed her online, the neighbor was prosecuted on "accessing protected computers without authorization" and one count of conspiracy, not the act of harassment itself (Associated Press, 2008). This ruling was overturned on August 28, 2009 by Judge Wu, because it was deemed to criminalize anyone who signed up for an online service using inaccurate or fake information, a common online practice. This confrontation between societal freedoms and individual safety is frequently repeated in debates on the subject. Lack of proximity is also an issue, because parties involved in solicitation and harassment online may reside in different states or countries, and are frequently anonymous to the end-user, making it difficult to discern where laws are being broken. The culture of the Internet does not place an emphasis on legal boundaries, as is evident with the current acceptability to youth of illegally downloading videos and movies. Finally, these problematic interactions unfold over time, and may not start online. For instance, some cyberbullying incidents overflow online from the schoolyard and involve student peers or offline friends. Other activities, such as the exchange of erotic pictures or text communiqués, may be lurid and uncomfortable for parents to talk about, yet take place between two willing (albeit underage) parties.

Perpetrator Characteristics

The focus of researchers interested in online harassment and solicitation has historically been on the individuals who initiate these encounters. Offending parties are frequently anonymous, and include both adults and youths. When youths could identify a perpetrator, it was often a similarly aged individual doing the solicitation (Wolak, Mitchell, & Finkelhor, 2006) and online harassment (Hinduja & Patchin, 2008a; McQuade & Sampat, 2008; Smith, Mahdavi, Carvalho, Fisher, Russell, & Tippett, 2008). There was also often an overlap between cyberbullying offenders and victims (Beran & Li, 2007; Kowalski & Limber, 2007; Ybarra & Mitchell, 2004a), meaning offenders and victims are not mutually exclusive groups. The minority of Internet solicitations that lead to offline sexual encounters took place between adults in their 20s and post-pubescent adolescents, in a model similar to that of statutory rape (Hines & Finkelhor, 2007).

Adults who solicit or commit sexual offenses against teenagers are a widely disparate group with few commonalities in psychology and motivations for offending. Contrary to the media's claim that pedophilia is at the root of most online sexual abuse, sexual attraction is only one of many reasons behind why adults perpetrate these crimes. Other factors include mental disorders (depression, poor impulse control), goals (desire for power and to engage in deviant acts), impulse control, and a generally anti-social character (Salter, 2004). Moreover, despite a frequent misuse of the term, adults who solicit or molest *adolescents* are, by definition, not *pedophiles* (American Psychological Association, 2000; World Health Organization, 2007), because "[s]exual practices between an adult and an adolescent and sexual aggression against young majors do not fall within the confines of pedophilia" (Arnaldo, 2001, p. 45).

The overall prevalence of adult and youth offenders in the general population is unknown. These remain extremely difficult populations to research, as they are mostly anonymous, globally distributed, and may not participate in offline crimes. Similar to many crimes, large-scale quantitative data on offenders, outside of data obtained from those in various stages of incarceration or rehabilitation, does not exist. The challenge of collecting meaningful information on these incidents has been called a "tip of the iceberg" problem, where the number of reported offenses might be much lower than the actual number of offenders (Sheldon & Howitt, 2007, p. 43), leading to vastly differing estimations of population size.

Sexual Solicitation and Internet-Initiated Offline Encounters

One of parents' greatest fears concerning online safety is the risk of "online predators" that entice youth to offline encounters. This topic is the center of tremendous public discourse (Marwick, 2008) and was the central theme in the popular TV show, "To Catch a Predator." In 2007, more than half (53%) of adults agreed with the statement that, "online predators are a threat to the children in their households" (Center for the Digital Future, 2008). Parents are particularly concerned that adults will coerce their children into offline sexual encounters, abduct them, or worse.

The reality is that few online solicitations lead to offline encounters, as many of these contacts are merely harassing or teasing, and physical abductions of children following from online meetings are nearly nonexistent. When online meetings develop into offline sexual encounters, they are most common between pairs of adolescents and between older adolescents and 20-somethings (Wolak et al. 2006; Wolak, Finkelhor, Mitchell, & Ybarra, 2008b). A sizeable minority (roughly 10–16%) of American youths make connections online that lead to in-person meetings, but they are primarily non-sexual and related to friendship (Wolak et al. (2006).

Fear of strangers sexually abusing children pre-dates the Internet (Glassner, 1999). While this "stranger danger" rhetoric is pervasive, it is not effective at keeping kids safe (McBride, 2005), because 95% of offline sexual assault cases reported to authorities are committed by family members or known acquaintances (Snyder & Sickmund, 2006). Dire predictions about the threat of Internet-initiated sex crimes committed by strangers appear to be exaggerated, as the vast majority of sexual abuse of youths still occurs offline between known parties (Finkelhor & Ormrod, 2000; Wolak, Finkelhor, & Mitchell, 2009).

Solicitation

An online sexual solicitation is broadly defined as an online communication where "someone on the Internet tried to get [a minor] to talk about sex when they did not want to," an offender asked a minor to "do something sexual they did not want to," or other sexual overtures coming out of online relationships (Finkelhor, Mitchell, & Wolak, 2000). The first and second Youth and Internet Safety Surveys (YISS) indicated that 13–19% of youths had experienced some form of online sexual solicitation in the past year. Given the anonymity of communication, it is often impossible to objectively assess the age of solicitors. Nonetheless, youths reported that they believed that 43% of solicitors were under 18, 30% were between 18 and 25, 9% were over 25, and 18% were completely unknown (Wolak et al., 2006).

The definition of solicitation encompasses a range of sexualized online contact, including taunting emails, lascivious text messages, fake social networking site profiles that use sexual imagery, and the distribution of digital photos. While some solicitations are designed to lead to an offline sexual encounter, very few actually do. Some of this behavior is "flirting" between minors (McQuade & Sampat, 2008, Smith, 2007), while other solicitations are simply meant to be upsetting (Biber, Doverspike, Baznik, Cober, & Ritter, 2002; Finn, 2004; Wolfe & Chiodo, 2008). This umbrella term of "online solicitation" can be considered similar in dimensionality to offline conceptions of sexual harassment, but is confounded by physical and temporal distance offered by new technologies. For instance, Fitzgerald, Gelfand, and Drasgow (1995) found support for a three-dimensional model of offline sexual harassment: gender harassment, unwanted sexual attention, and sexual coercion. Barak (2005) applied this model to sexual harassment on the Internet and argued that online disinhibition, openness, venturesome attitudes, and a masculine atmosphere were instrumental in online sexual harassment.

There are relatively few large-scale quantitative studies concerning the prevalence of online sexual solicitation (Fleming & Rickwood, 2004; McQuade & Sampat, 2008) and even fewer national U.S.-based studies (Wolak et al., 2006). The experiences of key stakeholders, such as school counselors and medical personnel, remain poorly understood. For instance, there have only been two studies that collected law enforcement data on Internet-initiated sex crimes against minors, called the National Juvenile Online Victimization (N-JOV) series of studies (Wolak, Finkelhor, & Mitchell, 2004; Wolak et al., 2009). Key points of data collection are as-yet untapped, such as nonprofits and rape crisis centers, which would yield data from sources other than incarcerated offenders and youths who experience this contact.

Online sexual solicitations by adults are of great concern because coercive communications are thought to "groom" youth (Berson, 2003) in a manner similar to offline child molesters (Lang & Frenzel, 1988). There are several reasons why online solicitations, although some are designed to entice youth into offline sexual relationships, are quite dissimilar to those of the media-propagated image of the pedophile enticing children to participate in either offline or online sexual encounters: neither online solicitations nor Internet-initiated relationships particularly tend to target pre-pubescent children; when offline sexual encounters occur, they happen multiple times; and significant deception is uncommon (Wolak et al., 2008b). While adults may shave off a few years from their real age, only 5% of offenders claimed to be the same age as the youth victim (Wolak et al., 2004). Wolak et al. (2008b) concluded that, "when deception does occur, it often involves promises of love and romance by offenders whose intentions are primarily sexual" (p. 113).

Online solicitations are not disturbing to a majority of recipients, as most youths (66–75%) who were solicited were not psychologically distressed by this type of contact (Wolak et al., 2006). A minority of all youths surveyed (4%) reported *distressing* sexual solicitations online which made them feel "very upset or afraid" (Wolak et al., 2006, p. 15), or *aggressive* online sexual solicitations (4%), where the offender "asked to meet the youth in person; called them on the telephone; or sent them offline mail, money, or gifts" (Wolak et al., 2006, p. 15). A small number (2%) of youths reported both aggressive and distressing solicitations. The researchers concluded that while some of the solicitations were problematic, "close to half of the solicitations were relatively mild events that did not appear to be dangerous or frightening" (Wolak et al., 2006, p. 15). Online solicitations were also concentrated in older adolescents; youths 14–17 years old reported 79% of aggressive incidents and 74% of distressing incidents (Wolak et al., 2006, p. 15).

Offline Connections

Most offline meetings between youths and adults who met on the Internet were not of a sexual nature. Between 9–16% of youths reported Internet-initiated offline meetings in the United States, across various locations, sample sizes, administration dates, and wording of surveys (Berrier, 2007; Berson & Berson, 2005; McQuade & Sampat, 2008; Rosen, Cheever, & Carrier, 2008; Wolak et al., 2006). Studies in Europe, New Zealand, and Singapore show a wider range (8–26%) of Internet-initiated offline encounters (Berson & Berson, 2005; Gennaro & Dutton, 2007; Liau, Khoo, & Ang, 2005; Livingstone & Bober, 2004; Livingstone & Haddon, 2008), likely due to social norms, varying Internet usage habits overseas, and other international differences. Out of these youths, a minority were involved in sexual contact. In the first Youth and Internet Safety Survey (YISS-1), administered in 2000, no instances of Internet-initiated sex were reported. In the second Youth and Internet Safety Survey (YISS-2), administered in 2005, 0.03% (4 in 1,500) of youths reported physical sexual contact with an adult they met online, and all were 17-year-olds engaging in sexual acts with adults. Two youths out of 1,500 (one 15-year-old girl and one 16-year-old girl) surveyed reported an offline sexual assault resulting from online solicitation. In the small number of offline meetings between minors and adults that involved sex, the offense typically followed a model of statutory rape: a post-pubescent minor had non-forcible sexual relations with an adult in their 20s (Hines & Finkelhor, 2007; Wolak et al., 2004; Wolak et al., 2008b).

Data from law enforcement similarly describe how the offense is similar to statutory rape, and involved sexually-themed chat over time, often leading to multiple meetings. Most (80%) online sex offenders brought up sex in online communication with youths, meaning that, "the victims knew they were interacting with adults who were interested in them sexually" (Wolak et al., 2004, p. 424.e18) before the first meeting. Most (73%) of Internet-initiated sexual relationships developed between an adult and a minor involved multiple meetings (Wolak et al., 2004), indicating that the minor was aware of the ongoing sexual nature of the relationship. Internet-initiated sexual encounters between an adult and adolescent were also unlikely to be violent (5% of cases) and none involved "stereotypical kidnappings in the sense of youths being taken against their will for a long distance or held for a considerable period of time" (Wolak et al., 2004, p. 424.e17).

This does not diminish the illegal nature of statutory sex crimes, but signals an opportunity to re-orient the messages provided to youth about how these crimes typically unfold. They are certainly not benign relationships, and some are psychologically harmful to youths (Hines & Finkelhor, 2007). At the same time, it is important to recognize the role that teens play in these types of relationships, because, "if some young people are initiating sexual activities with adults they meet on the Internet, we cannot be effective if we assume that all such relationships start with a predatory or criminally inclined adult" (Hines & Finkelhor, 2007, p. 301).

Victims and Perpetrators

The focus of research has shifted over the last several years, from offenders to characteristics of adolescents who are solicited online (Peter, Valkenburg, & Schouten, 2005; Ybarra & Mitchell, 2004a; Ybarra, Mitchell, Wolak, & Finkelhor, 2006). If online solicitations are relatively common, and offline sexual relationships typically necessitate an awareness of the youth of the nature of the relationship, what risk factors are related to this escalation? Adolescents are most likely to be solicited online, while the solicitation of pre-pubescent children by strangers (including solicitations leading to an offline sexual encounter) is rare (Wolak et al., 2006). Youth victims of online solicitation also tend to be female (Wolak et al., 2006) and experiencing difficulties offline, such as physical or sexual abuse (Mitchell, Wolak, & Finkelhor, 2007a).

Teens who reported online solicitations tended to be of the age that it is developmentally normal to be curious about sex (Ponton & Judice, 2004). Older youths (teenagers) are more likely to be solicited online and also to respond to these solicitations with real-world encounters, confirmed by both arrests for Internet-initiated sex crimes (Wolak et al., 2004) and youths' self-reports in surveys (Berson & Berson, 2005; McQuade & Sampat, 2008; Rosen et al., 2008; Wolak et al., 2006). Nearly all (99%) victims of Internet-initiated sex crime arrests in the N-JOV study were aged 13–17 years old, with 76% being high school aged, 14–17 (Wolak, Ybarra, Mitchell, & Finkelhor, 2007c), and none younger than 12 years old. Far from being naïve with regard to technology, these adolescents are thought to be more at-risk because they "engage in more complex and interactive Internet use. This actually puts them at greater risk than younger, less experienced youths" (Wolak et al., 2008b, p. 114). This is a perspective that is at odds with studies and programs that have equate younger age with greater risk due to a lack of understanding of how new technologies work (Fleming, Greentree, Cocotti-Muller, Elias, & Morrison, 2006; Brookshire & Maulhardt, 2005).

A typical scenario for Internet solicitation and Internet-initiated sexual encounters is a Caucasian male soliciting a teenage girl. According to interviewed youths, girls received the majority (70–75%) of online solicitations, and 73% of those solicited reported that the perpetrator was male (Wolak et al., 2006). In arrest records for Internet-initiated sex crimes, 99% of offenders were male and 75% of cases involved female victims (Wolak et al., 2004). In the N-JOV study, records showed that adult offenders who were arrested for Internet-initiated relationships online with minors tended to be male (99%), non-Hispanic white (81%), and communicated with the victim for 1–6 months (48%).

Online Harassment and Cyberbullying

Online harassment or "cyberbullying" is defined as "an overt, intentional act of aggression towards another person online" (Ybarra & Mitchell, 2004a, p. 1308) or a "willful and repeated harm inflicted through the use of computers, cell phones, and other electronic devices" (Hinduja & Patchin, 2008a, p. 5). This contact threatens, embarrasses, or humiliates youths (Lenhart, 2007), and involves private (such as chat or text messaging), semi-public (such as posting a harassing message on an email list), or public communications (such as creating a website devoted to making fun of the victim). The reach of cyberbullying is "magnified" when mediated through the Internet (Lenhart, 2007, p. 5) because the actual location of bullying may be away from the school setting, and is thus more pervasive (Ybarra, Diener-West, & Leaf, 2007a). "Cyberbullying" and "online harassment" have much conceptual similarity, and are frequently used interchangeably both within and outside of academic dialogue (Finkelhor, 2008, p. 26).

The problem of online harassment of minors is more widespread than solicitation, with 4–46% of youths reporting being cyberbullied (Agatston, Kowalski, & Limber, 2007; Fight crime sponsored studies: Opinion research corporation, 2006a; Fight crime sponsored studies: Opinion research corporation, 2006b; Finkelhor, Mitchell, & Wolak, 2000; Hinduja & Patchin, 2008a; Kowalski & Limber, 2007; Kowalski, Limber, & Agatston, 2007; McQuade & Sampat, 2008; National Children's Charity, 2005; Patchin & Hinduja, 2006; Smith et al., 2008; Williams & Guerra, 2007; Wolak et al., 2006), depending on how it is defined; date and location of data collection; and the time population under investigation. Despite its prevalence, cyberbullying is not reported to occur at higher overall rates than offline bullying (Lenhart, 2007; Li, 2007; Nansel, Overpeck, Pilla, June, Simons-Morton, & Scheidt, 2001). Cyberbullying also frequently lacks characteristics of "schoolyard bullying" such as aggression, repetition, and an imbalance of power (Wolak, Mitchell, & Finkelhor, 2007a), leading to dispute among researchers about the similarity between online and offline bullying (Burgess-Proctor, Patchin, & Hinduja, 2009; Hinduja & Patchin, 2008a; Wolak et al., 2007a).

Victims

About a third of cyberbullying involved "distressing harassment" (Wolak et al., 2006). Distress stemming from cyberbullying victimization can lead to negative effects similar to offline bullying such as depression, anxiety, and having negative social views of themselves (Hawker & Boulton, 2000). As Patchin and Hinduja describe it, "the negative effects inherent in cyberbullying... are not slight or trivial and have the potential to inflict serious psychological, emotional, or social harm" (Patchin & Hinduja, 2006, p. 149). Wolak et al. (2006) found that youths (10–17 year olds) who were bullied may feel upset (30%), afraid (24%), or embarrassed (22%). Similarly, Patchin and Hinduja (2006) found that 54% of victims were negatively affected in some way, such as feeling frustrated, angry, or sad. This is of concern not just because the youths had emotional responses, but also because negative emotions may be improperly resolved by adolescents through self-destructive behaviors, interpersonal violence, and various forms of delinquency (Borg, 1998; Ericson, 2001; Rigby, 2003; Roland, 2002; Seals & Young, 2003). Negative school-based effects of online harassment have been shown to occur, such as lower grades and absenteeism in school (Beran & Li, 2007).

Perpetrators

Online harassers and their victims are frequently both underage (Kowalski & Limber, 2007; Slonje & Smith, 2008; Wolak et al., 2006; Wolak et al., 2007a). Between 11–33% of minors admitted to harassing others online (Kowalski & Limber, 2007; McQuade & Sampat, 2008; National Children's Charity, 2005; Patchin & Hinduja, 2006; Wolak et al., 2006;). Half of the victims reported that cyberbullies were in their same grade (Stys, 2004), and 44% of victims reported that the perpetrator was an offline friend (Wolak et al., 2006).

Offline Connections

Distinguishing between victims and perpetrators can be challenging because some victims of online harassment are also perpetrators. Between 3 and 12% of youths have been found to be both victims and perpetrators of online harassment (Ybarra

& Mitchell, 2004a; Beran & Li, 2007; Kowalski & Limber, 2007). These aggressor-victims experience combinations of risks and are, "especially likely to also reveal serious psychosocial challenges, including problem behavior, substance use, depressive symptomatology, and low school commitment" (Ybarra & Mitchell, 2004a, p. 1314). The overlap between online perpetrators and victims shares conceptual similarities to offline "bully-victims" (those who are both bullies and the victims of bullying), which are reported to involve between 6–15% of U.S. youth (Haynie, Nansel, Eitel, Crump, Saylor, Yu, 2001; Nansel et al., 2001). Although these studies conceive of the victim-perpetrator overlap as being related to individual psychosocial qualities, victims may also respond to online harassment in a more direct manner. In a recent study, 27% of teenaged girls were found to "cyberbully back" in retaliation for being bullied online (Burgess-Proctor et al., 2009). Those who are engaged in online harassment but are offline victims may see the Internet as a "place to assert dominance over others as compensation for being bullied in person" or "a place where they take on a persona that is more aggressive than their in-person personality" (Ybarra & Mitchell, 2004a).

The connection between online and offline harassment is complex. Online bully and victim populations partly overlap, and sometimes involve entirely unknown harassers. Due to its apparent similarity to schoolyard bullying (which some researchers dispute), the most frequent way used to determine an overlap with offline bullying is whether it was experienced in a school setting, although this is sometimes difficult to determine, giving the range of technologies involved (an email could be sent at home and read in school, for example). By this measure, less than half of online harassment is related to school bullying, either through location (occurring at school) or peers (offender or target is a fellow student) (Beran & Li, 2007; Ybarra et al., 2007a). In other studies, over half of known bullies (or around 25% of the total number of cyberbullies) were identified as being from school (Slonje & Smith, 2008). Other studies demonstrated connections between online and offline bully perpetration (Hinduja & Patchin, 2007; Raskauskas & Stoltz, 2007), and online and offline bully victimization (Beran & Li, 2007; Hinduja & Patchin, 2007; Kowalski & Limber, 2007; Slonje & Smith, 2008, p. 152; Ybarra et al., 2007a). Social and academic performance is also hindered by cyberbullying. For example, those bullied outside of school were four times more likely to carry a weapon to school (Nansel, Overpeck, Haynie, Ruan, & Scheidt, 2003), and youths who experience cyberbullying were more likely to report alcohol and drug use, cheating at school, truancy, assaulting others, damaging property, and carrying a weapon (2007).

Solicitation

There is a small overlap between online harassment and solicitation, both in victims and perpetrators (Ybarra, Espelage, & Mitchell, 2007b), although little research has been performed on the topic. Youth who are "perpetrator-victims" (both perpetrators and victims of Internet harassment and unwanted sexual solicitation) comprise a very small minority of youths, but may be particularly troubled. They reported extremely high responses for offline perpetration of aggression (100%), offline victimization (100%), drug use such as inhalants (78%), and number of delinquent peers (on average, 3.2). This group was also particularly likely to be more aggressive offline, be victimized offline, and have a history of substance abuse.

Risk Factors

An ongoing body of research details demographic, environmental, habitual, and psychosocial risk factors that moderate the likelihood of youths being bullied or solicited online. David Finkelhor proposed a theory of "poly-victimization" that describes how certain youths are victimized in a multitude of ways by different parties and environmental situations, making them a vulnerable group for harm online and offline (Finkelhor, 2008). Several nontechnical means to combat solicitation and harassment have been identified, particularly a strong home environment and family life.

Demographics

Certain demographic factors have been correlated with increased risks for these types of harmful communication. Girls tend to be more at risk for being victimized by online solicitation (Wolak et al., 2006) and harassment (Agatston et al., 2007; DeHue, Bolman, & Völlink, 2008; Kowalski & Limber, 2007; Lenhart, 2007; Li, 2004,2006, 2007; Smith et al., 2008). Youths who are questioning their sexuality also face increased risks, as well. In a study where about 25% of cases of Internet solicitation in a nationwide survey were found to involve a male youth and a male adult, "most of the Internet-initiated cases involving boys had elements that made it clear victims were gay or questioning their sexual orientations (e.g., meeting offenders in gay-oriented chatrooms)" (Wolak et al., 2008b, p. 118). While all of the youths involved in these online activities may not identify as lesbian, gay, bisexual, or transgender in adult life, these studies do identify teens that are questioning their sexuality. The Internet may be a useful place to "come out" and try on new identities, but it appears to also be a place where gay, bisexual, or questioning teens are at a greater risk than their peers.

Teenagers and adolescents (aged 13–17 years) are more at-risk than prepubescent children (12 years of age or younger) for most threats, such as online solicitation (Beebe, Asche, Harrison, & Quinlan, 2004; Mitchell, Finkelhor, & Wolak, 2001; Mitchell et al., 2007a; Wolak et al., 2004, 2008b; Ybarra et al., 2007b). Online harassment also occurs less frequently among the youngest adolescents (Lenhart, 2007; Ybarra & Mitchell, 2004a) and children (McQuade & Sampat, 2008), and peaks around 13–14 years of age (Kowalski & Limber, 2007; Lenhart, 2007; McQuade & Sampat, 2008; Slonje & Smith, 2008; Williams & Guerra, 2007). Race is generally not a significant risk factor in crimes such as cyberbullying and online harassment (Hinduja & Patchin, 2008a; Nansel et al., 2001; Ybarra et al., 2007a).

Online Contact with Strangers

Communicating with anonymous individuals online is a common activity. Online social media are moving away from "walled gardens," and do not enforce age restrictions, making it nearly certain that youth with encounter postings or messages from people they do not know. Between 45% and 79% of U.S. youths were found to chat with strangers (McQuade & Sampat, 2008; Stahl & Fritz, 1999; Wolak, Mitchell, & Finkelhor, 2006), which was also correlated with receiving online solicitations (Beebe et al., 2004; Liau et al., 2005; Mitchell et al., 2001; Ybarra, Mitchell, Finkelhor, & Wolak, 2007c). Even more specifically, there is a correlation between youths who talk with strangers about *sexual topics* and those who are victimized (Wolak, Finkelhor, & Mitchell, 2008a).

At present, this link is merely correlational. In other words, there is no consensus on whether youths are more at-risk because they talk with strangers, or at-risk youths are more likely to talk with strangers. Some youths may be curious about sexual topics (Hines & Finkelhor, 2007, p. 301), particularly online where they can try out identities and new types of behavior. Other adolescents who are also involved in other risky behaviors (such as making rude or nasty comments, using file-sharing software to download images, visiting x-rated web sites, or talking about sex to people online) in addition to chat are more likely to receive *aggressive* solicitations, as well (Ybarra et al., 2007c; Wolak et al., 2008a).

Sharing of Personal Information

Posting personal or identifying information is often viewed as a risky behavior, although this by itself does not appear to be a significant risk factor. One reason is that posting information is common (or even required) on Internet sites, and "behaviors manifested by large numbers of people fail to predict events that are relatively uncommon" (Wolak et al., 2008b, p. 117). Other risky habits may be better predictors, and more related to why youths are at risk. In other words, the same psychosocial factors that place youths at risk for online solicitation and bullying are more significant risk factors than that of posting personal information online. For instance, "talking with people known only online ('strangers') under some conditions is related to interpersonal victimization, but sharing of personal information is not" (Ybarra et al., 2007c, p. 138). Despite anecdotal reports (Quayle & Taylor, 2001), cyberstalking, a crime where offenders locate victims offline using information found online (Jaishankar, Halder, & Ramdoss, 2008), only appears in youth online solicitation cases after the offender and victim meet offline (Wolak et al., 2009). Researchers consider cyberstalking to be driven by a desire to harass or control others, or as an online extension of offline stalking (Adam, 2002; Ogilvie, 2000; Philips & Morrissey, 2004; Sheridan & Grant, 2007).

Youths frequently post information of all sorts (text, images, video) online through social media such as SNSs. While investigation in this area is quite new, it appears that only a small number of teens are posting the most sensitive contact information such as a phone number, address, or full name (Hinduja & Patchin, 2008c; Lenhart & Madden, 2007; Pierce, 2007b). Jones et al. concluded that, "the inclusion of offline contact information was an anomaly in user profiles" (Jones, Millermaier, Goya-Martinez, & Schuler, 2008). Males were found to more frequently post personal information, while females posted images (Ybarra, Alexander, & Mitchell, 2005). More males were also found to have public profiles while females were more likely to have private profiles (Burgess-Proctor et al., 2009).

Passwords are particularly problematic to share, because they allow youths to impersonate others or hijack accounts. Pew Internet research from 2001 found that 22% of teens 12–17 had shared a password with a friend or someone they knew (Lenhart, Rainie, & Lewis, 2001), and McQuade and Sampat (2008) found that 13% of 4th–6th graders and 15% of 7th–9th graders experienced someone using their password without their permission and a slightly smaller percentage of youths had someone else impersonate them online. This again signals that the "friends" a youth has may indeed be the same people later harassing them or using their accounts for nefarious purposes.

The number of youths revealing personal information increased from 2000 (11%) to 2005 (35%) (Wolak et al., 2006), and still appears to be on the increase (Burgess-Proctor et al., 2009), in spite of efforts by advocacy groups (National Center for Missing and Exploited Children, 2006; Brookshire & Maulhardt, 2005). During this time there was no increase in Internet-instigated abductions or forcible Internet-initiated sexual encounters between adults and youths. Still, during this time of rapid technological change and transition, it remains to be seen how the risk of transmission of personal information interacts with or mediates other risk factors. In YISS-2, researchers concluded that, "it is not clear what kinds of information are particularly problematic, or exactly what the risks are with respect to the different situations in which youths disclose personal information online" (Wolak et al., 2006, p. 50).

Depression, Physical Abuse, and Substance Abuse

Depression, physical abuse, and substance abuse are all correlated with various risky behaviors that lead to poor choices with respect to online activities. Depressed youths were more likely to report increased unwanted exposure to online pornography (Wolak, Mitchell, & Finkelhor, 2007b), online harassment (Ybarra, 2004; Mitchell, Ybarra, & Finkelhor, 2007b; Ybarra, Leaf, & Diener-West, 2004), and solicitation (Mitchell et al., 2007b). Risk for online harassment was particularly pronounced among depressed male youths, who were 8 times more likely to be victimized than non-depressed male youths (Ybarra, 2004). Suicidal ideation has also been significantly correlated with online harassment victimization among adolescents (Hinduja & Patchin, 2008a). Self-harm, often a physical manifestation of depression, is also correlated with other risky behaviors that increase the likelihood of risk (Mitchell & Ybarra, 2007; Mitchell, Finkelhor, & Wolak, 2005). Depressed youths were also prone to a host of other risk factors, and were more likely to be heavy Internet users and talk with strangers online (Ybarra et al., 2005), making it difficult to untangle where the risk lies. Adolescents who have been sexually or physically abused offline are more likely to be solicited or harassed (Mitchell et al., 2007a; Mitchell et al., 2007b; Wells & Mitchell, 2008). Both youths who harass others (Ybarra & Mitchell, 2004b) and are solicitation victims online (Mitchell et al., 2007b) were more frequent drug users. Youths who were *both* perpetrator-victims of Internet harassment and unwanted online sexual solicitation were the heaviest drug users (Ybarra et al., 2007b). Offline, bullies tend to have used alcohol or other substances (Ybarra & Mitchell, 2007), paralleling the online environment.

Home Environment

The vast majority of parents (90%) are concerned about their child's online safety (Wolak et al., 2006), and, "a warm and communicative parent-child relationship is the most important nontechnical means that parents can use to deal with the challenges of the sexualized media environment" (Greenfield, 2004, p. 741). Home is where nearly all (91%) of youths reported using the Internet (Wolak et al., 2006), and adolescents who live in a poor home environment are at higher risk for online sexual victimization (Wolak, Mitchell, & Finkelhor, 2003a) and harassment (Ybarra & Mitchell, 2004b). Low parental monitoring was correlated with a host of negative offline consequences, such as increased likelihood of violence over time (Brendgen, Vitaro, Tremblay, & Lavoie, 2001), police contact (Pettit, Laird, Dodge, Bates, & Criss, 2001), and traditional offline bullying (Patterson, 2002; Steinberg & Silk, 2002).

A positive and communicative home environment inoculates youths against a host of dangers. Parents were generally responsible about their children going to real-world meetings resulting from online contact; 73% of parents were aware of real-world meetings and 75% accompanied the minor to the meeting (Wolak et al., 2006). About half of parents discussed related topics (such as online sexualized talk, adult pictures, and harassment) with their children, who were more safety-conscious as a result (Fleming et al., 2006). Similarly, parenting style was related to the techniques used to restrict access of minors to the Internet (Eastin, Greenberg, & Hofschire, 2006), more family rules about the Internet were correlated with less risk of a face-to-face meeting with someone met online (Liau et al., 2005), and a positive parental relationship improved effects from poverty and other socioeconomic factors (Barnow, Lucht, & Freyberger, 2001).

Despite an interest in the topic, parents held inaccurate beliefs about the risks of Internet communication (Livingstone & Bober, 2004; DeHue et al., 2008).

Parents under-estimated the amount of information adolescents posted online (Rosen et al., 2008), how frequently their children posted online personals, and corresponded with strangers (Computer Science and Telecommunications Board National Research Council, 2002, p. 165). This under-estimation of incidents may be due to the infrequent reporting of harassment or solicitation incidents (or even general Internet habits) to parents or other adults. Only around a third of youths who were harassed reported the occurrence to a parent or guardian (DeHue et al., 2008; Fight crime sponsored studies: Opinion research corporation, 2006b; National Children's Charity, 2005; Patchin & Hinduja, 2006; Wolak et al., 2006) and even less frequently told another adult such as a teacher. Youths may not report these incidents because they frequently (69%) felt that the incident was "not serious enough" to warrant discussion with an adult (Wolak et al., 2006, p. 26). Lines of communication should remain open between parents and their children, for mutual benefit.

Privacy Settings and Blocking

Youths benefit from an awareness of features that can be used to combat harassing contact. Between a third and half of SNS users employ privacy settings on SNSs. In 2006, Lenhart and Madden (2007) found that 66% of youths 12–17 had limited access to their SNS profiles, while Hinduja and Patchin reported slightly lower rates of setting profiles to private (Hinduja & Patchin, 2008c). Young SNS members also employed blocking features when sexually solicited online. Of the 7–9% of SNS members that were "approached for a sexual liaison," almost all immediately blocked the user (Rosen, 2006).

Youth have also developed a range of informal practices to deal with unwanted online contact. In a qualitative study, youths who are asked about such encounters draw parallels to spam or peculiar comments from strangers in public offline settings, noting that ignoring such solicitations typically makes them go away (boyd, 2008). There are situations where youths do not perceive the setting of media as private to be necessary. Lange (2007) described the awareness of YouTube users of the difference between being anonymous to a large number of people ("privately public"), or known but viewable by a smaller group ("publicly private"). Similarly Ben-Ze'ev (2003) states, "with complete strangers, the issue of privacy [online] is of little concern since we are in a sense anonymous" (p. 454). Of course, if a person's identity is disclosed, it can lead to a violation of a participant's expectancies. The revealing of personal information online and how youth view privacy is a controversial and evolving topic.

Mediating Technologies

Technologies that mediate online communication, such as chat rooms, message boards, and social network sites, attract different groups of youths and provide varying features to communicate. Youths tend to congregate online where their peers are, so many troubling interactions take place through popular technologies. Youths are also quite adept at using online technologies to suit their own goals. In some types of environments, it is more normative for youths to interact with people they don't know, such as on a message board where aliases are employed. At-risk youths are more attracted to some environments, elevating their levels of risk, as is demonstrated when depressed teens more heavily use online chat.

Chatrooms and Instant Messaging

Chatrooms and Instant Messaging (IM) have been the most prevalent media in online solicitation (Wolak et al., 2006) and harassment (Wolak et al., 2006; Kowalski & Limber, 2007; Fight crime sponsored studies: Opinion research corporation, 2006a; Fight crime sponsored studies: Opinion research corporation, 2006b) of minors. Chat and IM played a role in 77–86% of solicitation attempts and Internet-instigated relationships leading to offline sexual encounters (Wolak et al., 2004). This and other literature suggests the possibility that, "the nature of chat rooms and the kinds of interactions that occur in them create additional risk" (Wolak et al., 2007c, p. 329). Authorities have used these technologies extensively and effectively for "sting" arrests (Wolak, Mitchell, & Finkelhor, 2003b).

Several explanations exist for why chat and instant messaging are particularly prevalent in harassment and solicitation. Synchronous media may be particularly effective to get immediate feedback on if a youth is interested in a sexual topic, or irritate a target through a constant stream of messages. In addition to being popular, these technologies are used by youths for locating partners (Šmahel & Subrahmanyam, 2007) and general socialization (Leung, 2001). Youths who have a poor home environment or engage in other risky behaviors are more likely to use online chat frequently (Beebe et al., 2004), and chatroom use is correlated with increased depression (Ybarra et al., 2005), suggesting chat could be a particularly attractive mode of communication for youths who are not getting the emotional support they need.

Blogging

A sizeable minority of youths (28%) has created a blog (Lenhart, Madden, Macgill, & Smith, 2007b), but despite some suggestions that it is potentially dangerous (Huffaker, 2006), youth bloggers do not appear to have a higher level of interaction with strangers online nor are they more likely to be sexually solicited (Mitchell, Wolak, & Finkelhor, 2008). That said, they have been found to be more likely to experience online harassment (Mitchell et al., 2008).

Social Network Sites

Social network sites, or SNSs (boyd & Ellison, 2007), such as MySpace and Facebook, are among the most popular and controversial types of social media. As of 2007, over half of youths have used them to create profiles (Lenhart, Madden,

Macgill, & Smith, 2007a) and develop new friends (Smith, 2007). Young people are frequently members of SNSs (Lipsman, 2007), and use them to communicate and maintain social bonds (Granovetter, 1973, 1983), and as a base for online communities (Rheingold, 2001; Smith, 1999). Despite a strong public interest, research is inconclusive on the extent to which these sites present a risk.

Parents are justifiably concerned about youth interactions in these online spaces, as they know little about them, and what they do hear through the media tends to be negative. In 2007, 85% of adults were uncomfortable with their children participating in online communities (Center for the Digital Future, 2008) and in 2006, 63% of parents thought there were "quite a few sexual predators" on MySpace (Rosen, 2006). This worry was not carried by the youths who use these sites; 83% of teenagers felt this type of website is generally safe (Rosen, 2006), despite that a sizeable minority (19–22%) of youths reported being upset by harassment or solicitation on these sites (Rosen et al., 2008). Several other researchers have reported correlations between SNSs and either solicitation or harassment. Lenhart (2007) found that, "social network users are also more likely to be cyberbullied" (p. 4), particularly certain types of online harassment, such as spreading of rumors and harassing email (Lenhart, 2007). Girls appear to be more prone to receiving unwanted messages on SNSs (Smith, 2007), because harassers and solicitors generally target girls. Studies suggest SNS membership is slightly more female (51–54%) (Hinduja & Patchin, 2008b; Jones et al., 2008; Schrock, 2006; Thelwall, 2008), but this may be due to a minority of males self-identifying as female online.

Although certain SNS members (those who posted a picture and those who flirted online) were more likely to receive online contact from strangers, Smith concluded that, "despite popular concerns about teens and social networking, our analysis suggests that social network sites are not inherently more inviting to scary or uncomfortable contacts than other online activities" (Smith, 2007, p. 2). Similarly, Ybarra and Mitchell (2008) concluded that, "[b]road claims of victimization risk, at least defined as unwanted sexual solicitation or harassment, associated with social networking sites do not seem justified" (p. e350). The popularity of SNSs may be the attraction, because those seeking to victimize youths migrate to where they are; "When a medium becomes used by a huge portion of the population… it inevitably becomes a venue for deviant activity by some, but it is not necessarily a risk promoter" (Wolak et al., 2009, p. 8). Several troubling areas arise on SNSs, such as how, "the greatest exposure of children and adolescents to sex crimes is at the hands of people who are already a part of their families and social networks" (Wolak et al., 2009, p. 8). Much more research is needed in this area to provide a complete picture of how this technology is being adopted and used by youths, particularly in the context of sexualized communication.

Multiplayer Online Games and Environments

Nearly all American youths play games daily (Lenhart, Kahne, Middaugh, Macgill, Evans, & Vitak, 2008), half (47%) play with friends they know offline and 27% with people they met online. Contrary to stereotypes, females do play online games, but in lower numbers than males for most genres (Entertainment Software Association, 2008; Griffiths, Davies, & Chappell, 2003; Lenhart et al., 2008; Yee, 2006). The percentage of youth players may vary greatly between games, even within the same genre (Williams, Yee, & Caplan, 2008). The research is split on whether players of games such as MMOGs (Massively Multiplayer Online Games), are more at-risk than other youths for psychosocial factors such as depression, substance abuse, difficulties with self-regulation, trouble at school, and increased aggression (Ducheneaut, Yee, Nickell, & Moore, 2006; Ng & Wiemer-Hastings, 2005; Seay & Kraut, 2007; Williams & Skoric, 2005; Williams et al., 2008).

Online gaming environments frequently have multimedia capabilities and interactive possibilities well beyond web-based social media (such as SNSs). As concerns online communication, many games offer real-time multimedia chat during gameplay through text, voice, or video, and may encounter aggressive behavior (Anderson & Bushman, 2002; Funk, Baldacci, Pasold, & Baumgardner, 2004; Williams & Skoric, 2005). For instance, nearly half of game-playing teens report seeing or hearing "people being hateful, racist, or sexist while playing" at least sometimes, and 63% reported "people being mean and overly aggressive" (Lenhart et al., 2008).

In addition to more familiar modes of communication, 3-dimensional environments offer a new way for harassment to occur: "griefing." This is defined as when a player, "utilizes aspects of the game structure or physics in unintended ways to cause distress for other players" (Warner & Ratier, 2005, p. 47) and disrupts the gaming experience (Foo & Koivisto, 2004; Lin & Sun, 2005). For instance, players may be virtually confined in a cage, repeatedly killed, or teleported against their will. It is unclear if harassment on virtual worlds is inherently more distressing than other online technologies. Gamers may have a greater connection with their avatars, and may even feel that an avatar is physically their own body (Ehrsson, 2007; Lenggenhager, Tadi, Metzinger, & Blanke, 2007), raising the question of if people playing MMOs are more susceptible to psychological harm through griefing.

Multimedia Communications

Multimedia used in online harassment is not yet as widely prevalent as text forms, although this may change with increased adoption of mobile devices. For instance, 6% of youths reported having an embarrassing picture of them posted online without their permission (Lenhart, 2007) and 8% reported being a victim of images transmitted over a cell phone (Raskauskas & Stoltz, 2007). These multimedia communications include images and movies created by victims and then modified to poke fun at them (British Broadcasting Corporation, 2006), "mash-ups" that combine user-generated content with other imagery or videos (Jenkins, 2006), or content unrelated to the victim that is designed to disgust or offend (such as a sexually-themed picture). The role of technologies such as cell phones and webcams, which are being adopted by youths (Rainie, 2005), is not yet known, but multimedia communications can be more distressing (Smith et al., 2008).

Multimedia is also used in coercive online solicitation. Youths have been sent inappropriate images (such as of genitalia or sexual situations), or images were requested from youths. In the N-JOV study, arrested Internet-initiated sex offenders were found to send adult pornography (10%) or child pornography (9%) to victims (Wolak et al., 2004). One in five online child molesters took "sexually suggestive or explicit photographs of victims or convinced victims to take such photographs of themselves or friends" (Wolak et al., 2008b, p. 120). In a national survey, 4% of youths who use the Internet reported receiving a request for a sexual picture of themselves (Mitchell, Wolak, & Finkelhor, 2007c) (but only one youth in 1,500 complied), and 7% of students in grades 7–9 in the Rochester, New York, area received an online request for a nude picture (McQuade & Sampat, 2008).

Sexualized images and videos of youths, regardless of how they are created, constitute a troubling source of illegal, underage pornographic material for adults (if released on the Internet), and serve as fodder for future harassment or bullying (if stored on devices or computers). Despite low rates of compliance among youths, this is a serious issue, as, "[even] if only a small percentage cooperate, considering such requests flattering, glamorous, adventuresome, or testament of their love and devotion, this could be a major contribution to the production of illegal material" (Mitchell et al., 2007c, p. 201). Once these videos and images are uploaded, it is nearly impossible to keep them from being traded, downloaded, and viewed by third parties. Taylor and Quayle describe the way this content can never be deleted as "a permanent record of crime, and serves to perpetuate the images and memory of that abuse" (Taylor & Quayle, 2003, p. 24).

Future Research

In addition to the topics discussed here, some areas of youth safety are critically under-researched, particularly (1) conceptual clarity, (2) minor-minor solicitation and sexual relations, (3) the role of digital image and video capture devices, and (4) the impact of mobile technologies. New methodologies and standardized measures that can be compared across populations and studies are also needed to illuminate these under-researched topics. Finally, because these risks to youth are rapidly developing, there is a dire need for ongoing large-scale national surveys to synchronously track and quickly report these complex dynamics as they unfold.

Conceptual Clarity

Online solicitation and cyberbullying are investigated from a variety of disciplines, including sociology, child development, computer science, and communication. This inclusive approach led to vital findings, but not unified terminology. For instance, the phrase "online solicitation" on the surface level appears to specifically describe coercive communication, but its definition includes much communication that falls under sexual harassment, such as gender harassment and unwanted sexual attention. When statistics, such as the finding that one in five youths has been solicited online (Wolak et al., 2006), are quoted in the lay press, it leads to misunderstandings by the public, who may incorrectly assume that coercion is the motive and the perpetrators are sexually attracted to children. The time frame under investigation also varies between studies, and is a factor in the lack of consensus on how frequently these crimes occur. For instance, the widely varying statistics on the prevalence of cyberbullying come from surveys that ask youths about their experiences in the last few months, year, or lifetime. A theoretically rigorous multi-dimensional construct for cyberbullying and online solicitation, one that describes the areas of overlap as well as functional and time-based differences, would lead to more fruitful research and less confusion.

Minor-Minor Solicitation and Sexual Relations

Most research to date on solicitation has focused on solicitation by an adult, frequently a "stranger," who coerces a minor into a sexual relationship. Yet, research on online harassment and solicitation has revealed that this mode of operation isn't the norm, and the online coercion of children by adults into offline sexual relationships is scarce. More importantly, this focus on adult-minor coercive communication draws attention away from more prevalent and complex forms of online solicitation and harassment. It remains unclear how sexualized Internet solicitations (coercive and otherwise) are integrated with offline relationships among similar-aged youths, and where harmful encounters occur. Some of this contact is consensual, such as "sexting" or the writing of erotic emails between willing minors. However, there is the possibility that Internet-mediated communication is playing a role in undesirable offline outcomes such as rape, or that other offline crimes integrate new technologies in unforeseen ways. We need to consider a more holistic perspective when analyzing how sexual relationships and friendships are created, maintained, and terminated, and the emotional implications these have on teens. Finally, relying on the term "stranger" is problematic, because two people are not necessarily strangers after interacting together online, and the creation of online friendships (primarily non-sexual) is common. The concept may not be useful when considering online harm to youths.

Multimedia Communication

As more children and teenagers engage in the production or reception of amateur content in the process of harassment or solicitation, questions emerge about the content they are producing and integrating into daily life. Multimedia-capable devices are gaining in popularity (Center for the Digital Future, 2008; Hinduja & Patchin, 2008a), which offer multimedia recording through an "always on" connection direct to the Internet. Images and movies may be particularly distressing to victims of online harassment (Smith et al., 2008) or increase the initial attraction (Walther, Slovacek, & Tidwell, 2001). A similar charge can be leveled against research on multimedia harassment as was made against multimedia computer-mediated communication (CMC) in 2000 (Soukup, 2000): more research is required to overcome the "text-only bias" of online harassment. Harassment and solicitations are increasingly complex and multi-modal, and offenders may integrate, process, and post photographs and videos in ways we don't yet understand. Special care should be taken to assess the impact of and track this new form of cyberbullying over the next several years.

The rates of the use of multimedia for consensual sexual relations among minors is currently poorly understood, but seems likely, given the use of images to develop relationships online (Walther et al., 2001), the wide variety of amateur content created and distributed online both privately and publicly (Jacobs, 2007), and the presence of sexualized pictures on SNSs such as MySpace (Pierce, 2007a). These movies and images may be created during consensual sexual relationships between similar-aged adolescents, for instance, during flirting, which is common (Lenhart, 2007; Schiano, Chen, Ginsberg, Gretarsdottir, Huddleston, & Isaacs, 2002) or as an outlet for sexual thoughts and development (Atwood, 2006; Subrahmanyam & Greenfield, 2008). In one of the first surveys to include questions on the topic, 3% of 7th–9th graders asked for "naked pictures from another Internet user" (McQuade & Sampat, 2008). Finally, as previously discussed, video game and virtual world experiences may offer immersive new modes of harassment, such as griefing.

Intersection of Different Mobile and Internet-Based Technologies

The majority (77%) of Internet-initiated sex crimes against youths used multiple modes of communication (Wolak et al., 2004), but little is understood about the interplay between them. Most research has focused on the role of the Internet, but mobile phones are increasingly playing a role in sexual solicitation and harassment. It is already known that mobile phone use is a risk factor for receiving aggressive sexual solicitations online (Mitchell et al., 2007a) and online harassment (Hinduja & Patchin, 2008a). Mobile communication also provides a pervasive personal space where youths may be contacted at any time. How mobile devices are used in the United States for harassment and solicitation requires further examination over the next several years as these increasingly powerful devices are used by younger demographics.

The most recent online solicitation-related term the media have adopted is "sexting," or the sending of sexual multimedia content between minors on mobile devices or cell phones. According to LexisNexis, the term was not used in an indexed article in the United States until May 2008, and 75% of the articles (on a search performed March 21, 2009) were published in the first three months of

2009. Studies on the creation or transmission of sexual media on mobile devices are few, and statistics on how many youths receive solicitations or respond to them vary widely. A 2008 study conducted in the United Kingdom by TRU found that 20% of teens aged 13–19 sent nude or semi-nude photos or videos (The National Campaign, 2008). In this study, the majority of "sexting" communications appear to be from one individual to another; 71% of teenaged girls and 67% of teenaged males sent them to a significant other (The National Campaign, 2008).

References

Adam, A. (2002). Cyberstalking and Internet pornography: Gender and the gaze. *Ethics and Information Technology*, 4(2), 133–142.

Agatston, P. W., Kowalski, R., & Limber, S. (2007). Students' perspectives on cyber bullying. *Journal of Adolescent Health*, *41*, S59–S60.

American Psychological Association. (2000). *Diagnostic and statistical manual of mental disorders* (4th ed.). Arlington, VA: American Psychiatric Publishing.

Anderson, C. A., & Bushman, B. J. (2002). Effects of violent video games on aggressive behavior, aggressive cognition, aggressive affect, physiological arousal, and prosocial behavior: A meta-analytic review of the scientific literature. *Psychological Science*, *12*(5), 353–359.

Arnaldo, C. A. (2001). *Child Abuse on the Internet: Ending the Silence.* New York: Berghahn, Paris: UNESCO.

Associated Press. (2008). Mom indicted in MySpace suicide case: Computer charges against woman whose daughter feuded with victim. Retrieved August 10, 2010, from http://www.msnbc.msn.com/ id/24652422

Atwood, J. D. (2006). Mommy's little angel, Daddy's little girl: Do you know what your preteens are doing? *The American Journal of Family Therapy*, 34, 447–467.

Barak, A. (2005). Sexual harassment on the Internet. *Social Science Computer Review*, *23*(1), 77–92.

Barnow, S., Lucht, M., & Freyberger, H.-J. (2001). Influence of punishment, emotional rejection, child abuse, and broken home on aggression in adolescent: An examination of aggressive adolescents in Germany. *Psychopathology*, *34*(4), 167–173.

Beebe, T. J., Asche, S. E., Harrison, P. A., & Quinlan, K. B. (2004). Heightened vulnerability and increased risk-taking among adolescent chat room users: Results from a statewide school survey. *Journal of Adolescent Health*, *35*, 116–123.

Ben-Ze'ev, A. (2003). Privacy, emotional closeness, and openness in cyberspace. *Computers in Human Behavior*, *19*, 451–467.

Beran, T., & Li, Q. (2007). The relationship between cyberbullying and school bullying. *Journal of Student Wellbeing*, *1*(2), 15–33.

Berrier, T. (2007). *Sixth-, Seventh-, and Eighth-Grade Students' Experiences With the Internet and Their Internet Safety Knowledge.* Johnson City: East Tennessee State University.

Berson, I. R. (2003). Grooming cybervictims: The psychosocial effects of online exploitation for youth. *Journal of School Violence*, *2*(1), 5–18.

Berson, I. R., & Berson, M. J. (2005). Challenging online behaviors of youth: Findings from a comparative analysis of young people in the United States and New Zealand. *Social Science Computer Review*, *23*(1), 29–38.

Biber, J. K., Doverspike, D., Baznik, D., Cober, A., & Ritter, B. A. (2002). Sexual harassment in online communications: Effects of gender and discourse medium. *CyberPsychology & Behavior*, *5*(1), 33–42.

Borg, M. G. (1998). The emotional reaction of school bullies and their victims. *Educational Psychology, 18*(4), 433–444.

boyd, d. (2007). Why youth (heart) social network sites: The role of networked publics in teenage social life. In D. Buckingham (Ed.), *MacArthur Series on Digital Learning–Youth, Identity, and Digital Media Volume* (pp. 119–142). Cambridge, MA: MIT Press.

boyd, d. (2008). *Taken out of context: American teenage socialization in networked publics.* PhD Dissertation. University of California-Berkeley, School of Information, Berkeley: CA.

boyd, d., & Ellison, N. (2007). Social network sites: Definition, history, and scholarship. *Journal of Computer-Mediated Communication, 13*(1), 210-230.

Brendgen, M., Vitaro, F., Tremblay, R. E., & Lavoie, F. (2001). Reactive and proactive aggression: Predictions to physical violence in different contexts and moderating effects of parental monitoring and caregiving behavior. *Journal of Abnormal Child Psychology, 29*(4), 293–304.

British Broadcasting Corporation. (2006). Star Wars kid is top viral video. *BBC News* Retrieved August 10, 2009, from http://news.bbc.co.uk/2/hi/entertainment/6187554.stm

Brookshire, M., & Maulhardt, C. (2005). *Evaluation of the effectiveness of the netsmartz program: A study of Maine public schools.* Retrieved August 10, 2009, from http://www.netsmartz.org/pdf/gw_evaluation.pdf

Burgess-Proctor, A., Patchin, J., & Hinduja, S. (2009). Cyberbullying and online harassment: Reconceptualizing the victimization of adolescent girls. In V. Garcia & J. Clifford (Eds.), *Female crime victims: Reality reconsidered.* Upper Saddle River, NJ: Prentice Hall.

Cassell, J., & Cramer, M. (2007). High tech or high risk: Moral panics about girls online. In T. McPherson (Ed.), *MacArthur Foundation Series on Digital Media and Learning: Digital Youth, Innovation, and the Unexpected* (pp. 53–75). Cambridge, MA: MIT Press.

Center for the Digital Future. (2008). *Annual Internet Survey by the Center for the Digital Future Finds Shifting Trends among Adults about the Benefits and Consequences of Children Going Online.* Retrieved August 10, 2009, http://www.digitalcenter.org/pages/current_report.asp? intGlobalId=19

Computer Science and Telecommunications Board National Research Council. (2002). *Youth, pornography, and the Internet.* Washington DC: National Academies Press.

DeHue, F., Bolman, C., & Völlink, T. (2008). Cyberbullying: Youngsters' experiences and parental perception. *CyberPsychology & Behavior, 11*(2), 217–223.

Ducheneaut, N., Yee, N., Nickell, E., & Moore, R. J. (2006). *"Alone Together?" Exploring the social dynamics of massively multiplayer online games.* Paper presented at the Conference on Human Factors in Computing Systems, Montreal, Canada.

Eastin, M. S., Greenberg, B. S., & Hofschire, L. (2006). Parenting the Internet. *Journal of Communication, 56*(3), 486–504.

Ehrsson, H. (2007). The experimental induction of out-of-body experiences. *Science, 317*(5841), 1048.

Entertainment Software Association. (2008). *2008 Sale, Demographic and Usage Data: Essential Facts about the Computer and Video Game Industry.* Retrievd August 10, 2009, from http://www.theesa.com/facts/pdfs/ ESA_EF_2008.pdf

Ericson, N. (2001). Addressing the problem of juvenile bullying. *OJJDP Fact Sheet, 27.*

Fight crime sponsored studies: Opinion research corporation. (2006a). *Cyberbully Pre-Teen.* Retrieved August 10, 2009, from http://www.fightcrime.org/cyberbullying/cyberbullyingpreteen.pdf

Fight crime sponsored studies: Opinion research corporation. (2006b). *Cyberbully Teen.*

Finkelhor, D. (2008). *Childhood Victimization: Violence, Crime, and Abuse in the Lives of Young People.* New York: Oxford University Press.

Finkelhor, D., & Ormrod, R. (2000). Kidnapping of juveniles: Patterns from NIBRS. *OJJDP: Juvenile Justice Bulletin*. Retrieved August 10, 2009, from http://www.ncjrs.org/pdffiles1/ojjdp/181161.pdf

Finkelhor, D., Mitchell, K. J., & Wolak, J. (2000). *Online Victimization: A Report on the Nation's Youth*. Retrieved August 10, 2009, from http://eric.ed.gov/ERICWebPortal/recordDetail?accno=ED442039

Finn, J. (2004). A survey of online harassment at a university campus. *Journal of Interpersonal Violence, 19*(4), 468–483.

Fitzgerald, L. F., Gelfand, M. J., & Drasgow, F. (1995). Measuring sexual harassment: Theoretical and psychometric advances. *Basic and Applied Social Psychology, 17*(4), 425–445.

Fleming, M. J., Greentree, S., Cocotti-Muller, D., Elias, K. A., & Morrison, S. (2006). Safety in cyberspace: Adolescents' safety and exposure online. *Youth & Society, 38*(2), 135–154.

Fleming, M., & Rickwood, D. (2004). Teens in cyberspace: Do they encounter friend or foe? *Youth Studies Australia, 23*(3), 46–52.

Foo, C. Y., & Koivisto, E. M. I. (2004). *Defining grief play in MMORPGs: Player and developer perceptions*. Paper presented at the 2004 ACM SIGCHI International Conference on Advances in computer entertainment technology.

Funk, J. B., Baldacci, H. B., Pasold, T., & Baumgardner, J. (2004). Violence exposure in real-life, video games, television, movies, and the Internet: Is there desensitization? *Journal of Adolescence, 27*, 23–39.

Gennaro, C. D., & Dutton, W. H. (2007). Reconfiguring friendships: Social relationships and the Internet. *Information, Communication & Society, 10*(5), 591–618.

Glassner, B. (1999). *The Culture of Fear*. New York: Penguin.

Granovetter, M. S. (1973). The strength of weak ties. *The American Journal of Sociology, 78*(6), 1360–1380.

Granovetter, M. S. (1983). The strength of weak ties: A network theory revisited. *Sociological Theory, 1*, 201–233.

Greenfield, P. M. (2004). Inadvertent exposure to pornography on the Internet: Implications of peer-to-peer file-sharing networks for child development and families. *Journal of Applied Developmental Psychology, 25*(6), 741–750.

Griffiths, M. D., Davies, M. N. O., & Chappell, D. (2003). Breaking the stereotype: The case of online gaming. *CyberPsychology & Behavior, 6*(1), 81–91.

Gross, E. F. (2004). Adolescent Internet use: What we expect, what teens report. *Applied Developmental Psychology, 25*, 633–649.

Hawker, D. S. J., & Boulton, M. J. (2000). Twenty years' research on peer victimization and psychosocial maladjustment: A meta-analytic review of cross-sectional studies. *Journal of Child Psychology and Psychiatry, 41*(4), 441–455.

Haynie, D. L., Nansel, T., Eitel, P., Crump, A. D., Saylor, K., & Yu, K. (2001). Bullies, victims, and bully/victims: Distinct groups of at-risk youth. *The Journal of Early Adolescence, 21*(1), 29–49.

Hinduja, S., & Patchin, J. (2007). Offline consequences of online victimization: School violence and delinquency. *Journal of School Violence, 6*(3), 89–112.

Hinduja, S., & Patchin, J. (2008a). *Bullying beyond the Schoolyard: Preventing and Responding to Cyberbullying*. Thousand Oaks, CA: Sage.

Hinduja, S., & Patchin, J. (2008b). Cyberbullying: An exploratory analysis of factors related to offending and victimization. *Deviant Behavior, 29*(2), 129-156.

Hinduja, S., & Patchin, J. (2008c). Personal information of adolescents on the Internet: A quantitative content analysis of MySpace. *Journal of Adolescence, 31*, 125–146.

Hines, D. A., & Finkelhor, D. (2007). Statutory sex crime relationships between juveniles and adults: A review of social scientific research. *Aggression and Violent Behavior, 12*, 300–314.

Huffaker, D. (2006). *Teen blogs exposed: The private lives of teens made public.* Paper presented at the American Association for the Advancement of Science, St Louis, MO.

Ito, M., Baumer, S., Bittanti, M., boyd, d., Cody, R., Herr-Stephenson, B., Horst, H., Lange, P., Mahendran, D., Martinez, K., Pascoe, C., Perkel, D., Robinson, L., Sims, C., & Tripp, L. (2009). *Hanging out, messing around and geeking out: Living and learning with new media.* Cambridge, MA: MIT.

Jacobs, K. (2007). *Netporn: DIY web culture and sexual politics (Critical Media Studies).* Lanham, MD: Rowman & Littlefield.

Jaishankar, K., Halder, D., & Ramdoss, S. (2008). Pedophilia, pornography, and stalking: Analyzing child victimization on the Internet. In F. Schmallenger & Pittaro, M. (Ed.), *Crimes of the Internet* (pp. 28–42). Upper Saddle River, NJ: Prentice Hall.

Jenkins, H. (2006). *Convergence Culture.* New York: New York University Press.

Jones, S., Millermaier, S., Goya-Martinez, M., & Schuler, J. (2008). Whose space is MySpace? A content analysis of MySpace profiles. *First Monday, 13*(9).

Kowalski, R. M., & Limber, S. P. (2007). Electronic bullying among middle school students. *Journal of Adolescent Health, 41*, S22–S30.

Kowalski, R. M., Limber, S. P., & Agatston, P. W. (2007). *Cyber Bullying: Bullying in the Digital Age.* Malden, MA: Blackwell.

Lang, R. A., & Frenzel, R. R. (1988). How sex offenders lure children. *Annals of Sex Research, 1*(2), 303–317.

Lange, P. G. (2007). Publicly private and privately public: Social networking on YouTube. *Journal of Computer-Mediated Communication, 13*(1), 361-380.

Lenggenhager, B., Tadi, T., Metzinger, T., & Blanke, O. (2007). *Science, 317*(5841), 1096–1099.

Lenhart, A. (2007). *Pew Internet & American Life Project memo: Cyberbullying and online teens.* Retrieved August 10, 2009, from http://www.pewinternet.org/PPF/r/216/report_display.asp

Lenhart, A., Rainie, L., & Lewis, O. (2001). Teenage Life Online: The Rise of the Instant-message Generation and the Internet's Impact on Friendships and Family Relationships.

Lenhart, A., Kahne, J., Middaugh, E., Macgill, A. R., Evans, C., & Vitak, J. (2008). *Teens, video games, and civics.* Retrieved August 10, 2009, from http://www.pewinternet.org/PPF/r/263/report_display.asp

Lenhart, A., & Madden, M. (2007). Teens, privacy, & online social networks. Retrieved August 10, 2009, from http://www.pewinternet.org/Reports/2007/Teens-Privacy-and-Online-Social-Networks.aspx

Lenhart, A., Madden, M., Macgill, A. R., & Smith, A. (2007a). *Writing, technology and teens.* Retrieved August 10, 2009, from http://www.pewinternet.org/pdfs/PIP_Teens_Social_Media_Final.pdf

Lenhart, A., Madden, M., Macgill, A. R., & Smith, A. (2007b). *Teens and social media: The use of social media gains a greater foothold in teen life as they embrace the conversational nature of interactive online media.* Retrieved August 10, 2009, from http://www.pewinternet.org/PPF/r/230/report _display.asp

Leung, L. (2001). College student motives for chatting on ICQ. *New Media & Society, 3*(4), 483–500.

Li, Q. (2004). Cyberbullying in schools: Nature and extent of adolescents' experience. Retrieved August 10, 2009, from http://www.ucalgary.ca/~qinli/publication/cyberbully_ aera05 %20.html

Li, Q. (2006). Cyberbullying in schools: A research of gender differences. *School Psychology International, 27*(2), 157–170.

Li, Q. (2007). New bottle but old wine: A research of cyberbullying in schools. *Computers in Human Behavior, 23*, 1777–1791.

Liau, A. K., Khoo, A., & Ang, P. H. (2005). Factors influencing adolescents engagement in risky Internet behavior. *CyberPsychology & Behavior, 8*(6), 513–520.

Lin, H., & Sun, C.-T. (2005). *The 'White-eyed' player culture: Grief play and construction of deviance in MMORPGs.* Paper presented at the DiGRA 2005 Conference: Changing Views–Worlds in Play.

Lipsman, A. (2007). *Social networking goes global: Major social networking sites substantially expanded their global visitor base during past year.* Retrieved August 13, 2009, from http://www.comscore.com/press/release.asp?press=1555

Livingstone, S., & Bober, M. (2004). *UK children go online : Surveying the experiences of young people and their parents.* Retrieved from http://eprints.lse.ac.uk/395/

Livingstone, S., & Haddon, L. (2008). Risky experiences for children online: Charting European research on children and the Internet. *Children & Society, 22*, 314–323.

Madden, M. (2006). *Internet penetration and impact.* Retrieved August 13, 2009, from http://www.pewinternet .org/PPF/r/182/report_display.asp

Marwick, A. (2008). To catch a predator? The MySpace moral panic. *First Monday, 13*(6), article 3.

McBride, N. A. (2005). *Child safety is more than a slogan: "Stranger-Danger" warning not effective at keeping kids safer.* Retrieved August 13, 2009, from http://www.missingkids.com/en_US/publications/ StrangerDangerArticle.pdf

McQuade, S. C., & Sampat, N. M. (2008). *Survey of Internet and at-risk behaviors: Undertaken by school districts of Monroe county New York.* Retrieved from http://www.rrcsei.org/RIT%20Cyber%20Survey%20Final%20Report.pdf

Mitchell, K. J., Finkelhor, D., & Wolak, J. (2001). Risk factors for and impact of online sexual solicitation of youth. *Journal of the American Medical Association, 285*(23), 3011–3014.

Mitchell, K., Finkelhor, D., & Wolak, J. (2005). Police posing as juveniles online to catch sex offenders: Is it working? *Sexual Abuse: A Journal of Research and Treatment, 17*(3), 241–267.

Mitchell, K. J., Wolak, J., & Finkelhor, D. (2007a). Youth Internet users at risk for the most serious online sexual solicitations. *American Journal of Preventative Medicine, 32*(6), 532–537.

Mitchell, K. J., Wolak, J., & Finkelhor, D. (2007c). Online requests for sexual pictures from youth: Risk factors and incident characteristics. *Journal of Adolescent Health, 41*, 196–203.

Mitchell, K. J., Wolak, J., & Finkelhor, D. (2008). Are blogs putting youth at risk for online sexual solicitation or harassment? *Child Abuse & Neglect, 32*, 277–294.

Mitchell, K. J., Ybarra, M., & Finkelhor, D. (2007b). The relative importance of online victimization in understanding depression, delinquency, and substance use. *Child Maltreatment, 12*(4), 314–324.

Mitchell, K. J., & Ybarra, M. (2007). Online behavior of youth who engage in self-harm provides clues for preventive intervention. *Preventative Medicine, 45*, 392–396.

Nansel, T. R., Overpeck, M. D., Haynie, D. L., Ruan, W. J., & Scheidt, P. C. (2003). Relationships between bullying and violence among US youth. *Archives of Pediatrics & Adolescent Medicine, 157*(4), 348–353.

Nansel, T. R., Overpeck, M., Pilla, R. S., June, R. W., Simons-Morton, B., & Scheidt, P. (2001). Bullying behaviors among US youth: Prevalence and association with psychosocial adjustment. *Journal of the American Medical Association, 16*, 2094–2100.

National Center for Missing and Exploited Children. (2006). *2006 Annual Report.*

National Children's Charity. (2005). *Putting U in the picture.* Retrieved August 13, 2009, from http://www.nch.org. uk/uploads/documents/Mobile_bullying_%20report.pdf

Ng, B. D., & Wiemer-Hastings, P. (2005). Addiction to the Internet and online gaming. *CyberPsychology & Behavior*, 8(2), 110–113.

Ogilvie, E. (2000). *The Internet and cyberstalking.* Paper presented at the Criminal Justice Responses Conference, Sydney.

Palfrey, J., & Gasser, U. (2008). *Born digital: Understanding the first generation of digital natives.* New York: Basic.

Patchin, J., & Hinduja, S. (2006). Bullies move beyond the schoolyard: A preliminary look at cyberbullying. *Youth Violence and Juvenile Justice*, 4(2), 148–169.

Patterson, G. R., & Fisher, P. A. (2002). Recent developments in our understanding of parenting: Bidirectional effects, causal models, and the search for parsimony. In M. H. Bornstein (Ed.), Handbook of parenting: Vol. 5. Practical issues in parenting (pp. 59-88). Mahwah, NJ: Erlbaum.

Peter, J., Valkenburg, P., & Schouten, A. (2005). *Characteristics and motives of adolescents: Talking with strangers on the Internet and its consequences.* Paper presented at the 55th Annual Conference of the International Communication Association (ICA), New York.

Pettit, G. S., Laird, R. D., Dodge, K. A., Bates, J. E., & Criss, M. M. (2001). Antecedents and behavior-problem outcomes of parental monitoring and psychological control in early adolescence. *Child Development*, 72(2), 583–598.

Philips, F., & Morrissey, G. (2004). Cyberstalking and cyberpredators: A threat to safe sexuality on the Internet. *Convergence: The International Journal of Research into New Media Technologies*, 10(1), 66–79.

Pierce, T. A. (2007a). *Teens' use of MySpace & the type of content posted on the sites.* Retrieved August 13, 2009, from http://www.fresno.k12.ca.us/divdept/cfen/Flyer/mySpace.pdf

Pierce, T. A. (2007b). X-Posed on MySpace: A content analysis of "MySpace" social networking sites. *Journal of Media Psychology, 12*(1).

Ponton, L. E., & Judice, S. (2004). Typical adolescent sexual development. *Child and Adolescent Psychiatric Clinics of North America, 13*, 497.

Quayle, E., & Taylor, M. (2001). Child seduction and self-representation on the Internet. *CyberPsychology & Behavior*, 4(5), 597–608.

Rainie, L. (2005). *16% of Internet users have viewed a remote person or placing using a webcam.* Retrieved August 13, 2009, from http://www.pewinternet.org/pdfs/PIP_webcam_use.pdf

Raskauskas, J., & Stoltz, A. D. (2007). Involvement in traditional and electronic bullying among adolescents. *Developmental Psychology*, 43(3), 564–575.

Rheingold, H. (2001). The virtual community. In D. Trend (Ed.), *Reading digital culture (keyworks in cultural studies)* (pp. 272–281). Reading, MA: Wiley-Blackwell.

Rideout, V. (2007). *Parents, children & media: A Kaiser Family Foundation survey.* Retrieved August 10, 2009, from http://www.kff.org/entmedia/7638.cfm

Rigby, K. (2003). Consequences of bullying in schools. *Canadian Journal of Psychiatry*, 48, 583–590.

Roland, E. (2002). Bullying, depressive symptoms and suicidal thoughts. *Educational Research*, 44, 55–67.

Rosen, L. (2006). Adolescents in MySpace: Identity formation, friendship and sexual predators. Retrieved August 13, 2009, from http://www.csudh.edu/psych/lrosen.htm

Rosen, L. D., Cheever, N. A., & Carrier, L. M. (2008). The association of parenting style and child age with parental limit setting and adolescent MySpace behavior. *Journal of Applied Developmental Psychology*, 29(6):459–471.

Salter, A. (2004). *Predators: Pedophiles, rapists, and other sex offenders.* New York: Basic.

Schiano, D. J., Chen, C. P., Ginsberg, J., Gretarsdottir, U., Huddleston, M., & Isaacs, E. (2002). *Teen use of messaging media.* Paper presented at the CHI, Minneapolis, Minnesota.

Schrock, A. (2006). *MySpace or Yourspace: A media system dependency view of MySpace.* Orlando, FL: University of Central Florida.

Seals, D., & Young, J. (2003). Bullying and victimization: Prevalence and relationship to gender, grade level, ethnicity, self-esteem and depression. *Adolescence, 38*, 735–747.

Seay, A. F., & Kraut, R. E. (2007). *Project massive: Self-regulation and problematic use of online gaming.* Proceedings of the ACM conference on human factors in computing systems (pp. 829–838). New York: ACM.

Sheldon, K., & Howitt, D. (2007). *Sex offenders and the Internet.* New York: Wiley.

Sheridan, L. P., & Grant, T. (2007). Is cyberstalking different? *Psychology, Crime & Law, 13*(6), 627–640.

Slonje, R., & Smith, P. K. (2008). Cyberbullying: Another main type of bullying? *Scandinavian Journal of Psychology, 49*, 147–154.

Šmahel, D., & Subrahmanyam, K. (2007). Any girls want to chat press 911: Partner selection in monitored and unmonitored teen chat rooms. *CyberPsychology & Behavior, 10*(3), 346–353.

Smith, A. (2007). *Pew Internet & American Life Project memo: Teens and online stranger contact.* Retrieved August 13, 2009, from http://www.pewinternet.org/PPF/r/223/report_display.asp

Smith, M. A. (1999). *Communities in cyberspace.* London: Routledge.

Smith, P. K., Mahdavi, J., Carvalho, M., Fisher, S., Russell, S., & Tippett, N. (2008). Cyberbullying: Its nature and impact in secondary school pupils. *Journal of Child Psychology and Psychiatry, 49*(4), 376–385.

Snyder, H. N., & Sickmund, M. (2006). *Juvenile offenders and victims: 2006 national report.* Retrieved August 10, 2009, from http://ojjdp.ncjrs.gov/ojstatbb/nr2006/index.html

Soukup, C. (2000). Building a theory of multimedia CMC. *New Media & Society, 2*(4), 407–425.

Stahl, C., & Fritz, N. (1999). Internet safety: Adolescents' self-report. *Journal of Adolescent Health, 31*, 7–10.

Steinberg, L., & Silk, J. S. (2002). Parenting adolescents. In M. H. Bornstein (Ed.), *Handbook of parenting: Volume 1. Children and parenting.* Mahwah, NJ: Erlbaum.

Stys, Y. (2004). *Beyond the schoolyard: Examining bullying among canadian youth.* Unpublished. Masters thesis, Carleton University, Ottawa.

Subrahmanyam, K., & Greenfield, P. (2008). Online communication and adolescent relationships. *The Future of Children, 18*(1), 119–146.

Taylor, M., & Quayle, E. (2003). *Child pornography–An Internet crime.* Hove, UK: Brunner-Routledge.

The National Campaign. (2008). *Sex and tech: Results from a survey of teens and young adults.* Retrieved August 10, 2009, from http://www.thenationalcampaign.org/sextech/PDF/SexTech_Summary.pdf

Thelwall, M. (2008). Social networks, gender, and friending: An analysis of MySpace member profiles. *Journal of the American Society for Information Science and Technology, 59*(8), 1523–1527.

Walther, J. B., Slovacek, C. L., & Tidwell, L. C. (2001). Is a picture worth a thousand words? Photographic images in long-term and short-term computer-mediated communication. *Communication Research, 28*(1), 105–134.

Warner, D. E., & Ratier, M. (2005). Social context in massively-multiplayer online games (MMOGs): Ethical questions in shared space. *International Review of Information Ethics, 4*, 7.

Wells, M., & Mitchell, K. J. (2008). How do high-risk youth use the Internet? Characteristics and implications for prevention. *Child Maltreatment, 13*(3): 227-234.

Williams, D., & Skoric, M. (2005). Internet fantasy violence: A test of aggression in an online game. *Communication Monographs, 72*(2), 217–233.

Williams, D., Yee, N., & Caplan, S. E. (2008). Who plays, how much, and why? Debunking the stereotypical gamer profile. *Journal of Computer Mediated Communication, 13*, 993–1018.

Williams, K. R., & Guerra, N. G. (2007). Prevalence and predictors of Internet bullying. *Journal of Adolescent Health, 41*, S14–S21.

Wolak, J., Finkelhor, D., & Mitchell, K. J. (2004). Internet-initiated sex crimes against minors: Implications for prevention based on findings from a national study. *Journal of Adolescent Health, 35*(5), 424.e11-424.e20.

Wolak, J., Finkelhor, D., & Mitchell, K. (2008a). Is talking online to unknown people always risky? Distinguishing online interaction styles in a national sample of youth Internet users. *CyberPsychology & Behavior, 11*(3), 340–343.

Wolak, J., Finkelhor, D., & Mitchell, K. (2009). *Trends in arrests of "Online Predators"*. Retrieved August 9, 2009, from http://www.unh.edu/news/NJOV2.pdf

Wolak, J., Finkelhor, D., Mitchell, K., & Ybarra, M. (2008b). Online "Predators" and their victims: Myths, realities, and implications for prevention and treatment. *American Psychologist, 63*(2), 111–128.

Wolak, J., Mitchell, K. J., & Finkelhor, D. (2003a). Escaping or connecting? Characteristics of youth who form close online relationships. *Journal of Adolescence, 26*, 105–119.

Wolak, J., Mitchell, K. J., & Finkelhor, D. (2003b). *Internet sex crimes against minors: The response of law enforcement.* Retrieved August 9, 2009, from http://www.ncjrs.gov/App/publications/Abstract.aspx?id=202909.

Wolak, J., Mitchell, K., & Finkelhor, D. (2006). *Online victimization of youth: Five years later.* Retrieved August 9, 2009, from http://www.unh.edu/ccrc/pdf/CV138.pdf

Wolak, J., Mitchell, K. J., & Finkelhor, D. (2007a). Does online harassment constitute bullying? An exploration of online harassment by known peers and online-only contacts. *Journal of Adolescent Health, 41*, S51–S58.

Wolak, J., Mitchell, K. J., & Finkelhor, D. (2007b). Unwanted and wanted exposure to online pornography in a national sample of youth Internet users. *Pediatrics, 119*(2), 247–257.

Wolak, J., Ybarra, M., Mitchell, K. J., & Finkelhor, D. (2007c). Current research knowledge about adolescent victimization on the Internet. *Adolescent Medicine, 18*, 325–241.

Wolfe, D. A., & Chiodo, D. (2008). Sexual *harassment and related behaviors reported among youth from grade 9 to grade 11.* Retrieved August 9, 2009, from http://www.camh.net/News_events/Media_centre/CAMH%20harassment%20paper.pdf

World Health Organization. (2007). International statistical classification of diseases and related health problems, 10th revision. Geneva, Switzerland: WHO.

Ybarra, M. (2004). Linkages between depressive symptomology and Internet harassment among young regular Internet users. *CyberPsychology & Behavior, 7*(2), 247–257.

Ybarra, M., Alexander, C., & Mitchell, K. J. (2005). Depressive symptomatology, youth Internet use, and online interactions: A national survey. *Journal of Adolescent Health, 36*(1), 9–18.

Ybarra, M., Diener-West, M., & Leaf, P. J. (2007a). Examining the overlap in Internet harassment and school bullying: Implications for school intervention. *Journal of Adolescent Health, 41*, S42–S50.

Ybarra, M., Espelage, D. L., & Mitchell, K. J. (2007b). The co-occurrence of Internet harassment and unwanted sexual solicitation victimization and perpetration: Associations with psychosocial indicators. *Journal of Adolescent Health, 41*, S31–S41.

Ybarra, M., Leaf, P. J., & Diener-West, M. (2004). Sex differences in youth-reported depressive symptomatology and unwanted Internet sexual solicitation. *Journal of Medical Internet Research*, *6*(1).

Ybarra, M., & Mitchell, K. J. (2004a). Online aggressor/targets, aggressors, and targets: A comparison of associated youth characteristics. *Journal of Child Psychology and Psychiatry*, *45*(7), 1308–1316.

Ybarra, M., & Mitchell, K. J. (2004b). Youth engaging in online harassment: Associations with caregiver-child relationships, Internet use, and personal characteristics. *Journal of Adolescence*, *27*, 319–336.

Ybarra, M., & Mitchell, K. J. (2007). Prevalence and frequency of Internet harassment instigation: Implications for adolescent health. *Journal of Adolescent Health*, *41*, 189–195.

Ybarra, M., & Mitchell, K. J. (2008). How risky are social networking sites? A comparison of places online where youth sexual solicitation and harassment occurs. *Pediatrics*, *121*(2), e350–e357.

Ybarra, M., Mitchell, K., Finkelhor, D., & Wolak, J. (2007c). Internet prevention messages: Targeting the right online behaviors. *Archives of Pediatrics & Adolescent Medicine*, *161*, 138–145.

Ybarra, M., Mitchell, K., Wolak, J., & Finkelhor, D. (2006). Examining characteristics and associated distress related to Internet harassment: Findings from the second Youth Internet Safety Survey. *Pediatrics*, *118*(4), e1169–e1177.

Yee, N. (2006). The demographics, motivations, and derived experiences of users of massively multi-user online graphical environments. *Presence*, *15*(3), 309–329.

LIST OF CONTRIBUTORS

Patricia Amason (Ph.D., Purdue University, 1993) is an Associate Professor and Associate Chair, Department of Communication, University of Arkansas. Her research examines provision of social support in interpersonal relationships and the communication surrounding sexual issues among romantic partners. Her research appears in *Journal of Applied Communication Research, Communication Studies, Communication Yearbook 10, Southern Communication Journal, Sex Roles, Health Communication,* and *Journal of Family Communication.*

Theodore A. Avtgis (Ph.D., Kent State University) is Associate Professor of Communication Studies, West Virginia University. He has published five books, numerous book chapters, and over 40 peer reviewed research articles appearing in *Communication Education* and *Management Communication Quarterly,* among others. He is co-founder of *Medical Communication Specialists* and current editor of *Communication Research Reports.* Among numerous awards, he was named a Centennial Scholar by the Eastern Communication Association.

Deborah Ballard-Reisch (Ph.D., 1983, Bowling Green State University) is Kansas Health Foundation Distinguished Chair in Strategic Communication, Professor, Elliott School of Communication, Wichita State University. Dr. Ballard-Reisch has authored over 40 published essays in scholarly journals/edited volumes. She co-edited *Communication and Sex Role Socialization* and special issues of the *Journal of Family Communication* (research methodology) and *Women and Language* (globalization and feminism). Her current research examines health promotion in underserved populations, community-based participatory research, and social media.

Jennifer A. H. Becker (Ph.D., 2005, University of Oklahoma) is Instructor at the University of Alabama. Her research focuses on challenges in interpersonal communication in relational and organizational contexts. Her research appears in *Communication Monographs, Journal of Applied Communication Research,* and *Journal of Social and Personal Relationships.*

danah boyd (Ph.D., 2009, University of California, Berkeley) is a researcher at Microsoft Research New England. Her research examines social media, youth practices, and social network sites. Her research appears in *New Media and Society, Journal of Children and Media, International Journal of Media and Cultural Politics,* and *Journal of Computer-Mediated Communication.*

Erin M. Bryant (M.A., 2008, Washington State University) is a doctoral student studying interpersonal and computer-mediated communication in the Hugh Downs School of Human Communication, Arizona State University. She has co-authored several manuscripts examining social networking sites in the context of uses and gratifications, relational maintenance, friendship rules, and relational reconnection. Other recent research has examined modality switching, textual harassment, and workplace deception. Her work was honored at the National Communication Association with "Top 4 Paper" and "Top Student Paper" awards.

Jeffrey T. Child (Ph.D., 2007, North Dakota State University) is an Assistant Professor in the School of Communication Studies, Kent State University. His research examines relational communication, privacy, and new communication technology. His work related to the latter line of research appears in *Computers in Human Behavior, Journal of the American Society for Information Science and Technology,* and *Management Communication Quarterly.*

Stacey L. Connaughton (Ph.D., 2002, University of Texas at Austin), is Associate Professor in Communication, Purdue University. Her research examines identification and leadership in geographically distributed contexts and is funded by NSF, the Carnegie Foundation, and the Russell Sage Foundation. Her research appears in *Journal of Communication, Management Communication Quarterly, Small Group Research,* and *Knowledge Management Review.* She has facilitated workshops for corporate, governmental, and educational groups on virtual teams and leadership.

Kathryn Dindia (Ph.D., 1981, University of Washington) is Professor of Communication at the University of Wisconsin-Milwaukee. Her research examines sex differences, self-disclosure, and relational maintenance. She authored over 50 essays published in scholarly journals and edited volumes; her research appears in *Psychological Bulletin, Human Communication Research, Communication Research, Journal of Social and Personal Relationships,* and *Personal Relationships.*

Norah E. Dunbar (Ph.D., 2000, University of Arizona) is Associate Professor of Communication and Faculty Associate at the Center for Applied Social Research, University of Oklahoma. Her research examines nonverbal communication, deception detection, and the development of interpersonal power in relationships. Her research appears in *Communication Monographs* and the *Journal of Social and Personal Relationships.*

Paige P. Edley (Ph.D., 1997, Rutgers University) is Associate Professor at Loyola Marymount University. Her research examines the intersections of power, gender, and identity in organizations, work-life balance, and women-owned businesses. Her research appears in *Management Communication Quarterly, Communication Yearbook, Electronic Journal of Communication, Women and Language,* and *Argumentation and Advocacy.* She is committed to issues of social justice and blends activism with her teaching and scholarship.

Michal Frenkel (M.A., 2009, University of Haifa). Frenkel's research examines the effects of new media on family communication and cohesion as well as new media in organizations. Her research appears in the *Scandinavian Journal of Management.*

Jeffrey T. Hancock (Ph.D., 2002, Dalhousie University) is Associate Professor in Communication, Cornell University. His research examines social interactions mediated by information and communication technology, with an emphasis on how people produce and understand language in these contexts. His research appears in *Communication Research, Journal of Communication, Human Communication Research*, and *Media Psychology*.

Mark L. Hans (B.S., 1989, Liberty University) is a graduate teaching assistant in communication at the University of Arkansas. He is a former television engineer with Disney/ABC; a sound engineer; and a former consultant in the marketing, advertising, and financial disciplines. His research emphases are organizational, small group, and interpersonal communication.

Lou Heldman (B.A., 1972, Ohio State University) is Distinguished Senior Fellow in Media Management and Journalism at Wichita State University. He is the creator and Executive Producer of the television series *Wichita State & the World.* He is retired president and publisher of *The Wichita Eagle* and Kansas.com. His research focuses primarily on social media.

Andrew C. High (M.A., 2006, University of Delaware) is a graduate teaching and research assistant in the Department of Communication Arts and Sciences at Penn State University. His research examines the process of social support and computer-mediated communication, especially in the ways communication technologies alter the processes of support provision and reception. His research appears in *Computers in Human Behavior, Small Group Research,* and *Communication Research Reports.*

Renée Houston (Ph.D., Florida State University) is Associate Professor of Communication Studies, University of Puget Sound. Her research and teaching interests include organizational communication and systems theory as well as information and communication technologies. Her most recent research examines the intersections of public and organizational policy on women, the working poor, and the homeless. Her research appears in numerous book chapters and *Communication Theory, Management Communication Quarterly, Journal of Research Practice, International Journal of Learning,* and *World Futures.*

Matthew Jensen (Ph.D., 2007, University of Arizona) is Assistant Professor of Management Information Systems and Faculty Associate at the Center for Applied Social Research, University of Oklahoma. His research examines deception and credibility in online interaction, and appears in *Group Decision and Negotiation* and the *Journal of Management Information Systems.*

Amy Janan Johnson (Ph.D., 1999, Michigan State University) is Associate Professor in Communication, University of Oklahoma. Her research examines long-distance relationships and computer-mediated communication, friendships, stepfamilies, and interpersonal argument. She has published 18 journal articles and 7 book chapters in venues including *Communication Monographs*, *Journal of Communication*, *Journal of Social and Personal Relationships*, and *Personal Relationships*.

David Kamerer (Ph.D., 1989, Indiana University) is accredited in public relations (APR) by the Universal Accreditation Board. He serves as Assistant Professor in the School of Communication, Loyola University Chicago, where he teaches courses in social media, public relations, and digital marketing. He has previously taught at Wichita State University and has worked in marketing, public relations, and corporate communication for non-profit agencies including Big Brothers Big Sisters of Kansas, Envision, and Via Christi Health.

Jinsuk Kim (M.A., 2007, University of Wisconsin-Milwaukee) is a doctoral candidate in the Department of Communication at Michigan State University. Her research examines self-disclosure, impression management, and relationship development in online environments. Her research appears in the edited volume *A Networked Self* and in journals including *Asian Journal of Social Psychology* and *Human Communication Research.*

Jennifer Marmo (M.A., 2006, Eastern Michigan University) is a doctoral student and teaching associate in the Hugh Downs School of Human Communication, Arizona State University. Her research examines interpersonal relationships and issues of motivation, health, and conflict. Her research examining relational maintenance and friendship rules on Facebook was awarded "Top Student Paper" by the National Communication Association's Interpersonal Communication Division.

Katheryn C. Maguire (Ph.D., 2001, University of Texas) is Assistant Professor in the Department of Communication at Wayne State University. Her research examines how individuals use both mediated and face-to-face communication to maintain relationships and cope with stressful situations in contexts including military deployments and long distance romances. Her research appears in *Communication Monographs, Journal of Applied Communication Research,* and *Communication Quarterly.*

Gustavo S. Mesch (Ph.D., 1993, Ohio State University) is Associate Professor of Sociology at the University of Haifa and Chair of the Communication and Information Technologies Section of the American Sociological Association. His research examines privacy in the information society as well as youth and the Internet. He is co-author of *Wired Youth: The Social World of Adolescence in the Information Age* (Routledge, 2010). His research appears in *Human Communication Research* and *Journal of Computer Mediated Communication.*

Ahlam Muhtaseb (Ph.D., 2004, University of Memphis) is an assistant professor in the Department of Communication Studies at California State University, San

Bernardino. Her research examines on-line communities as alternative spaces for political activism and social support for marginalized groups.

Sandra Petronio (Ph.D., 1979, University of Michigan) is Professor of Communication and Senior Faculty Associate, Indiana School of Medicine as well as Fairbanks Medical Ethics Center, IUPUI. She published numerous articles in interdisciplinary journals and five books on her evidenced-based theory, *Communication Privacy Management*, testing the theory over the last 30 years. Her research examines the intersection of privacy, disclosure, and confidentiality across contexts like social networking. She is past president of the International Association of Relationship Research and the Western Communication Association.

Makenzie Phillips (M.A., 2009, San Diego State University) is Special Lecturer in Communication at Boise State University. Her research examines the effects of computer-mediated communication on romantic relationships, nonverbal communication, and the dark-side of communication.

E. Phillips Polack (M.D., West Virginia University) is Clinical Professor of Surgery at West Virginia University and a practicing plastic surgeon at Wheeling Hospital in Wheeling, West Virginia. Dr. Polack holds certifications from the American Board of Plastic Surgery and the American Board of Surgery, among others. He has published several articles on communication in both communication and health science journals. He is senior author of two books including *Medical Communication: Defining the Discipline* (Kendall-Hunt) and co-founder of *Medical Communication Specialists.*

Artemio Ramirez, Jr. (Ph.D., University of Arizona, 2000) is Assistant Professor in the Hugh Downs School of Human Communication, Arizona State University. His research examines interpersonal aspects of computer-mediated communication, particularly the role and effect of social information seeking on impressions and relationships. His research appears in *Communication Monographs*, *Communication Research*, *Human Communication Research*, *Journal of Communication*, *Journal of Social and Personal Relationships*, and *New Media & Society*.

N. Lamar Reinsch, Jr. (Ph.D., 1973, University of Kansas) is Professor of Management at Georgetown University. His research examines the human impact of communication technologies, crisis management, and leadership development.

Bobby Rozzell (B.A., 2007; M.A., 2010, Wichita State University) is Graduate Teaching Assistant in Communication, University of Oklahoma. His research examines social media, particularly the Twitter community in Wichita, Kansas. He has made numerous presentations on the use and potential of social media at academic conferences including the National Communication Association Conference, the Southwest Texas Popular Culture Association Conference, and the Sooner Conference.

W. Scott Sanders (M.A., 2007, Purdue University) is a doctoral candidate in the Annenberg School for Communication & Journalism at the University of Southern California. His research examines the development of trust and interpersonal relationships in online communities.

Andrew R. Schrock (M.A., 2006, University of Central Florida) is a second-year Ph.D. student at the Annenberg School for Communication and Journalism, University of Southern California, where he is a research assistant for the Metamorphosis Project directed by Sandra Ball-Rokeach. His research examines risky behaviors, group collaboration and communities, particularly those involving globalization and technological change.

Brittney D. Selvidge (B.A, 2009, Ouachita Baptist University) is Graduate Teaching Assistant in Communication, University of Arkansas. Her research examines gender and on-line communication, especially blogging.

Denise H. Solomon (Ph.D., Northwestern University) is Professor of Communication Arts and Sciences and Associate Dean for Research and Graduate Studies at Pennsylvania State University. Her research examines the causes and consequences of turbulence in romantic associations and culminated in the relational turbulence model, describing how transitions in romantic relationships promote relationship qualities that polarize cognitive, emotional, and communicative reactions to both ordinary and extraordinary experiences.

Brian H. Spitzberg (Ph.D., 1981, University of Southern California) is Professor of Communication at San Diego State University. His research examines interpersonal competence, conflict, jealousy, and stalking. Dr. Spitzberg is co-author of several books, many scholarly articles, and frequently serves as reviewer for numerous scholarly journals. He also serves in an advisory capacity for the San Diego City Attorney's Domestic Violence Unit, San Diego District Attorney's Stalking Strike Force, and is a member of the Association of Threat Assessment Professionals.

Sydney M. Staggers (M.A., Emerson College) is a doctoral student in the Department of Communication Studies, West Virginia University. Her research examines health communication and interpersonal communication. Ms. Staggers has co-authored several articles and convention papers examining health communication and health organizations.

Katie A. Tinker (M.A., 2010, University of Arkansas) was a graduate teaching assistant in communication at the University of Arkansas. Her research examines the intersection of race, class, and gender as constructed through rhetoric.

Catalina L. Toma (Ph.D., 2010, Cornell University) is Assistant Professor in Communication Arts, University of Wisconsin-Madison. Her research examines the effect of communication technology on interpersonal processes, such as impression management, impression formation, deception, trust and social connectivity. Her research appears in *Communication Research, Journal of Communication,* and *Personality and Social Psychology Bulletin*.

Stephanie Tom Tong (M.A., 2008, Michigan State University) is completing doctoral studies in communication at Michigan State. Her research examines impression formation and interpersonal relations online, including the effects of social media systems, interface cues, sociometric cues, and stereotypes in online conversations, social perceptions, and social influence. Her research appears in *Journal of Computer-Mediated Communication, Human Communication Research,* and *Journal of Social and Personal Relationships*.

Jeanine Warisse Turner (Ph.D., 1996, The Ohio State University) is Associate Professor in Communication, Culture, and Technology at Georgetown University. Her research examines human factors and management issues involved in the introduction and use of new communication technologies. Her research appears in *Health Communication, Journal of Business Communication,* and *Journal of Business and Technical Communication*.

Joseph B. Walther (Ph.D., 1990, University of Arizona) is Professor of Communication and Telecommunication, Information Studies and Media, Michigan State University. His research examines the interpersonal dynamics of computer-mediated communication in personal relationships, groups, and social support—areas in which he has published original theories and numerous empirical studies. His professional honors include twice receiving the National Communication Association's Woolbert Award for articles that have stood the test of time and had major impacts on the discipline.

Lynne M. Webb (Ph.D., 1980, University of Oregon) is Professor of Communication at the University of Arkansas. She previously served as a tenured faculty member at the Universities of Florida and Memphis. Her research examines young adults' interpersonal communication in romantic and family contexts. Her research appears in over 50 essays published in scholarly journals and edited volumes, including *Computers in Human Behavior, Communication Education, Health Communication,* and *Journal of Family Communication*.

Susan M. Wieczorek (M.A., University of Pittsburgh) is a public speaking and speaking enhancement coordinator for general education at the University of Pittsburgh, Johnstown. Ms. Wieczorek has presented research in both the communication and health science disciplines; her research examines patient-doctor interactions, proxemics, and classroom pedagogical techniques.

Kevin B. Wright (Ph.D., University of Oklahoma) is Professor in Communication at the University of Oklahoma. His research examines interpersonal communication, social support related to health outcomes, and computer-mediated relationships. He coauthored *Health Communication in the 21st Century*, and his research appears in over 45 book chapters and journal articles, including the *Journal of Communication*, *Communication Monographs*, the *Journal of Social and Personal Relationships*, *Communication Quarterly*, *Journal of Applied Communication Research*, *Health Communication*, and the *Journal of Computer-Mediated Communication*.

INDEX

G

H

I

Q

R

U

V

W

Y